GUIDE TO MARITIME SECURITY AND THE ISPS CODE

2012 EDITION

IMO INTERNATIONAL MARITIME ORGANIZATION
London, 2012

Published in 2012 by the
INTERNATIONAL MARITIME ORGANIZATION
4 Albert Embankment, London SE1 7SR
www.imo.org

Printed by Polestar Wheatons (UK) Ltd, Exeter, EX2 8RP

ISBN: 978-92-801-1544-4

IMO PUBLICATION
Sales number: IA116E

This publication has been prepared from official documents of IMO, and every effort has been made to eliminate errors and reproduce the original text(s) faithfully. Readers should be aware that, in case of inconsistency, the official IMO text will prevail.

075556

Guide to maritime security and the ISPS Code

Guidance for port facilities, ports and ships

Disclaimer

This Guide has been developed to consolidate existing IMO maritime security-related material into an easily read companion guide to SOLAS chapter XI-2 and the International Ship and Port Facility Security (ISPS) Code in order to assist States in promoting maritime security through development of the requisite legal framework, the associated administrative practices and procedures, and the necessary material, technical and human resources. It is intended both to assist SOLAS Contracting Governments in the implementation of, verification of, compliance with and enforcement of the provisions of SOLAS chapter XI-2 and the ISPS Code, and to serve as an aid and reference for those engaged in delivering capacity-building activities in the field of maritime security.

While the guidance in this Guide was developed with contributions from international maritime security practitioners and is based on generally recognized maritime security practices and procedures, the suggested practices and procedures are not the only means of implementing the Maritime Security Measures of SOLAS chapter XI-2 and the ISPS Code. Other methods of meeting the requirements may be equally appropriate and effective. It should be noted that, because of the diversity of legal and administrative structures within individual States, the practices and procedures proposed in the text may need to be modified to fit within such structures.

Guidance contained in this Guide is not intended to replace, supersede, contradict, nullify or override any international or domestic law of any State or any further instrument developed by the Organization and should not be interpreted or applied in a manner which may undermine compliance with the maritime security requirements of individual States, which should, in any case, take precedence over any non-mandatory guidance included herein which may not have been adopted by the State concerned.

Users of this Guide should be aware that references to IMO instruments may become out of date by the adoption of more recent instruments. Users are therefore invited to consult their national Administration or the IMO website for information on the status of referenced instruments.

Contents

Page

Guide to maritime security and the ISPS Code

Section 4 – Security responsibilities of ship operators

Section 5 – Framework for conducting security assessments

Foreword

In the wake of the tragic events of 11 September 2001 in the United States of America, a Diplomatic Conference on maritime security was held at the London Headquarters of the International Maritime Organization (IMO) from 9 to 13 December 2002. This Conference adopted a number of amendments to the International Convention for the Safety of Life at Sea, 1974, the most far-reaching of which enshrined the new International Ship and Port Facility Security (ISPS) Code. The Conference also adopted a series of resolutions designed to add weight to the amendments, encourage the application of the Measures to ships and port facilities not covered by the ISPS Code and pave the way for future work on the subject.

The ISPS Code was produced in just over a year by IMO's Maritime Safety Committee (MSC) and its maritime security Working Group. It contains detailed security-related requirements for Governments, port authorities and shipping companies in a mandatory section (part A), together with a series of guidelines about how to meet these requirements in a second, non-mandatory section (part B).

Due to the urgent need to have security measures in place, the ISPS Code came into effect on 1 July 2004, just 18 months after its adoption. To assist Contracting Governments in exercising their implementation responsibilities, particularly those in lesser developed countries, one of the resolutions at the Diplomatic Conference invited IMO to develop training materials and, if necessary, further guidance on various aspects of the ISPS Code. This was accomplished in the 2003–2008 period through the development of model training courses; the issuance of specific guidance, mainly in the form of MSC circulars; the organization of over 100 regional and national workshops; and the conduct of several advisory and assessment missions in response to requests from individual Governments.

In 2009, as IMO's focus was shifting to other pressing security issues (notably piracy and armed robbery) and the implementation of long-range identification and tracking systems, there was a growing recognition of the need to reinforce implementation of the ISPS Code and to strengthen linkages with other IMO initiatives. In responding to this need, IMO took stock of the training and guidance materials that it had issued over the preceding six years. It found that, while some of the materials had become outdated, much remained relevant but was situated in an array of documentation that was not easily accessible by maritime security practitioners.

This Guide has been prepared as a practical way of providing Government and industry practitioners responsible for implementing the ISPS Code with a consolidated and up-to-date source of guidance material with appropriate linkages to other ongoing IMO initiatives.

Section 1 – Introduction

1.1 Purpose of the Guide

1.1.1 This Guide is intended to provide consolidated guidance on the implementation of the security-related amendments to the International Convention on the Safety of Life at Sea, 1974 (SOLAS Convention) which were adopted in December 2002. These amendments included a new chapter XI-2 in the SOLAS Convention, "Special measures to enhance maritime security", which enshrined the International Ship and Port Facility Security (ISPS) Code. Throughout this Guide, these are collectively referred to as the Maritime Security Measures.

1.1.2 The guidance in the Guide is addressed primarily to:

.1 Government officials who exercise the responsibilities that the Maritime Security Measures place on Contracting Governments;

.2 port facility employees who exercise the responsibilities that the Maritime Security Measures place on port facilities; and

.3 shipping company employees, including shipboard personnel, who exercise the responsibilities that the Maritime Security Measures place on shipping companies and their ships.

1.1.3 The guidance may also be relevant to those responsible for, or undertaking, any security-related responsibility at port facilities, in ports and on ships.

1.2 Structure

1.2.1 The Guide is presented in five sections:

.1 section 1 describes the purpose and content of the Guide and provides an overview of the Maritime Security Measures, outlines the benefits and challenges in their implementation and the need to maintain security awareness;

.2 section 2 provides guidance on the security responsibilities that the Maritime Security Measures place on Governments and those who may be authorized to undertake a Government's security responsibilities;

.3 section 3 provides guidance on the security responsibilities that the Maritime Security Measures place on port facilities and those undertaking these responsibilities at port facilities;

.4 section 4 provides guidance on the security responsibilities that the Maritime Security Measures place on shipping companies and those undertaking these responsibilities within companies and on their ships; and

.5 section 5 describes a security assessment methodology for port facilities and ports.

1.2.2 Each section contains a series of subsections corresponding to the main areas of security responsibility. Each subsection can be further broken down to address specific responsibilities. Where appropriate, the text in each subsection reflects the experience of Contracting Governments in implementing the Maritime Security Measures; appendices are used to supplement the short narrative by providing references, templates, checklists, practices and methodologies that have been adopted by Contracting Governments.

1.2.3 In order to achieve a clear distinction between the mandatory provisions of the Maritime Security Measures and supporting guidance material, attention has been paid throughout the Guide to the consistent use of verbs as follows:

- **.1** mandatory text uses 'must' or 'is/are required to', as appropriate; and
- **.2** guidance text uses 'should', 'could' or 'may', as appropriate.

1.2.4 The Guide is to be reviewed and updated on a regular basis.

1.2.5 Many aspects of the Maritime Security Measures have responsibilities for Governments, port facility operators and ship operators. To assist with understanding how these responsibilities complement each other, appendix 1.1 identifies their location in sections 2–4 of the Guide.

1.3 Sources

1.3.1 The guidance in the Guide is mainly drawn from IMO sources. In addition to part B of the ISPS Code, they include a variety of resolutions, circulars and circular letters. A full list is provided in appendix 1.2 – IMO Guidance material on Maritime Security Measures, 1986–2011. These documents are on IMO's website and may be accessed at: http://www.imo.org/OurWork/Security/Pages/MaritimeSecurity.aspx

1.3.2 Other sources of guidance material include:

- **.1** the ILO/IMO Code of practice on security in ports;
- **.2** presentations at IMO regional and national workshops;
- **.3** internet sites of Contracting Governments and their multilateral organizations; and
- **.4** information made available to IMO by Contracting Governments on their organizational structures, practices and procedures; the guidance issued to their port facilities and shipping companies; and their implementation experience.

1.3.3 To a lesser extent, elements of the guidance in the Guide are derived from material on the internet sites of non-governmental organizations representing the ports and shipping industries, and of individual port authorities and shipping companies.

1.3.4 To the extent possible, the Guide's contents include illustrative examples drawn from the sources described above.

1.4 Overview of the Maritime Security Measures

Origins

1.4.1 After the 1985 attack on the *Achille Lauro*, the Maritime Safety Committee (MSC) issued guidance on the security of cruise ships and the ports that they use. The guidance covered:

- **.1** the appointment within Government of a Designated Authority responsible for cruise ship and cruise port security;
- **.2** the appointment of an operator security officer by shipping companies operating cruise ships;
- **.3** the appointment of a ship security officer for each cruise ship;
- **.4** undertaking a ship security survey of each cruise ship;
- **.5** preparation of a ship security plan for each cruise ship and its approval by a Designated Authority within Government;
- **.6** appointment of facility security officers at cruise ports;
- **.7** undertaking a facility security survey for each cruise port; and
- **.8** preparation of a facility security plan for each cruise port and approval by the Designated Authority.

1.4.2 Some Governments imported elements of this guidance into their national legislation.

1.4.3 In 1996, the MSC extended the application of the above guidance to international passenger ferry services and the ports that they use. This further guidance recommended the use of three threat levels:

.1 background;

.2 moderate; and

.3 high.

1.4.4 In November 2001, IMO issued a resolution which called for a review of the existing international legal and technical measures to prevent and suppress terrorist acts against ships at sea and in port, and to improve security aboard and ashore. The aim was to:

.1 reduce risks to passengers, crew and port personnel on board ships and in port areas as well as to ships and their cargoes;

.2 enhance ship and port security; and

.3 prevent shipping from becoming a target of international terrorism.

1.4.5 In December 2002, a Diplomatic Conference on maritime security was held at the London headquarters of IMO. It was attended by 109 Contracting Governments to the SOLAS Convention (see below) and observers from other United Nations agencies, intergovernmental organizations and non-governmental international associations. Its work resulted in the adoption of the SOLAS amendments 2002 (see below).

The SOLAS Convention

1.4.6 The 1974 SOLAS Convention is one of 32 international conventions and agreements that have been adopted by IMO. It is the premier international treaty dealing with the safety of ships and specifies minimum standards for the construction, equipment and operation of ships. Since its adoption in 1974, the SOLAS Convention has been amended on numerous occasions.

The SOLAS amendments 2002

1.4.7 In December 2002, IMO adopted security-related amendments to the SOLAS Convention aimed at enhancing the security of ships and the port facilities that they use. The amendments include thirteen mandatory regulations in chapter XI-2, "Special measures to enhance maritime security", and the linked International Ship and Port Facility Security (ISPS) Code, collectively referred to as the Maritime Security Measures throughout this Guide.

1.4.8 The ISPS Code has a mandatory section (part A) and a recommendatory section (part B). The guidance given in part B of the ISPS Code should be taken into account when implementing the SOLAS chapter XI-2 regulations and the mandatory provisions in part A. IMO has published the ISPS Code, including chapter XI-2, in English, French, Spanish and Arabic; an electronic version is also available in English, French, Spanish and Russian. Both versions may be obtained by accessing IMO's website at: www.imo.org/Publications

Conference resolutions

1.4.9 In addition to adopting the SOLAS amendments 2002, the Diplomatic Conference considered a range of maritime security issues and adopted nine Conference resolutions addressing:

.1 further work by the International Maritime Organization pertaining to the enhancement of maritime security;

.2 future amendments to chapters XI-1 and XI-2 of the 1974 SOLAS Convention on special measures to enhance maritime safety and security;

.3 promotion of technical co-operation and assistance;

.4 early implementation of the special measures to enhance maritime security;

.5 establishment of appropriate measures to enhance the security of ships, port facilities, mobile offshore drilling units on location and fixed and floating platforms not covered by chapter XI-2 of the 1974 SOLAS Convention;

.6 enhancement of security in co-operation with the International Labour Organization;

.7 enhancement of security in co-operation with the World Customs Organization;

.8 early implementation of long-range ship's identification and tracking; and

.9 human element-related aspects and shore leave for seafarers.

The Maritime Security Measures in brief

1.4.10 Most Governments have to enact national legislation to give full effect to the Maritime Security Measures. While Governments have the discretion to extend provisions from the Maritime Security Measures to ships and port facilities that the Measures do not apply to, they cannot adopt legislative provisions whose effect would be to apply lower requirements to ships and port facilities regulated under the Maritime Security Measures than those specified in the Measures.

1.4.11 The following paragraphs outline some of the key features of the Maritime Security Measures.

Organizations within Government

1.4.12 Contracting Governments can establish Designated Authorities within Government to undertake their port facility security responsibilities. Governments or their Designated Authorities and Administrations may delegate the undertaking of certain responsibilities to recognized security organizations outside Government.

Security levels

1.4.13 The setting of the security level applying at any particular time is the responsibility of Governments and will apply to ships flying their flag and to their port facilities. The ISPS Code defines three security levels for international use:

.1 security level 1, normal;

.2 security level 2, lasting for the period of time when there is a heightened risk of a security incident; and

.3 security level 3, lasting for the period of time when there is the probable or imminent risk of a security incident.

Information to IMO

1.4.14 The Maritime Security Measures require certain information to be provided to IMO and information to be made available to allow effective communication between company/ship security officers and the port facility security officers responsible for the port facility and the ships that they serve.

Risk management

1.4.15 In essence, the Maritime Security Measures were developed with the basic understanding that ensuring the security of ships and port facilities was a risk-management activity and that, to determine what security measures are appropriate, an assessment of the risks must be made in each particular case. The purpose of the ISPS Code is to provide a standardized, consistent framework for evaluating risk, enabling Governments to offset changes in threat levels with changes in vulnerability for ships and port facilities.

1.4.16 This concept of risk management is embodied in the Maritime Security Measures through a number of functional security requirements for ships and port facilities, including (but not limited to) security assessments, security plans and access control.

1.4.17 Any shipping company operating ships to which the Maritime Security Measures apply must appoint at least one company security officer for the Company and a ship security officer for each of its ships.

1.4.18 Governments are required to undertake a port facility security assessment (PFSA) on each port facility within the scope of the Maritime Security Measures. The results have to be approved by the Government and are to be used to help determine which port facilities are required to appoint a port facility security officer (PFSO). Each PFSA should be reviewed regularly. When completed, the PFSA has to be provided to the PFSO.

Milestones

1.4.19 The Maritime Security Measures entered into force internationally on 1 July 2004.

1.4.20 Certain elements of the information that Governments must provide to IMO must be updated and returned to IMO at five-year intervals, first by 1 July 2009 and again by 1 July 2014. The required information is identified in subsection 2.19.

1.5 Benefits of, and challenges in, implementing the Maritime Security Measures

1.5.1 Following adoption of the Maritime Security Measures in December 2002, Governments had until 1 July 2004 to implement the Maritime Security Measures in their national legislation and to make the necessary administrative and organizational alterations to facilitate their implementation.

1.5.2 Many Governments achieved this target, although a number of interim arrangements were required. In many cases, enhancements were made later in the light of experience.

1.5.3 A number of Governments have also applied security requirements to port facilities, port areas and ships not covered by the Maritime Security Measures. This has included extending the application to ships operating domestic services and the application of provisions taken from the ILO/IMO Code of practice on security in ports.

1.5.4 Since the entry into force of the Maritime Security Measures, a number of port facilities have reported a marked reduction in both the incidence of thefts and the number of accidents in security-restricted areas. In addition, it has been reported that, during the first six months since the introduction of the Measures, there was a significant reduction in stowaway cases in US ports.

1.5.5 A review of the statistics published by regional Memoranda of Understanding on Port State Control indicated that security-related deficiencies found on ships to which the Maritime Security Measures apply also showed a positive trend, albeit after some difficulties in the period immediately following their introduction.

1.5.6 Maritime security measures were developed in response to perceived terrorist threats. However, to varying degrees, the Measures are applicable to countering other forms of security threats, notably piracy and armed robbery in international and territorial waters; and unlawful activities such as drug smuggling at ports. Thus, the fundamental purpose of the ISPS Code can be considered to be to reduce the vulnerability of the maritime industry to security threats, regardless of their nature.

1.5.7 As with all other aspects of shipping regulated through multilateral treaty instruments, the effectiveness of the requirements is dependent on the degree to which the relevant provisions are universally implemented and enforced. Thus, the success of the Maritime Security Measures is in the hands of Governments and the shipping and port industries.

1.5.8 When the Maritime Security Measures are implemented and enforced proportionally (i.e., ensuring that the action taken 'fits' the seriousness of the contravention) and effectively, they have proved to be successful in protecting ships and port facilities from unlawful acts. However, although the Maritime Security Measures came into effect on 1 July 2004, gaps in their implementation and application persist.

1.5.9 Many Governments are still striving to fully implement the Maritime Security Measures, particularly those pertaining to port facilities, due to a variety of factors, including:

- **.1** competing priorities for funds – these may include anti-piracy and armed robbery measures, maritime safety and environmental protection, and security measures for the other modes of transportation;

.2 the high cost of implementing security measures at port facilities – estimated in a 2007 study by the UN Conference on Trade and Development to average US$287,000 in investment costs and US$105,000 in annual running costs per port facility;

.3 difficulty in quantifying the effectiveness of security measures other than by means of anecdotal evidence – although a focus on such factors as fewer deaths, theft-related infractions and unauthorized entries into restricted areas may provide empirical measures of success;

.4 difficulty in estimating the probability and consequences of each type of potential threat and integrating it with known vulnerabilities, particularly for port facilities;

.5 the lack of the legal and policy instruments required to achieve compliance with the Maritime Security Measures and to resolve jurisdictional issues between Government agencies;

.6 limitations in the training received by security practitioners – training programmes should be designed by qualified personnel to meet the specific implementation responsibilities of each type of practitioner (e.g., Government officials, security officers, guards, managers); and

.7 limitations in the guidance readily accessible to security practitioners, particularly on the implementation experience of Governments and the industry.

1.5.10 As the Maritime Security Measures have become an accepted part of the shipping and port industries, there have been reports of varying levels of diligence in their implementation. New patterns of security threats and incidents can emerge, and have emerged.

1.5.11 From their inception, it has been repeatedly emphasized that those implementing the Maritime Security Measures should give due regard to the welfare of seafarers, particularly with reference to seafarers' access to shore and shore leave and allowing access to ships by representatives of organizations committed to the welfare of seafarers. Problems in these respects can still arise, and this Guide re-emphasizes IMO's collective view that the Maritime Security Measures should not be used to impose unnecessary restrictions or additional costs on seafarers.

1.5.12 This Guide is a response to these challenges by providing practitioners with a consolidated and up-to-date source of guidance material on port facility and ship security. In doing so, it recognizes the need to refocus implementation efforts and to strengthen linkages with other ongoing IMO initiatives, notably:

.1 benefits of the Measures in efforts to counter piracy and armed robbery at sea;

.2 utility of long-range identification and tracking systems for enhanced maritime situational awareness;

.3 role of seafarers in a security regime; and

.4 balance between the facilitation of trade and security.

1.6 Maintaining security awareness

Introduction

1.6.1 Historically, the port and shipping industries experienced high levels of criminal activity, particularly smuggling and pilferage, which impeded the development of a positive security culture in the maritime industries.

1.6.2 Effective implementation of the Maritime Security Measures has given port and ship users greater confidence that their cargoes will arrive intact and without tampering. This has resulted in economic benefits to port facilities that maintain high security standards.

1.6.3 Despite this, the promotion of security awareness and the continued development of a security culture across the port and shipping industries remains a continuing challenge for all those involved in port and ship security. In order to play a leadership role, Government organizations need to co-ordinate their efforts in meeting this challenge.

1.6.4 Security awareness is part of the training required under the Maritime Security Measures for PFSOs, those undertaking port facility security-related duties and other port facility personnel.

1.6.5 Similarly, security awareness is part of the training required for company security officers, ship security officers and all shipboard personnel.

Security awareness programmes

1.6.6 Promoting security awareness is vital to the security and safety and health of all port facility, port and ship personnel. For this reason, it is important that those responsible for implementing or overseeing the implementation of the Maritime Security Measures take the steps necessary to maintain and enhance security awareness among their stakeholders and employees.

1.6.7 Typically, this is achieved through awareness programmes. To be successful, the designers of such programmes should ask themselves the following questions:

.1 what message(s) needs to be conveyed?

.2 who should receive it?

.3 how should it be communicated?

.4 is follow-up required?

1.6.8 Governments generally convey broad messages to wide audiences either directly or through their national authorities, e.g., information on the Government's security policy, threat levels and effective security measures, as well as requesting the public to exercise continuing vigilance and to report security concerns.

1.6.9 Law-enforcement services issue similar messages but directed to stakeholders at the regional or local level.

1.6.10 Port facility operators, port administrators and shipping companies are likely to focus on advising their personnel about:

.1 their security policy;

.2 information received on specific security threats (as is appropriate to release);

.3 available training courses;

.4 the need to continually exercise vigilance;

.5 the need and the procedures to be followed for reporting unusual incidents or behaviour; and

.6 the actions that should be taken into account in the event of a security incident, including taking part in security drills or exercises.

1.6.11 The means of communication can take various forms, depending on the message and the intended audience.

1.6.12 General messages to wide audiences are likely to use the media whereas more specific messages are more likely to use avenues such as presentations to security committees, the delivery of customized awareness training and the issuance of promotional material (e.g., posters, pamphlets, magazine articles and DVDs).

1.6.13 It should be noted that effective communication with local communities, land-holders and operators of small boats, whose past rights of access into and around port areas may now be affected by new security measures, remains a challenge for many Designated Authorities and operators of port facilities.

1.6.14 Follow-up and repetition are recommended practice, with examples including security awareness being a standing item at security committees and the aim of regular drills and employee training days.

1.6.15 It is important to recognize that successful security-awareness programmes tend to be tailored to the particular needs and concerns of each group of stakeholders; conversely, multi-stakeholder programmes may not be effective if the message becomes blurred or is not available in the local language.

1.6.16 A list of websites with material on security awareness is in appendix 1.3 – Websites showing security awareness programmes.

1.7 Abbreviations

1.7.1 The following abbreviations (acronyms) are used in the Guide:

AFA – Armed forces authority

AIS – Automatic identification system

APEC – Asia Pacific Economic Cooperation

ASA – Alternative security agreement

CCTV – Closed-circuit television

CSO – Company security officer

CSP – Continuous service provider

DoS – Declaration of Security

EMSA – European Maritime Safety Agency

ESA – Equivalent security arrangement

FAL – IMO's Facilitation Committee

FPSO – Floating production storage and offloading vessel

GISIS – Global Integrated Shipping Information System

GMDSS – Global Maritime Distress and Safety System

GNSS – Global Navigation Satellite System

ID – Identification document

IDE – International Data Exchange

ILO – International Labour Organization

IMO – International Maritime Organization

ISM Code – International Safety Management Code

ISPS Code – International Ship and Port Facility Security Code

ISSC – International Ship Security Certificate

LRIT – Long-range identification and tracking of ships

MODU – Mobile offshore drilling unit

MOU – Memorandum of Understanding

MRCC – Maritime Rescue Co-ordination Centre

MSC – IMO's Maritime Safety Committee

OAS – Organization of American States

PFSA – Port facility security assessment

PFSO – Port facility security officer

PFSP – Port facility security plan

PSA – Port security assessment

PSAC – Port security advisory committee

PSC – Port security committee

PSO – Port security officer

PSP – Port security plan

RO – Recognized organization

RSO – Recognized security organization

SAFE – WCO SAFE Framework of Standards to secure and facilitate global trade

SAR – Search and rescue

SOC – Statement of Compliance

SOLAS – International Convention for the Safety of Life at Sea

SSA – Ship security assessment

SSAS – Ship security alert system

SSO – Ship security officer

SSP – Ship security plan

STCW – International Convention on Standards of Training, Certification and Watchkeeping for Seafarers

SUA – Convention for the Suppression of Unlawful Acts against the Safety of Maritime Navigation

UN – United Nations

WCO – World Customs Organization

1.8 Definitions

1.8.1 The following definitions apply to this Guide:

Administration means the Government of the State whose flag the ship is entitled to fly. In the Maritime Security Measures and this Guide, "Administration" is used to describe the organization within Government responsible for ship security.

Alternative security agreement (ASA) means a bilateral or multilateral agreement between Governments covering short international voyages on fixed routes between dedicated port facilities, allowing the security measures and procedures applied to the port facilities and ships to differ from those required under the Maritime Security Measures.

Application of the Measures means determining the port facilities covered by the Maritime Security Measures, i.e., those required to appoint a PFSO and submit a PFSP, and communicating their location along with the identity and title of their PFSO and the date of approval of the PFSP. In cases where port facilities are occasionally used by ships on international voyages, undertaking a PFSA to decide the extent of application of the Maritime Security Measures.

Armed forces authority (AFA) means the organization within Government responsible for co-ordinating the response by themilitary or security forces to a security incident.

Certification means issuing International Ship Security Certificates (ISSCs), Interim ISSCs and Statements of Compliance for port facilities (optional).

Chapter means a chapter of the SOLAS Convention.

Clear grounds means reasons for believing that a ship does not comply with requirements of the Maritime Security Measures.

Company means the owner of the ship or any other organization or person, such as the manager or the bareboat charterer, who has assumed the responsibility for operation of the ship from the owner of the ship and who, on assuming such responsibility, has agreed to take over all the duties and responsibilities imposed by the International Safety Management (ISM) Code.

Company security officer (CSO) means the person designated by the Company for ensuring that a ship security assessment is carried out; that a ship security plan is developed, submitted for approval, and thereafter implemented and maintained; and for liaison with port facility security officers and the ship security officer.

Competent authority means an organization designated by an Administration to receive and act on a ship-to-shore security alert.

Compliance verifications means undertaking intermediate and renewal verifications of compliance for ISSC issuance.

Continuous Synopsis Record (CSR) is a record maintained and updated throughout a ship's life and issued by the ship's Administration under SOLAS chapter XI-I, "Special measures to enhance maritime safety", containing information, including the name of the Administration or Contracting Government who issued the ship's current ISSC or Interim ISSC and the name of the body who carried out the verification of which the Certificate was issued if not the Administration or Contracting Government. The original names of those who issued previous International Ship Security Certificates have to remain in the CSR.

Contracting Government means a Government that has agreed to be bound by the SOLAS Convention. In this Guide the simpler term "Government" is generally used in place of "Contracting Government" unless there is a direct quotation from SOLAS chapter XI-2 or from the ISPS Code part A or part B. Depending on the context, "Government" can also be used in IMO Maritime Security Measures with either the term "Administration" or "Designated Authority", or with both, or in place of either or both.

Control and compliance measures mean actions that can be taken by a duly authorized officer when it is believed that clear grounds exist that a foreign-flagged ship does not comply with the requirements of the Maritime Security Measures; notifying the relevant Government when such measures have been applied to a ship, designating the contact point to receive communication from Governments exercising control and compliance measures, and communicating the contact details to IMO.

Declaration of Security (DoS) means an agreement reached between a ship and either a port facility or another ship with which it interfaces, specifying the security measures each will implement.

Deficiency means a failure to comply with the requirements of the Maritime Security Measures.

Designated Authority means the organization(s) or the Administration(s) identified, within the Contracting Government, as responsible for ensuring the implementation of the provisions of the Maritime Security Measures pertaining to port facility security and ship/port interface, from the point of view of the port facility. In the ILO/IMO Code of practice on security in ports the term is used to describe the organization within Government responsible for port security.

Duly authorized officer means a Government official given specific authorization to undertake official duties, usually associated with inspection and enforcement activities. Such duties under the Maritime Security Measures include undertaking control and compliance measures in respect of foreign-flagged vessels under the Maritime Security Measures, and the use of the term in this Guide is usually associated with that activity.

Emergency response services includes police, military, fire and ambulance services responding to a security incident or to an accident.

Equivalent security arrangements (ESA) means a Designated Authority or Administration allowing a port facility, a group of port facilities or a ship to implement security measures other than those in the Maritime Security Measures but equivalent to those in the Maritime Security Measures.

Government is used in this Guide in place of "Contracting Government". Depending on the context, the term may be used in the Guide with "Administrations" or "Designated Authority", or in their place.

Government official means any Government employee who has security-related responsibilities under the Maritime Security Measures and includes duly authorized officers undertaking control and compliance measures in respect of foreign-flagged vessels using the Maritime Security Measures.

ILO/IMO Code of practice means the ILO/IMO Code of practice on security in ports.

Interim International Ship Security Certificate (Interim ISSC) is a Certificate issued by, or on behalf of, a ship's Administration for a ship without an ISSC:

- on delivery or prior to entry into service,
- following transfer between Contracting Governments to the SOLAS Convention,
- following transfer to a Contracting Government from a non-Contracting Government, or
- following a change of the company operating the ship.

International Safety Management (ISM) Code means the International Management Code for the Safe Operation of Ships and for Pollution Prevention required to be carried by all SOLAS ships under SOLAS chapter IX, "Management for the safe operation of ships".

International Ship and Port Facility Security (ISPS) Code means the International Code for the Security of Ships and of Port Facilities, consisting of part A (the provisions of which shall be treated as mandatory) and part B (the provisions of which shall be treated as recommendatory).

International Ship Security Certificate (ISSC) is a Certificate issued following verification by, or on behalf of, the ship's Administration that the ship complies with the requirements in SOLAS chapter XI-2 and the ISPS Code.

International voyage means a voyage from a country to which the SOLAS Convention applies to a port outside such a country, or conversely (SOLAS chapter I, "General provisions").

Maritime Security Measures means SOLAS chapter XI-2, "Special measures to enhance maritime security", and the ISPS Code, parts A and B.

Member State means a Member State of the International Maritime Organization or International Labour Organization.

Non-SOLAS port facilities means port facilities to which the SOLAS Convention does not apply or which occasionally handle ships to which the Maritime Security Measures apply but which do not have to appoint a PFSO or submit a PFSP.

Non-SOLAS ship is a ship to which the SOLAS Convention does not apply – see *Ship*.

Port means the geographic area defined by the Government or Designated Authority, including port facilities as defined in the ISPS Code, in which maritime and other activities occur.

Port facility means a location, as determined by the Contracting Government or by the Designated Authority, where the ship/port interface takes place. This includes areas such as anchorages, waiting berths and approaches from seaward, as appropriate.

Port facility security assessment (PFSA) means a risk assessment undertaken by, or for, a Designated Authority which is provided to port facility security officers as a prelude to the preparation of a port facility security plan or the review, or amendment, of an approved port facility security plan. A port facility security assessment also has to be undertaken by, or for, the Designated Authority for port facilities occasionally used by SOLAS ships that have not had to appoint a port facility security officer.

Port facility security officer (PFSO) means the person designated as responsible for the development, implementation, revision and maintenance of the port facility security plan and for liaison with the ship security officers and company security officers.

Port facility security plan (PFSP) means a plan developed to ensure the application of measures designed to protect the port facility and ships, persons, cargo, cargo transport units and ship's stores within the port facility from the risks of a security incident.

Port security officer (PSO) means the person tasked to manage and co-ordinate security in the port.

Port security plan (PFP) means a plan developed to ensure the application of measures designed to protect the port and ships, persons, cargo, cargo transport units and ship's stores within the port from the risks of a security incident.

Recognized security organization (RSO) means an organization with appropriate expertise in security matters and with appropriate knowledge of ship and port operations that is authorized to carry out an assessment, or a verification, or an approval or a certification activity required by the Maritime Security Measures.

Regulation means a regulation of the SOLAS Convention.

Security advice and assistance: designating a national contact point to provide security advice or assistance to ships or to receive reports of security concerns from ships, and communicating contact details to IMO.

Security incident means any suspicious act or circumstance threatening the security of a ship, including a mobile offshore drilling unit and a high-speed craft, or of a port facility or of any ship/port interface or any ship-to-ship activity.

Security level means the qualification of the degree of risk that a security incident will be attempted or will occur.

Security level 1 means the level for which minimum appropriate protective security measures shall be maintained at all times.

Security level 2 means the level for which appropriate additional protective security measures shall be maintained for a period of time as a result of heightened risk of a security incident.

Security level 3 means the level for which further specific protective security measures shall be maintained for a limited period of time when a security incident is probable or imminent, although it may not be possible to identify the specific target.

Security plans: approving security plans submitted by port facilities (PFSPs) and shipping companies (SSPs), and any subsequent amendments.

Ship means a passenger ship carrying more than 12 passengers or a cargo ship engaged in an international voyage, and includes high-speed craft and mobile offshore drilling units (MODUs). Generally, the provisions of the SOLAS Convention apply to cargo ships of, or over, 500 gross tonnage (gt). The Maritime Security Measures apply to passenger ships, as above, and to cargo ships over 500 gt. However, certain provisions from chapter V, "Safety of navigation", of the SOLAS Convention also specifically apply to cargo ships of, or over, 300 gt, including mandatory fitting of equipment associated with automatic identification systems (AIS) and long-range identification and tracking (LRIT) systems.

Shipboard personnel means the master and the members of the crew or other persons employed or engaged in any capacity on board a ship in the business of that ship, including high-speed craft, special-purpose ships and mobile offshore drilling units not on location.

Shipping company: see *Company.*

Ship/port interface means the interactions that occur when a ship is directly and immediately affected by actions involving the movement of persons, goods or the provisions of port services to or from the ship.

Ship security alert system (SSAS): provides the means by which a ship can transmit a security alert to a competent authority on shore, indicating that the security of the ship is under threat or has been compromised.

Ship security assessment means a risk assessment undertaken by, or for, a company security officer as a prelude to the preparation of a ship security plan or the review, or amendment, of an approved ship security plan.

Ship security officer (SSO) means the person on board the ship, accountable to the master, who is designated by the Company as responsible for the security of the ship, including implementation and maintenance of the ship security plan, and for liaison with the company security officer and port facility security officers.

Ship security plan (SSP) means a plan developed to ensure the application of measures on board the ship designed to protect persons on board, cargo, cargo transport units, ship's stores or the ship from the risks of a security incident.

Ship-to-ship activity means any activity not related to a port facility that involves the transfer of goods or persons from one ship to another.

Short international voyage is an international voyage in the course of which a ship is not at any time more than 200 miles from a port or a place in which the passengers and crew could be placed in safety. Neither the distance between the last port of call in the country in which the voyage begins and the final port of destination, nor the return voyage, shall exceed 600 miles. The final port of destination is the last port of call in the scheduled voyage at which the ship commences its return voyage to the country in which the voyage began.

SOLAS Convention means the International Convention for the Safety of Life at Sea, 1974, as amended.

Threat is the likelihood that an unlawful act will be committed against a particular target, based on a perpetrator's intent and capability.

Appendix 1.1
Cross-reference of Government and industry responsibilities

Maritime Security Measure (with Government and industry responsibilities)	Reference in Guide to responsibilities for:		
	Government officials	**Port facility operators**	**Ship operators**
Recognized security organizations	2.5	3.2.5–3.2.8	4.2.6–4.2.8
Security levels	2.6	3.3	4.3
Declarations of Security	2.7	3.4	4.4
Designating port facilities	2.8.1– 2.8.9	3.2.1	–
Port facility boundaries	2.8.10–2.8.12	3.2.2–3.2.3	–
Non-SOLAS port facilities	2.8.14–2.8.16	3.10	–
Port security committees	2.8.17–2.8.18	–	4.2.5
Port facility security officers	2.8.19–2.8.24	3.5.1–3.5.6	–
Port facility security assessments	2.8.25–2.8.33	3.6	–
Port facility security plans	2.8.34–2.8.42	3.7	–
Appointment and qualifications of ship security personnel	2.9.1–2.9.11	–	4.5
Ship security assessments	2.9.12–2.9.14	–	4.7
Ship security plans	2.9.15–2.9.30	–	4.8.1–4.8.11
Reporting security incidents	2.9.37	3.8.8–3.8.10	4.8.34–4.8.37
Security records	2.9.38	–	4.8.38–4.8.39
Continuous Synopsis Records	2.9.42	–	4.10.8
International Ship Security Certificates	2.10	–	4.9
Ship security alert systems	2.12.4–2.12.15	–	4.6.1–4.6.11
Automatic identification systems	2.12.16–2.12.19	–	4.6.12– 4.6.15
Pre-arrival notification	2.12.20–2.12.24	–	4.6.13–4.6.15
Long-range identification and tracking systems	2.12.25–2.12.37	–	4.6.16–4.6.18
Alternative security agreements	2.13	3.2.9–3.2.10	4.2.9–4.2.11
Equivalent security arrangements	2.14	3.2.11	4.2.12
Control and compliance measures	2.11	–	4.10
Seafarer access considerations	2.17.5– 2.17.10	3.8.13–3.8.19	4.8.30–4.8.33
Non-SOLAS vessels	2.18.3–2.18.15	–	4.11
Port security	2.18.16–2.18.20	3.9	–

Appendix 1.2
IMO Guidance material on Maritime Security Measures, 1986–2011

Category and identifier	Full title	Adopted
AIS		
Resolution A.956(23)	Amendments to the Guidelines for the onboard operational use of shipborne automatic identification systems (AIS) (resolution A.917(22))	05-Dec-03
Continuous Synopsis Record (CSR)		
Resolution A.959(23)	Format and guidelines for the maintenance of the Continuous Synopsis Record (CSR)	05-Dec-03
Resolution MSC.198(80)	Adoption of amendments to the Format and Guidelines for the maintenance of the Continuous Synopsis Record (CSR) (resolution A.959(23))	20-May-05
Control and compliance measures		
MSC/Circ.1113	Guidance to port State control officers on the non-security-related elements of the 2002 SOLAS amendments	07-Jun-04
MSC.1/Circ.1191	Further reminder of the obligation to notify flag States when exercising control and compliance measures	30-May-06
Facilitating secure trade		
MSC-FAL.1/Circ.1	Securing and facilitating international trade	21-Oct-07
ILO/IMO Code of practice on security in ports		24-Dec-03
LRIT		
Resolution MSC.263(84)	Revised performance standards and functional requirements for the long-range identification and tracking of ships	16-May-08
Resolution MSC.211(81)	Arrangements for the timely establishment of the long-range identification and tracking system	19-May-06
Resolution MSC.242(83)	Use of the long-range identification and tracking information for maritime safety and marine environment protection purposes	12-Oct-07
Resolution MSC.243(83)	Establishment of International LRIT Data Exchange on an interim basis	12-Oct-07
Resolution MSC.254(83)	Adoption of amendments to the Performance standards and functional requirements for the long-range identification and tracking of ships	12-Oct-07
Resolution MSC.298(87)	Establishment of a distribution facility for the provision of LRIT information to security forces operating in waters of the Gulf of Aden and the western Indian Ocean to aid their work in the repression of piracy and armed robbery against ships (the distribution facility)	21-May-10
MSC.1/Circ.1259/Rev.3	Long-range identification and tracking system – Technical documentation (Part I)	21-May-10
MSC.1/Circ.1294/Rev.2	Long-range identification and tracking system – Technical documentation (Part II)	15-Feb-11
MSC.1/Circ.1295	Guidance in relation to certain types of ships which are required to transmit LRIT information, on exemptions and equivalents and on certain operational matters	08-Dec-08
MSC.1/Circ.1298	Guidance on the implementation of the LRIT system	08-Dec-08
MSC.1/Circ.1307	Guidance on the survey and certification of compliance of ships with the requirement to transmit LRIT information	09-June-09
MSC.1/Circ.1308	Guidance to search and rescue services in relation to requesting and receiving LRIT information	09-June-09
MSC.1/Circ.1376	Continuity of service plan for the LRIT system	03-Dec-10
MSC.1/Circ.1377	List of application service providers authorized to conduct conformance tests and issue LRIT conformance test reports on behalf of Administrations	06-Dec-10

Category and identifier	Full title	Adopted
MSC 86/INF.16	LRIT Data Distribution plan – Accessing and entering information – Guidance notes for Contracting Governments and Developmental and integration testing – Guidance notes for LRIT Data Centres	
Ship security alert system (SSAS)		
Resolution MSC.136(76)	Performance standards for a ship security alert system	11-Dec-02
Resolution MSC.147(77)	Adoption of the revised performance standards for a ship security alert system	29-May-03
MSC/Circ.1072	Guidance on provision of ship security alert systems	26-Jun-03
MSC/Circ.1109/Rev.1	False security alerts and distress/security double alerts	14-Dec-04
MSC/Circ.1155	Guidance on the message priority and the testing of ship security alert systems	23-May-05
MSC.1/Circ.1190	Guidance on the provision of information for identifying ships when transmitting ship security alerts	30-May-06
Maritime Rescue Co-ordination Centres (MRCCs)		
MSC/Circ.1073	Directives for Maritime Rescue Co-ordination Centres (MRCCs) on acts of violence against ships	10-Jun-03
Maritime terrorism		
Resolution A.924(22)	Review of measures and procedures to prevent acts of terrorism which threaten the security of passengers and crews and the safety of ships	20-Nov-01
Non-SOLAS ships		
MSC.1/Circ.1283	Non-mandatory guidelines on security aspects of the operation of vessels which do not fall within the scope of SOLAS chapter XI-2 and the ISPS Code	22-Dec-08
Port security		
MSC/Circ.1106	Implementation of SOLAS chapter XI-2 and the ISPS Code to port facilities	29-Mar-04
Recognized security organization		
MSC/Circ.1074	Interim Guidelines for the authorization of RSOs acting on behalf of the Administration and/or Designated Authority of a Contracting Government	10-Jun-03
Training of seafarers and port personnel		
Resolution A.955(23)	Amendments to the Principles of safe manning (resolution A.890(21))	05-Dec-03
MSC/Circ.1154	Guidelines on training and certification for company security officers	23-May-05
Resolution MSC.203(81)	Adoption of amendments to the International Convention on Standards of Training, Certification and Watchkeeping for Seafarers (STCW), 1978, as amended	18-May-06
Resolution MSC.209(81)	Adoption of amendments to the Seafarers' Training, Certification and Watchkeeping (STCW) Code	18-May-06
MSC.1/Circ.1188	Guidelines on training and certification for port facility security officers	22-May-06
STCW.6/Circ.9	Amendments to part B of the Seafarers' Training, Certification and Watchkeeping (STCW) Code	22-May-06
MSC.1/Circ.1235	Guidelines on security-related training and familiarization for shipboard personnel	21-Oct-07
MSC.1/Circ.1341	Guidelines on security-related training and familiarization for port facility personnel	27-May-10
IMO Model Course 3.19	Ship security officer	
IMO Model Course 3.20	Company security officer	
IMO Model Course 3.21	Port facility security officer	

Category and identifier	Full title	Adopted
Security-related information		
MSC.1/Circ.1305	Revised guidance to masters, companies and duly authorized officers on the requirements relating to the submission of security-related information prior to the entry of a ship into port	09-Jun-09
Shore leave		
MSC.1/Circ.1342	Reminder in connection with shore leave and access to ships	27-May-10
Smuggling of drugs		
Resolution A.985(24) Rev.1	Revision of the Guidelines for the prevention and suppression of the smuggling of drugs, psychotropic substances and precursor chemicals on ships engaged in international maritime traffic (resolution A.872(20))	01-Dec-05
Resolution FAL.9(34)	Revised Guidelines for the prevention and suppression of the smuggling of drugs, psychotropic substances and precursor chemicals on ships engaged in international maritime traffic	30-Mar-07
Special purpose ships		
MSC/Circ.1157	Interim scheme for the compliance of certain cargo ships with the special measures to enhance maritime security	23-May-05
MSC.1/Circ.1189	Interim scheme for the compliance of special purpose ships with the special measures to enhance maritime security	30-May-06
SOLAS chapter XI-2 and the ISPS Code		
MSC/Circ.1067	Early implementation of the special measures to enhance maritime security	28-Feb-03
MSC/Circ.1097	Guidance relating to the implementation of SOLAS chapter XI-2 and the ISPS Code	06-Jun-03
Circular letter No.2514	Information required from SOLAS Contracting Governments under the provisions of SOLAS regulation XI-2/13	08-Dec-03
MSC/Circ.1104	Implementation of SOLAS chapter XI-2 and the ISPS Code	15-Jan-04
Circular letter No.2529	Information required from SOLAS Contracting Governments under the provisions of SOLAS regulation XI-2/13.1.1 on communication of a single national contact point	12-Feb-04
MSC/Circ.1110	Matters related to SOLAS regulations XI-2/6 and XI-2/7	07-Jun-04
MSC/Circ.1111	Guidance relating to the implementation of SOLAS chapter XI-2 and the ISPS Code	07-Jun-04
MSC/Circ.1132	Guidance relating to the implementation of SOLAS chapter XI-2 and of the ISPS Code	14-Dec-04
Resolution MSC.194(80)	Adoption of amendments to the International Convention for the Safety of Life at Sea, 1974 as amended (relates to the amendments (related aspects only))	20-May-05
Resolution MSC.196(80)	Adoption of amendments to the International Code for the Security of Ships and of Port Facilities (International Ship and Port Facility Security (ISPS) Code)	20-May-05
MSC/Circ.1156	Guidance on the access of public authorities, emergency response services and pilots on board ships to which SOLAS chapter XI-2 and the ISPS Code apply	23-May-05
Resolution MSC.202(81)	Adoption of amendments to the International Convention for the Safety of Life at Sea, 1974, as amended	19-May-06
MSC.1/Circ.1194	Effective implementation of SOLAS chapter XI-2 and the ISPS Code	30-May-06
Stowaway cases		
Resolution A.871(20)	Guidelines on the allocation of responsibilities to seek the successful resolution of stowaway cases	25-Nov-97
FAL.2/Circ.50/Rev.2	Reports on stowaway incidents	29-Nov-10

Category and identifier	Full title	Adopted
Resolution A.1027(26)	Application and revision of the guidelines on the allocation of responsibilities to seek the successful resolution of stowaway cases (resolution A.871(20))	02-Dec-09
SUA Treaties (International Conference on the revision) – Final Act		19-Oct-05
	Protocol of 2005 to the Convention for the Suppression of Unlawful Acts against the Safety of Maritime Navigation	01-Nov-05
	Protocol of 2005 to the Protocol for the Suppression of Unlawful Acts against the Safety of Fixed Platforms Located on the Continental Shelf	01-Nov-05
Trafficking or transport of migrants by sea		
MSC/Circ.896/Rev.1	Interim Measures for combating unsafe practices associated with the trafficking or transport of migrants by sea	12-Jun-01
Unlawful acts		
Resolution A.584(14)	Measures to prevent unlawful acts which threaten the safety of ships and the security of their passengers and crews	20-Nov-85
MSC/Circ.443	Measures to prevent unlawful acts against passengers and crews on board ships	26-Sep-86
MSC/Circ.754	Passenger ferry security	05-Jul-96
Voluntary self-assessment		
MSC.1/Circ.1192	Guidance on voluntary self-assessment by SOLAS Contracting Governments and by port facilities	30-May-06
MSC.1/Circ.1193	Guidance on voluntary self-assessment by Administrations and for ship security	30-May-06
MSC.1/Circ.1217	Interim Guidance on voluntary self-assessment by companies and CSOs for ship security	14-Dec-06
Other		
Resolution MSC.160(78)	Adoption of IMO Unique Company and Registered Owner Identification Number Scheme	20-May-04
Resolution MSC.159(78)	Interim Guidance on control and compliance measures to enhance maritime security	21-May-04
Circular letter No.2554	Implementation of IMO Unique Company and Registered Owner Identification Number Scheme (resolution MSC.160(78))	24-Jun-04
MSC.1/Circ.1371	List of codes, recommendations, guidelines and other safety- and security-related non-mandatory instruments	30-Jul-10

Appendix 1.3
Websites showing security awareness programmes

1 America's Waterway Watch Program is a nationwide initiative that asks its members to report suspicious activity around maritime locations to local law-enforcement agencies.

The website for additional information is: www.americaswaterwaywatch.us/

2 Project Kraken is a regional initiative in the UK that asks local residents and maritime stakeholders to report suspicious activity around maritime locations to the local police force.

The website for additional information is: www.hampshire.police.uk/Internet/advice/kraken/

3 The Maritime and Port Authority of Singapore has produced a tri-lingual Harbour Craft Security Code poster which can be viewed at:

www.mpa.gov.sg/sites/circulars_and_notices/pdfs/maritime_security_notices/pc04-18.pdf

4 The International Merchant Marine Registry of Belize has developed a set of maritime security guidelines for shipping companies which use the registry as well as a wide range of security practitioners. The 34-page document summarizes the maritime security framework in Belize; outlines the respective responsibilities of the national authority and shipping companies for implementing the Maritime Security Measures; and provides guidance on the Measures to be considered in response to threats to ships and other incidents at sea.

The website can be accessed at: www.immarbe.com/maritimesecurity.html

Section 2 – Security responsibilities of Governments and their national authorities

2.1 Introduction

2.1.1 This section provides guidance on the security responsibilities of Governments under the Maritime Security Measures. The specific topics include:

- **.1** alternative security agreements (ASAs);
- **.2** application of the Measures;
- **.3** certification;
- **.4** compliance verifications;
- **.5** Continuous Synopsis Record (CSR);
- **.6** control and compliance measures;
- **.7** Declaration of Security (DoS);
- **.8** equivalent security arrangements (ESAs);
- **.9** non-SOLAS port facilities;
- **.10** port facility security assessments (PFSAs);
- **.11** recognized security organizations (RSOs);
- **.12** security advice and assistance;
- **.13** security levels;
- **.14** security plans; and
- **.15** ship security alert systems (SSASs).

2.1.2 This section also documents the experience to date of Governments in establishing their framework for implementing and overseeing the implementation of the Maritime Security Measures. Topics include:

- **.1** national legislation;
- **.2** organizations within Government;
- **.3** Government co-ordination mechanisms;
- **.4** port facility and ship inspections;
- **.5** ship security communications;
- **.6** enforcement actions;
- **.7** training of Government officials with security responsibilities;
- **.8** national oversight;
- **.9** non-SOLAS vessels;
- **.10** additional security-related instruments and guidance issued by IMO;

.11 information to IMO; and

.12 wider aspects of port security.

2.1.3 Several aspects of the Maritime Security Measures have responsibilities for both Governments and port facility/ship operators. To assist with understanding how these responsibilities complement each other, the chart below identifies their location within each section.

2.1.4 IMO has encouraged Governments to assess the effectiveness with which their national authorities have fulfilled, and continue to fulfil, their obligations in respect of port facility and ship security. Implementation questionnaires issued as guidance for Designated Authorities and Administrations to examine the status of implementing their security responsibilities under the Maritime Security Measures are shown in appendix 2.1 – Implementation questionnaire for Designated Authorities and in appendix 2.2 – Implementation questionnaire for Administrations, respectively.

2.1.5 IMO also requires Governments to provide information on their national contact points and other aspects of their responsibilities, as specified in subsection 2.19.

Maritime Security Measure	Reference in Guide to responsibilities for:		
	Government officials	Port facility operators	Ship operators
Recognized security organizations	2.5	3.2.5–3.2.8	4.2.6–4.2.8
Security levels	2.6	3.3	4.3
Declarations of Security	2.7	3.4	4.4
Designating port facilities	2.8.1– 2.8.9	3.2.1	–
Port facility boundaries	2.8.10–2.8.12	3.2.2–3.2.3	–
Non-SOLAS port facilities	2.8.14–2.8.16	3.10	–
Port security committees	2.8.17–2.8.18	–	4.2.5
Port facility security officers	2.8.19–2.8.24	3.5.1–3.5.6	–
Port facility security assessments	2.8.25–2.8.33	3.6	–
Port facility security plans	2.8.34–2.8.42	3.7	–
Appointment and qualifications of ship security personnel	2.9.1–2.9.11	–	4.5
Ship security assessments	2.9.12–2.9.14	–	4.7
Ship security plans	2.9.15–2.9.30	–	4.8.1–4.8.11
Reporting security incidents	2.9.37	3.8.8–3.8.10	4.8.34–4.8.37
Security records	2.9.38	–	4.8.38–4.8.39
Continuous Synopsis Records	2.9.42	–	4.10.8
International Ship Security Certificates	2.10	–	4.9
Control and compliance measures	2.11	–	4.10
Automatic identification systems	2.12.16–2.12.19	–	4.6.11– 4.6.12
Pre-arrival notification	2.12.20–2.12.24	–	4.6.13–4.6.15
Long-range identification and tracking systems	2.12.25–2.12.37	–	4.6.16–4.6.18
Alternative security agreements	2.13	3.2.9–3.2.10	4.2.9–4.2.11
Equivalent security arrangements	2.14	3.2.11	4.2.12
Seafarer access considerations	2.17.5– 2.17.10	3.8.13–3.8.19	4.8.30–4.8.33
Non-SOLAS vessels	2.18.3–2.18.15	–	4.11
Port security	2.18.16–2.18.20	3.9	–

2.2 National legislation

Introduction

2.2.1 Essential to the successful implementation and oversight of the Maritime Security Measures is the drafting and enactment of appropriate national legislation. As a minimum, this should provide for the full implementation and oversight of the Maritime Security Measures.

2.2.2 A Government has the discretion to extend the application of the Maritime Security Measures, or requirements drawn from them, to the following elements under its jurisdiction (IMO has encouraged Governments to consider such extensions to ships and port facilities and a number have done so):

.1 non-SOLAS ships;

.2 the port facilities used by non-SOLAS ships; and

.3 offshore activities.

2.2.3 The legislation should also specify the powers needed for Government officials to undertake their duties, including the inspection and testing of security measures and procedures in place at ports and port facilities and on ships, and the application of enforcement actions to correct incidents of non-compliance.

2.2.4 The term "legislation" encompasses all primary and secondary legislation promulgated to implement the Maritime Security Measures. "Primary legislation" refers to acts, laws and decrees while "secondary legislation" refers to regulations, instructions, orders and by-laws issued under powers granted in primary legislation.

Experience to date

2.2.5 Most Governments have enacted legislation to implement the Maritime Security Measures. The precise approach taken has depended on the specific constitutional and legislative arrangements in each country. A number of countries have yet to put in place the legal instruments needed to fully implement the Maritime Security Measures.

2.2.6 In some countries, international legal instruments and amendments such as the Maritime Security Measures automatically apply in national law. However, in most countries the Maritime Security Measures have been implemented through the amendment of existing security, port, or shipping legislation, or through the enactment of new legal instruments.

2.2.7 For port facilities, implementation of the Maritime Security Measures has involved amendments to existing national or local port-related legislation (often in the form of port regulations or port by-laws), which already applied provisions controlling or restricting access to port areas and regulating activities within port areas.

2.2.8 Security requirements may have already been specified for ports under existing legislation relating to national security and the protection of critical national infrastructure. A number of Governments amended such legislation to incorporate the requirements of the Maritime Security Measures; in some cases, incorporation could be achieved without the need for formal amendment.

2.2.9 For ships, incorporation of the requirements in the Maritime Security Measures has been achieved through amendments to existing merchant shipping legislation which has been the means of implementing the other mandatory requirements in the SOLAS Convention.

2.2.10 A number of Governments have enacted specific new legislation to apply the requirements of the Maritime Security Measures to both their port facilities and ships. A limited number of Governments had already enacted legislation which had imposed security requirements on cruise ships using their ports.

Legislating for the Maritime Security Measures

Introduction

2.2.11 The following paragraphs provide guidance on several aspects of national legislation that could be utilized to fully implement the Maritime Security Measures.

Part B of the ISPS Code

2.2.12 While the term "Maritime Security Measures", which is used throughout this Guide, encompasses both parts of the ISPS Code, national legislation has generally focussed on the mandatory requirements in part A. However, certain sections of part A of the ISPS Code include the statement: "... taking into account the guidance given in part B of this Code".

2.2.13 A significant number of Governments have enacted legislation making significant extracts from the guidance originally provided in part B of the ISPS Code mandatory. Some have made all the guidance in part B mandatory.

2.2.14 Governments have taken elements from the guidance in part B of the ISPS Code when defining the responsibilities of:

- **.1** national authorities and their officials, including duly authorized officers responsible for control and compliance measures;
- **.2** port facility operators and their security personnel;
- **.3** shipping companies and their security personnel, including ships' masters; and
- **.4** RSOs undertaking duties for, or on behalf of, national authorities.

2.2.15 The guidance in part B has also been used to define the processes involved when officials, port facility operators, shipping companies and their security officers are undertaking their responsibilities.

Provisions in national legislation

2.2.16 To fully implement the requirements in the Maritime Security Measures, the legislation could cover:

- **.1** definitions;
- **.2** application;
- **.3** Designated Authority and Administration;
- **.4** security levels;
- **.5** port facilities;
- **.6** port facility security assessments;
- **.7** ship;
- **.8** port facility security plans and ship security plans;
- **.9** retention of records and Declarations of Security;
- **.10** inspection of port facilities and ships;
- **.11** enforcement action;
- **.12** control and compliance measures; and
- **.13** offences relating to the Maritime Security Measures.

2.2.17 The powers required for Designated Authorities and Administrations to undertake their specific responsibilities are discussed in the following paragraphs.

Definitions in legislation

2.2.18 The definitions used in national legislation should, as far as appropriate, be similar to those used in the Maritime Security Measures. However, there are certain terms used in the Maritime Security Measures that are not defined within them, including:

.1 Administration;

.2 shipping company;

.3 competent authority (used in connection with ship security alert systems);

.4 international voyage;

.5 master; and

.6 restricted area.

2.2.19 It may be necessary to provide definitions for such terms in national legislation. Some are defined elsewhere in the SOLAS Convention. To the extent possible, any definition should reflect the context in which the term is found in the Maritime Security Measures. As an example, the term "restricted area" could be defined as: "Restricted area means an area in a port facility or a ship that is identified as such in a port facility security plan or a ship security plan."

Application of legislation

2.2.20 The Maritime Security Measures apply to port facilities within a State's jurisdiction, to its SOLAS ships and to its territorial sea. The Maritime Security Measures also apply to a State's overseas territories.

2.2.21 The national legislation implementing the Maritime Security Measures should define their territorial application, including the State's territorial sea and, when appropriate, their extension to any overseas territory which does not have its own legislative authority.

Legislation: Designated Authority and Administration

2.2.22 The legislation could specify which organization within Government is to regulate port facility security (i.e., the Designated Authority), and which organization is to regulate ship security (i.e., the Administration). Responsibility for port facility and ship security can be combined in a single organization (refer to paragraphs 2.3.1 and 2.3.2).

2.2.23 The legislation could also specify whether the organizations and their officials have delegated power to act on their own behalf, in the organization's name, or whether they act under the authority of the relevant Minister.

2.2.24 The term "Designated Authority" is new in the Maritime Security Measures and could be defined in national legislation. As most Governments have enacted legislation to implement earlier provisions in the SOLAS Convention and other IMO legal instruments, the term "Administration" may already have been defined in merchant shipping legislation.

Legislation: Security levels

2.2.25 Setting the security level is a Government responsibility. There are few examples of national legislation that identifies the organization within Government responsible for setting it, unless it is the Designated Authority or Administration. However, national legislation could specify who is responsible for communicating changes in security level and for receiving and responding to such changes.

2.2.26 National legislation could give the Designated Authority and Administration the power to establish the time allowed to implement a change in security level. It could also specify the action to be taken when those responsible for:

.1 communicating changes in security level fail to do so;

.2 initiating the response to such a change fail to do so within the specified time.

Legislation: Port facilities

2.2.27 Designated Authorities need the authority to designate a port facility as:

- **.1** one required to appoint a PFSO and prepare a PFSP; and/or
- **.2** one used occasionally by SOLAS ships where the Designated Authority appoints an organization or person ashore to be responsible for shore-side security.

2.2.28 In the second case identified above, the Designated Authority has to undertake a PFSA.

2.2.29 National legislation could establish the requirements relating to:

- **.1** notification to the owner or operator of a designated port facility that there is a requirement to appoint a PFSO and prepare a PFSP;
- **.2** notification of the appointment of an organization or person ashore responsible for communicating with SOLAS ships at port facilities occasionally used by such ships and the responsibilities of that organization or person; and
- **.3** the employment status of the appointed PFSO, who should either be an employee of the port facility operator or owner, or engaged on a contract or other basis by the port facility owner or operator.

2.2.30 National legislation could establish that the operator or owner of such a port facility is responsible for the actions of their PFSO and for the security of their facility.

Legislation: Port facility security assessment

2.2.31 Port facility security assessments are undertaken by Designated Authority officials or by recognized security organizations on their behalf. As part of the process of completing an assessment, national legislation could authorize those undertaking such assessments to:

- **.1** enter land or premises;
- **.2** inspect documents, records and plans; and
- **.3** inspect security equipment.

2.2.32 National legislation could include criteria as to how often and under what circumstances PFSAs should be reviewed and refreshed.

Legislation: Ships

2.2.33 The Maritime Security Measures require shipping companies operating SOLAS ships to appoint:

- **.1** at least one CSO with the responsibility to undertake an SSA and prepare an SSP for each SOLAS ship; and
- **.2** an SSO, accountable to the master, responsible for implementing the SSP.

2.2.34 National legislation could establish that the shipping company is responsible for the actions of their Company and SSOs and for the security of their ships.

Legislation: Port facility plans and ship security plans

2.2.35 The legislation could set out the requirements and the procedures applying to:

- **.1** the submission of PFSPs and SSPs;
- **.2** the approval of PFSPs and SSPs, with or without modification;
- **.3** the requirements to review an approved PFSP or SSP; and
- **.4** the submission of amendments to an approved PFSP or SSP.

Legislation: Retention of records and Declarations of Security

2.2.36 National legislation could specify the minimum time that security records and Declarations of Security have to be retained at the port facility or on a ship.

Legislation: Inspection of port facilities and ships

2.2.37 The legislation could give officials in Designated Authorities and Administrations, or those authorized to undertake inspection duties on their behalf, authority to enter port facilities or to board ships to assess their compliance with the requirements of the Maritime Security Measures.

2.2.38 These powers could include the authority to:

- **.1** inspect a port facility or ship to assess compliance;
- **.2** inspect security equipment;
- **.3** initiate a port facility or ship security drill;
- **.4** enter any premises associated with a port facility or shipping company;
- **.5** request and inspect documents, records and plans;
- **.6** interview individuals regarding the security of a port facility or ship; and
- **.7** obtain and retain evidence relating to a security deficiency found at a port facility or on a ship.

2.2.39 Inspections could relate to:

- **.1** the issue or verification of a port facility's Statement of Compliance;
- **.2** the issue or verification of a ship's International Ship Security Certificate (ISSC) or Interim ISSC; and
- **.3** inspection, review or audit to assess the compliance of a port facility or ship with the requirements of the Maritime Security Measures.

Legislation: Enforcement action

2.2.40 The legislation could specify the actions that a Designated Authority and Administration can take if a security deficiency is found at a port facility or on a SOLAS ship.

2.2.41 If a serious deficiency is found which compromises the ability of a port facility or ship to operate at security levels 1 to 3, the legislation should give officials the power to issue restriction or suspension notices applying to specific activities at the port facility or ships until the deficiency is corrected or until appropriate alternative security measures and procedures are in place.

2.2.42 If a security deficiency does not compromise the ability of a port facility or ship to operate at security levels 1 to 3 and the port facility or ship fails to take action to correct the deficiency, the legislation should give officials the power to issue an enforcement notice requiring the port facility or ship to correct the deficiency within a stated period.

2.2.43 Legislation could also establish the procedures covering withdrawal of an approved PFSP or SSP and the procedure to allow their reinstatement.

2.2.44 In their legislation, many Governments provide procedures allowing operators of port facilities and ships to appeal the service of an enforcement notice and for such appeals to be considered. Similar rights of appeal could be considered in respect of restriction and suspension notices and the withdrawal of approved PFSPs or SSPs.

2.2.45 The legislation could establish administrative, civil or criminal penalties when a port facility or ship fails to comply, for example, with an enforcement, restriction or suspension notice and procedures relating to the application of such penalties, including the right of appeal against the imposition of a penalty.

Legislation: Control and compliance measures

2.2.46 The Maritime Security Measures allow control measures to be taken when a foreign-flagged SOLAS ship is in port, or has indicated its intention to enter the port. These control measures can involve:

.1 inspection of the ship;

.2 delaying the ship;

.3 detention of the ship;

.4 restrictions on operation;

.5 expulsion from port;

.6 refusal of entry into port; and

.7 other lesser administrative or corrective measures.

2.2.47 The legislation could establish procedures relating to the imposition of such control measures.

2.2.48 The legislation could specify that control measures allowing expulsion from a port or refusal of entry into a port should only be applied when the ship is considered to pose an immediate security threat.

2.2.49 The Maritime Security Measures provide that compensation can be claimed if a ship is unduly detained or delayed. The legislation could establish procedures for submitting and considering claims for compensation in these circumstances.

Legislation: Offences relating to the Maritime Security Measures

2.2.50 The Maritime Security Measures do not themselves establish any offences. The criminal or terrorist acts that they seek to detect and deter are typically already offences under a State's criminal law or criminal code.

2.2.51 When implementing the Maritime Security Measures, a number of Governments have established offences in their legislation relating to:

.1 failure to comply with an enforcement notice;

.2 intentional obstruction or impersonation of a Government official, or other person acting on behalf of a Designated Authority or Administration;

.3 failure to provide information requested by a Government official, or other person acting on behalf of a Designated Authority or Administration;

.4 providing information known to be false to a Government official, or other person acting on behalf of a Designated Authority or Administration; and

.5 unauthorized presence in a restricted area of a port facility or ship.

Extending the application of the Maritime Security Measures

2.2.52 The Maritime Security Measures apply to port facilities serving SOLAS ships and to SOLAS ships. Governments have been recommended to consider extending their application, in appropriate circumstances, to port facilities and ships that are not covered by them.

2.2.53 A number of Governments have applied requirements drawn from the Maritime Security Measures to:

.1 passenger and cargo ships solely involved in domestic voyages, including vessels involved in domestic voyages involving significant distances to overseas territories;

.2 harbour craft and other craft that interact in ship-to-ship activities with ships covered by the Maritime Security Measures;

.3 offshore supply and support vessels;

.4 fishing vessels and recreational crafts; and

.5 facilities used by the above.

2.2.54 The Maritime Security Measures do not apply to port facilities that are used primarily for military purposes. A number of Designated Authorities have applied requirements from the Maritime Security Measures to such port facilities if regular commercial services operate from them.

2.3 Organizations within Government

Organizational structures

2.3.1 The Maritime Security Measures differentiate between the roles of the Designated Authority (as the organization within Government responsible for port facility security) and the Administration (with responsibility for ship security). It is a matter for each individual Government where the specific responsibilities of the Designated Authority and Administration are located within the Government's administrative structures.

2.3.2 Most commonly, the responsibilities of the Designated Authority and the Administration are undertaken within the Departments or Ministry responsible for port and shipping matters, often a Transport Department or Ministry, or within an independent Governmental organization reporting to a Transport Minister. A number of Governments maintain a distinction between their Designated Authority with responsibility for port facility security and their Administration responsible for ship security. Others have combined the security responsibilities of the Designated Authority and Administration in a single organization. Occasionally the responsibilities for port facility and ship security are combined with responsibility for the security of other transport modes, including aviation.

Delegation of responsibility

2.3.3 There are a limited number of circumstances under the Maritime Security Measures when a Designated Authority can appoint a recognized security organization (RSO) to undertake duties on its behalf on port facility security. Refer to subsection 2.5 for a list of responsibilities and conditions of delegation.

2.3.4 Delegation of ship security responsibilities is a more common practice either by Governments or their Administrations. Some Governments have chosen to delegate responsibilities to off-shore international registries, albeit with oversight provided by their Transport Department or Ministry. In other instances, Administrations have delegated many of their ship security responsibilities to RSOs. Paragraphs 2.5.6 and 2.5.7 provide a full list of responsibilities and conditions of delegation.

2.4 Government co-ordination mechanisms

Introduction

2.4.1 Enhanced port facility, port and ship security forms part of Governments' efforts to counter terrorism and combat threats and can involve many organizations in addition to the national authorities responsible for applying the Maritime Security Measures. The main ones are listed below.

2.4.2 National customs and immigration authorities undertake their own control duties at ports and on ships and have detailed knowledge of the criminal activities that they seek to detect and deter. Many of these authorities have adopted practices and procedures drawn from the WCO's Framework of Standards to Secure and Facilitate Global Trade (the SAFE Framework) applying to ports, port facilities and ships as part of the global cargo supply chain.

2.4.3 National authorities set appropriate security levels, particularly those relating to terrorist threats, based on essential input from intelligence services and security forces authorities. Police, coast guard and military services form a major part of a Government's response to a serious security incident and generally have their own intelligence on criminality and threats in their areas of jurisdiction. Law-enforcement authorities are involved in the prosecution of offenders.

2.4.4 Decisions made by national authorities responsible for ship and port security should be based on close co-ordination across Government and between Government organizations. This can be assisted by the establishment of an appropriate national maritime security committee structure and the development of a national maritime security framework or strategy. Development of such a framework or strategy can avoid the possible duplication of security procedures and measures required by different Government organizations at ports and on board ships.

National maritime security framework/strategy

2.4.5 A number of Governments have developed national maritime security frameworks or strategies and policy statements. When appropriate, such frameworks or strategies could be established through national legislation.

2.4.6 National maritime security frameworks or strategies provide an effective way of establishing the national context within which to understand security concerns and requirements; and provide direction and guidance on undertaking security assessments and plans. A national maritime security framework/strategy could meet the recommendation in the ILO/IMO Code of practice on security in ports that Governments should develop a ports security policy document.

2.4.7 A national maritime security framework or strategy could cover, in appropriate detail, the following:

- **.1** extent and significance of the country's maritime industries and infrastructure;
- **.2** perception of current maritime threats;
- **.3** roles and responsibilities of Government organizations;
- **.4** national security policies applying to ports and ships;
- **.5** security responsibilities of the port and shipping industries;
- **.6** co-ordination of Government and industry responses;
- **.7** short- and longer-term security priorities; and
- **.8** development of a security culture across the maritime industries.

National maritime security committees

2.4.8 The fostering and maintenance of effective linkages between Government and industry can significantly assist the effective application of the Maritime Security Measures.

2.4.9 The work of a national maritime security committee and the development, relevance and acceptability of a national maritime security framework or strategy is enhanced if appropriate arrangements are in place to consult, and involve, representatives of those regulated: the port and shipping industries, those working in ports or on ships, cargo and passenger interests.

2.4.10 Effective co-ordination at the national and port levels allows those responsible for port and ship security to gain an appreciation of the security issues and threats that they should consider in their security assessments and seek to detect and deter through the procedures and measures in their security plans. A balanced appreciation of the security risks and threats actually faced allows the development of effective, proportionate and sustainable security procedures and measures. The imposition of excessive or inappropriate security procedures and measures can reduce their acceptability and effectiveness and impose unnecessary delays, or restrictions, on passenger or cargo movements.

2.4.11 Many Governments established national committees or working groups to co-ordinate the initial implementation of the Maritime Security Measures. While some were subsequently disbanded, several Governments formalized the arrangements and established permanent national maritime security committees, or equivalents, covering the port and shipping sectors.

2.4.12 A national maritime security committee can undertake two essentially interlinked activities:

.1 assisting the co-ordination of port and ship security requirements across Government; and

.2 facilitating full consultation on security issues with those regulated – the port and shipping industries, those employed at ports and on ships and those using ports and ships.

2.4.13 There are as many possible committee structures as there are established committees. For co-ordination within Governments, the core membership could consist of senior level representation from:

.1 department(s)/Ministry(ies) responsible for ports and shipping;

.2 their national authorities;

.3 intelligence/security services;

.4 national customs and immigration authorities;

.5 national police/law-enforcement agencies;

.6 military (naval) forces;

.7 coastguards; and

.8 foreign service.

2.4.14 For consultation with all major stakeholders in the port and shipping industries, membership could be added at senior levels from the national representatives of the port and shipping industries, port workers and seafarers, and cargo and passenger interests.

2.4.15 Many of these committees have found it useful to establish specialist sub-committees or working groups to focus attention on particular security issues or initiatives. Often, stakeholder representation is to be found at the sub-committee level due to the more specific nature of the topics being addressed.

2.4.16 The terms of reference of a national maritime security committee could include:

.1 identifying security threats and vulnerabilities;

.2 establishing security priorities;

.3 planning, co-ordinating and evaluating security initiatives;

.4 developing or contributing to a national maritime security framework or strategy;

.5 developing or contributing to Government policy statements on maritime security;

.6 developing co-ordinated positions on meeting international obligations;

.7 addressing jurisdictional issues involving member organizations; and

.8 handling major security issues, with multi-organization implications, referred to the committee by high-level committees.

2.4.17 Few national maritime security committees have any executive authority (that rests with their member Government organizations), but their efforts have reshaped security strategies, enhancing their acceptability and effectiveness when implemented.

Participation in international and regional organizations

2.4.18 In addition to IMO, an international organization, there are several regional organizations that have committees or sub-committees with a mandate to address issues related to implementing the Maritime Security Measures within the broader concept of maritime security, including:

.1 the Asia-Pacific Economic Cooperation (APEC), which has representation from 21 Contracting Governments in the APEC Region. Its maritime security programme is administered by the Transportation Working Group, whose website may be accessed at:

www.apec-tptwg.org.cn/

.2 the Organization of American States (OAS), which has representation from 34 Contracting Governments throughout the Americas and the Caribbean. Its maritime security programme is administered by two committees – the Inter-American Committee for Counter-Terrorism and the Inter-American Committee for Ports. Their websites may be accessed respectively at:

www.cicte.oas.org/Rev/En/Programs/Port.asp and www.safeports.org/

.3 the European Maritime Safety Agency (EMSA), which has representation from 27 Contracting Governments. Its maritime security programme may be accessed at:

www.emsa.europa.eu

.4 the Secretariat of the Pacific Community (SPC), which has 26 members including 22 Pacific Island countries and territories. Its maritime security programme may be accessed at:

www.spc.int/maritime

2.5 Recognized security organizations

Introduction

2.5.1 Governments can authorize recognized security organizations (RSOs) to undertake certain of their responsibilities under the Maritime Security Measures. Delegation of responsibilities relating to port facility security to RSOs is usually through the Designated Authority, while delegation of responsibilities relating to ship security is usually through the Administration.

2.5.2 The scope for delegating port facility security responsibilities to RSOs is restricted under the Maritime Security Measures, and a limited number of Designated Authorities have authorized RSOs to undertake port facility security assessments on their behalf. A port authority or port facility operator may be appointed as an RSO undertaking duties relating to port facilities if it can demonstrate the appropriate competencies.

2.5.3 The scope for authorizing RSOs to undertake Government responsibilities for ship security is more extensive under the Maritime Security Measures than for port facility security. Many Administrations have authorized RSOs, particularly those which also have received RSO status, to conduct inspections, surveys, verifications, and approvals and to issue Certificates for ships flying their flag under provisions elsewhere in the SOLAS and other IMO Conventions.

2.5.4 While a number of Governments have authorized RSOs to undertake certain Government responsibilities for the security of ships and port facilities, some Governments have preferred not to authorize any RSO to undertake any of their port facility and ship security responsibilities.

2.5.5 Under the Maritime Security Measures, Governments are required to provide IMO with the name and contact details of any RSO authorized to act on their behalf as well as details of their specific responsibilities and conditions of authority delegated to such organizations. This information can be provided using IMO's web-based database: http://gisis.imo.org, and should be kept updated.

Permitted delegations

2.5.6 Governments may authorize an RSO to undertake the following duties:

- **.1** approval of SSPs and their subsequent amendments (provided that the RSO was not involved with their development or implementation);
- **.2** verification and certification of compliance of ships with the Maritime Security Measures;
- **.3** conduct of PFSAs;
- **.4** provision of advice and assistance on security matters, including the completion of PFSAs, PFSPs, SSAs and SSPs.

2.5.7 The Maritime Security Measures specify that Governments may not delegate any of the following duties to RSOs:

- **.1** setting the security level;
- **.2** establishing the requirements for a Declaration of Security;
- **.3** determining which port facilities have to appoint a PFSO and prepare a PFSP;
- **.4** approving a PFSA or subsequent amendments;
- **.5** approving a PFSP or subsequent amendments;
- **.6** exercising control and compliance measures in respect of foreign-flagged SOLAS ships;
- **.7** approval of SSPs and subsequent amendments if they assisted in their preparation; and
- **.8** issuing Certificates to shipboard personnel under the STCW Convention and STCW Code.

Authorization

2.5.8 Governments should satisfy themselves that RSOs have demonstrated the organizational effectiveness and technical capabilities necessary to undertake the specific duties that may be delegated to them. These competencies are identified in appendix 2.3 – Criteria for selecting recognized security organizations – in the form of selection criteria.

2.5.9 In keeping with sound business practices, there should be a formal written agreement signed by both parties. As a minimum, it should:

- **.1** specify the scope and duration of the delegation;
- **.2** identify the main points of contact within the national authority and the RSO;
- **.3** detail the procedures for communications between the national authority and the RSO;
- **.4** detail the oversight procedures to be used by the national authority to verify that the RSO is carrying out its delegated activities in a satisfactory manner;
- **.5** detail the procedures for assessing reports received from the RSO;
- **.6** detail the procedures to be followed by the RSO if a ship is found not to be in compliance with the regulatory requirements for which that RSO has been delegated authority;
- **.7** detail the procedures to be followed by the Administration and the RSO if another Government imposes control measures on a ship for which that RSO has been delegated authority for issuing the ISSC;
- **.8** detail the data to be provided to the national authority to assist with the authority's approval of SSPs, PFSAs and PFSPs;
- **.9** identify the legislation, policies, procedures and other work instruments to be provided to the RSO;

.10 specify the records to be maintained by the RSO and made available as necessary to the national authority;

.11 specify any reports to be provided on a regular basis, including changes in capability (e.g., loss of key personnel); and

.12 specify a process for resolving performance-related issues.

Oversight

2.5.10 In addition to the oversight procedures identified above, Governments should ensure the adequacy and consistency of the work performed by RSOs on their behalf by establishing an oversight system that includes:

.1 undertaking inspections and audits of port facilities and ships where RSOs have undertaken delegated activities; and

.2 establishing requirements for certifying the RSO's quality system by independent auditors acceptable to the national authority.

2.5.11 Governments retain ultimate responsibility for the work undertaken on their behalf by the RSOs that they appoint. They have the authority to modify or revoke their delegations to an RSO which fails to meet agreed performance standards.

Experience to date

2.5.12 Several Designated Authorities have authorized RSOs to:

.1 undertake PFSAs on their behalf;

.2 assist port facilities with preparing their PFSPs;

.3 be training providers for PFSOs and other port facility security personnel;

.4 approve training courses on port facility security by training providers or institutions other than the RSO itself.

2.5.13 Many Administrations have approved RSOs as training providers for CSOs and SSOs.

2.5.14 A number of Governments have adopted legislation requiring a review at least every five years of the performance and authorization of any RSOs to which port facility and ship security responsibilities have been delegated under the Maritime Security Measures.

2.6 Security levels

Introduction

2.6.1 The Maritime Security Measures require Contracting Governments to gather and assess information with respect to security threats which could occur at a port facility or on, or against, a SOLAS ship. This process is essential to allow their national authorities to set the appropriate security level applying to their port facilities and to ships flying their flag.

2.6.2 The term "security level" refers to the degree of risk that a security incident will occur or be attempted. The Maritime Security Measures identify three levels of risk which are now used internationally:

.1 *security level 1* means the level for which minimum appropriate protective security measures shall be implemented at all times.

.2 *security level 2* means the level for which appropriate additional protective security measures shall be maintained for a period of time as a result of the heightened risk of a security incident.

.3 *security level 3* means the level for which further specific protective security measures shall be maintained for a limited period of time when a security incident is probable or imminent, although it may not be possible to identify a specific target.

2.6.3 At security level 1, the security measures and procedures in port security plans (PSPs), PFSPs or SSPs should be sufficient to counter most forms of criminality associated with ports and ships, in particular trespass, pilferage and stowaways. The priority is to allow normal commercial operations.

2.6.4 At security level 2, the priority is also to allow the continued commercial operation of the port, port facility or ship but with increased security restrictions.

2.6.5 At security level 3, the strictest security restrictions will be in place and could lead to the eventual suspension of commercial activities, with control of the security response transferred to the Government organizations responding to a significant incident.

2.6.6 Some national authorities have established maximum time periods in which their ports, port facilities or ships have to put in place the additional or further security procedures and measures following a change of security level. Designated Authorities and Administrations should specify the time allowed to change to operate at a higher security level. The time period is variable, depending on the reason for the change, but is usually between 3 and 24 hours.

Setting the security level

2.6.7 Many Governments use and communicate national security levels to alert the public to the perceived risk of a terrorist attack. In such cases, they may consider that the security levels developed in the Maritime Security Measures apply only to the risk of a terrorist attack. This need not be the case. Governments may set higher security levels to advise of the risk of other threats, particularly attacks by pirates or armed robbers against ships.

2.6.8 When a Government sets a higher security level on grounds other than the risk of a terrorist attack, a brief statement describing the kind of threat that has led to the change could be included when it is being communicated or transmitted. Some Governments provide stakeholders with examples of the type of risks that could lead to security levels being raised to level 2 or 3.

2.6.9 In setting the security level applying to port facilities and ships, Governments should take account of general and specific threat information. The Maritime Security Measures consider that the factors to be considered when setting the appropriate security level are:

.1 the degree to which the threat information is credible;

.2 the degree to which the threat information is corroborated;

.3 the degree to which the threat information is specific or imminent; and

.4 the potential consequences of the threatened security incident.

2.6.10 Information on terrorist threats is likely to be held by intelligence or security services, and Governments set the appropriate security level on advice provided by such sources. In other cases the security level is set by the Designated Authority for ports and port facilities or by the Administration for ships, based on threat information received from intelligence or security services.

2.6.11 Governments or their national authorities should only set security level 3 for ports, port facilities or ships in exceptional circumstances when there is credible information that a security incident is probable or imminent. Security level 3 should only be set for the duration of the identified threat or the duration of an actual incident.

2.6.12 Governments can apply the same security level to all their ports and port facilities. They may also apply different security levels to groups of ports and port facilities or to parts of a port or a particular port facility. Similarly, Governments can apply the same security level to all its ships or apply different security levels to individual ships, types of ship, ships operating in specific sea areas or ships using specific foreign ports or port facilities.

2.6.13 Governments can apply the same security level over their territorial sea or apply different security levels to different parts of their territorial sea.

Communicating the security level

2.6.14 National authorities should establish robust communication procedures to ensure that updated information on changes (both increases and decreases) in security levels is provided without delay to their port facilities and ships and to foreign-flagged ships in or intending to enter their port facilities or intending to transit their territorial sea. The procedures should also ensure that the information reaches their own officials, particularly those that may be located in port areas.

2.6.15 If the applicable security level is set by a national authority other than the Designated Authority or Administration, the Designated Authority or Administration still retains responsibility for the effective notification of changes in security levels to port facilities and ships.

2.6.16 Practices for achieving robust communication procedures are:

.1 charting the communication process;

.2 creating and maintaining an accurate contact list for communications by means of fax, e-mail or text message; and

.3 regular testing.

2.6.17 Communication procedures can vary, but the general practice for port facilities is that the Designated Authority provides the information to:

.1 the port security officer, or equivalent officer in the port/harbour authority, in the relevant ports, who forwards the information to the PFSOs and masters/SSOs of ships in port or intending to enter port;

.2 individual PFSOs, who then pass the information to ships at or intending to use their port facility.

2.6.18 In the case of ships, some Administrations communicate changes in security level directly to their own flag ships. Use of NAVTEX and Inmarsat-C SafetyNET allows Administrations to issue security-related messages directly to ships which are received through the ship's Global Maritime Distress Safety System (GMDSS). Administrations may also use terrestrial or satellite-based fax transmission to communicate with ships.

2.6.19 Other Administrations provide the information to their CSOs, who are then responsible for its onward transmission to ships. Occasionally, the information can be issued to CSOs and ships through the national register of ships or through RSOs. Changes in security level have also been communicated through the issue of a Notice to Mariners.

2.6.20 Under the Maritime Security Measures, Governments should establish means of communicating information on security levels to foreign-flagged ships operating in their territorial sea or that have communicated their intention to enter their territorial sea. This can be done through NAVTEX, Inmarsat-C SafetyNET and sureFax. Changes to security levels applying to all, or part, of the territorial sea can also be transmitted by Maritime Rescue Co-ordination Centres (MRCCs).

2.6.21 When a security risk has been identified that results in the application of a higher security level to all or part of the territorial sea, ships that might be affected have to be able to communicate with a contact point ashore. This contact point should be available at all times to receive reports of security concerns from ships and to offer guidance to ships. The contact point may also receive reports from port facilities of their security concerns.

2.6.22 When a higher security level applies, the contact point should be in a position to:

.1 advise ships operating in, or intending to enter, the territorial sea of the security procedures and measures that the coastal State considers appropriate to protect the ship from attack; and

.2 inform the ship of the security measures that the coastal State has put in place to counter the identified security risk.

2.6.23 Depending on the circumstances, the contact point could offer advice to a ship, including:

.1 altering or delaying its intended passage;

.2 navigating on a specified course or proceeding to a specific location;

.3 the availability of security personnel or equipment that could be provided to the ship;

.4 co-ordinating the passage, port arrival or departure of the ship to allow escort by patrol craft or aircraft; and

.5 any restricted areas established by the coastal State in response to a security threat or incident.

2.6.24 Foreign-flagged ships receiving such advice have discretion as to the action they take, having regard to the provisions in their SSP and any guidance or instructions that they may receive from their Administration.

2.6.25 Some Administrations have specified that ships flying their flag should apply the same security level as the coastal State when transiting the State's territorial sea or operating within its Exclusive Economic Zone.

2.6.26 Contracting Governments have discretion on the extent to which they choose to exchange information on security threats with other Governments.

2.7 Declarations of Security

Introduction

2.7.1 A Declaration of Security (DoS) is an agreement between a port or port facility and a ship or between a ship and another ship. It confirms the security responsibilities of each party during a ship/port interface (refer to subsection 3.4) or a ship-to-ship activity (refer to subsection 4.4). As such, a DoS should detail what measures can be shared or additionally provided and by which party.

Establishing the requirement for a DoS

2.7.2 The Maritime Security Measures require Governments to determine when a DoS is required by assessing the risk that the ship/port interface or ship-to-ship activity poses to persons, property or the environment. These circumstances are usually specified by the Designated Authority or Administration for inclusion in PSPs, PFSPs and SSPs. They cannot be specified by RSOs.

2.7.3 The circumstances warranting a DoS can include the following scenarios:

.1 a ship is operating at a higher security level than the port facility with which it is interfacing;

.2 there has been a security threat or a security incident involving a port facility or a ship with which it is interfacing;

.3 a port facility or ship is operating at security level 3;

.4 there has been a change to the security level applying to a port facility or a ship with which it is interfacing;

.5 a specific ship/port interface could endanger local facilities or residents;

.6 a specific ship/port interface could pose a significant pollution risk;

.7 a ship/port interface involves embarking or disembarking passengers or handling of dangerous cargo;

.8 a ship is using a non-SOLAS port facility;

.9 a ship is undertaking a ship-to-ship activity while operating at a higher security level than the other ship;

.10 a ship is undertaking a ship-to-ship activity with a non-SOLAS ship;

.11 a ship-to-ship activity involves the transfer of passengers or dangerous cargo at sea;

.12 a ship-to-ship activity could involve the risk of significant marine pollution;

.13 there is a Government-to-Government agreement requiring a DoS covering specified international voyages and the ships engaged on such voyages or ship-to-ship activities during such voyages;

.14 a non-SOLAS ship proposes to use a SOLAS port facility.

.15 the need to do so is indicated by a port facility's Designated Authority or ship's Administration;

.16 a ship is not compliant with the Maritime Security Measures (e.g., without a valid ISSC).

2.7.4 The requirements to request a DoS, and those relating to the response to such requests, should be based on security considerations. Declarations of Security should never be the norm and should not normally be required when both the port facility and the ship are operating at security level 1.

2.7.5 Developing a matrix similar to the one shown below may be a useful way of ensuring consistency in determining when a DoS should be initiated by a port facility.

Situation	Port Facility at Security Level 1	Port Facility at Security Level 2	Port Facility at Security Level 3
Non-SOLAS ship entering port facility	Required Not required	Required Not required	Required Not required
Non-ISPS Code-compliant ship entering port facility	Required Not required	Required Not required	Required Not required
Ship at security level 1	Required Not required	Required Not required	Required Not required
Ship at security level 2	Required Not required	Required Not required	Required Not required
Ship at security level 3	Required Not required	Required Not required	Required Not required
Following a security incident at port facility or on ship	Required Not required	Required Not required	Required Not required
Following a threat to port facility or ship	Required Not required	Required Not required	Required Not required

2.7.6 The precise circumstances when a DoS is required by a port facility from a ship can be established through the PFSA. Similarly, the precise circumstances when a DoS is to be requested by a ship from a port facility or another ship can be established through the ship's security assessment.

2.7.7 Experience to date indicates that, in addition to identifying the circumstances when a DoS is to be requested, some national authorities have:

.1 specified the validity and retention periods;

.2 modified the model form issued by IMO (refer to appendix 3.1 – Declaration of Security Form); and

.3 permitted the use of a single DoS for multiple visits by a ship to the same facility.

Government-to-Government agreement

2.7.8 A DoS under a Government-to-Government agreement usually applies to specific voyages between two countries and to specific passenger and cargo movements between the countries when both Governments consider that the activity poses additional security risks but wish to avoid imposing a higher security level. It is distinct from an alternative security agreement, which applies to short, high-frequency international voyages (refer to subsection 2.13).

Continuous Declarations of Security

2.7.9 Continuous Declarations of Security mean that a DoS is not required for each ship/port interface or ship-to-ship activity involving the same ship with the same port facility or between the same ships. In such instances, the DoS would remain in force either for a specified time or until circumstances change.

2.7.10 The circumstances under which a continuous DoS can be applied, its duration and when the DoS becomes invalid need to be carefully defined by the relevant national authorities following security assessments of the interfaces or activities involved.

Exclusive Economic Zone and Continental Shelf

2.7.11 The Maritime Security Measures do not apply to the off-shore activities beyond a country's territorial sea but within its Exclusive Economic Zone or Continental Shelf. It is likely that SOLAS ships will operate in these waters and interface with off-shore installations and undertake ship-to-ship activities with a non-SOLAS ship. Governments have been encouraged to develop security regimes for these areas.

2.7.12 These security regimes should facilitate agreement of a DoS or equivalent agreement between a SOLAS ship and any offshore installation with which it is interfacing, including single-buoy moorings, and between a SOLAS ship and any non-SOLAS ship, particularly mobile offshore drilling units (MODUs) on location and floating production storage and offloading vessels (FPSOs).

Retention

2.7.13 Port facilities and ships should retain Declarations of Security for the period specified by their respective national authorities. In many cases, this is three to five years.

2.7.14 Ships should have any Declarations of Security established during the period covering the ship's last ten ports of call available for inspection by Government officials undertaking control and compliance measures under the Maritime Security Measures (refer to subsections 2.11 and 4.10). This includes any DoS for a ship/port interface or ship-to-ship activity.

Request by a port facility

2.7.15 If a port facility requests a ship to agree to a DoS, the ship must comply. The PFSP will indicate the circumstances, specified by the Designated Authority, when such a request should be made.

Request by a ship

2.7.16 A ship can request that a DoS be agreed by a port facility or another ship. Again, the circumstances when such a request should be made will be those specified by the Administration and incorporated in the SSP.

2.7.17 If a ship requests that a port facility agrees a DoS, the port facility has to acknowledge that the request was made. The port facility does not have to agree a DoS with the requesting ship unless the circumstances relating to the request conform to those in the PFSP.

2.8 Port facility security responsibilities

Designating port facilities

2.8.1 Fundamental to the successful implementation of the Maritime Security Measures is the identification by the Designated Authority of all the port facilities within its territory used by SOLAS ships. The Designated Authority has to determine whether:

.1 the port facility is required to appoint a PFSO and submit a PFSP; or

.2 the port facility is occasionally used by SOLAS ships and does not have to appoint a PFSO.

2.8.2 Some Designated Authorities consider that all their port facilities used by SOLAS ships, even if the use is occasional, should appoint a PFSO and prepare a PFSP.

2.8.3 Designated Authorities have wide discretion as to how they designate their port facilities.

2.8.4 Factors to be considered for determining whether a port facility occasionally used by SOLAS ships should appoint a PFSO and prepare a PFSP could include:

.1 the frequency of use;

.2 use by ships considered to pose a heightened security risk, e.g., cruise ships or ships carrying dangerous cargoes; or

.3 proximity to populated areas.

2.8.5 Most Designated Authorities have defined multiple port facilities within each of their port areas.

2.8.6 Others have defined an entire port area, often a significant area involving the entire range of shipping activities, as a single port facility.

2.8.7 One Designated Authority has determined that each port facility includes several port areas.

2.8.8 Some Designated Authorities who initially designated entire port areas as a single port facility have subsequently changed their approach to designate multiple port facilities within each port area.

2.8.9 Many Designated Authorities have categorized their port facilities based on the type of operation at the facility and consideration of the security risks that can be associated with the operation. Examples include facilities handling:

.1 cruise ships;

.2 ro–ro passenger ships;

.3 chemical, oil and gas shipments in bulk;

.4 containers and ro–ro shipments;

.5 general cargo shipments;

.6 bulk cargoes (e.g., ore, coal and grain).

Port facility boundaries

2.8.10 A port facility can include an area of land or water, or land and water; it may be used either wholly or partly for the embarkation or disembarkation of passengers, or with the loading or unloading of cargo, from SOLAS ships. Essential to the designation of individual port facilities is the delineation of a clear boundary within which the port facility is responsible for exercising its responsibilities under the Maritime Security Measures. A key factor in designating individual port facilities is identifying those responsible for the ship/port interface at the facility. Usually this will be the facility operator. In multiple-use facilities, where there are a number of operators, the Designated Authority has to determine who is responsible for the overall security of the facility. This may be the owner of the facility rather than any of the operators. In water areas where control often rests with the port authority, or other authority, regulating the movement of ships within the port area, their designation as a distinct port facility appears to be rare.

2.8.11 How Designated Authorities have defined the extent of individual port facilities varies. Experience to date includes:

.1 limiting port facilities to the land area immediately adjacent to the berth(s);

.2 including all the contiguous land area, including buildings, associated with the embarkation or disembarkation of passengers or the storage, loading and unloading of cargo at the berth(s);

.3 using physical features, such as tree lines, fences or lines where temporary barriers may be used;

.4 recording the boundary accurately on a map and including it in both the PFSA and the PFSP;

.5 taking responsibility for the security of water-side areas adjacent to their berth(s), particularly in relation to manoeuvring areas at oil or gas terminals where safety considerations also apply;

.6 including other water areas, e.g., anchorages, waiting areas and approaches from seaward;

.7 including berths within port areas where harbour craft, including tugs and pilot vessels, are berthed or from which they operate;

.8 including ship yards and ship repair yards;

.9 including fishing ports and marinas when they are within port areas including port facilities or are immediately adjacent to a designated port facility.

2.8.12 Designated Authorities have, on occasion, not designated port facilities regularly used by SOLAS ships on the grounds that the port facility is owned and operated by a Government-appointed ports' authority. However, such a practice does not conform to the requirements in the Maritime Security Measures.

Notification

2.8.13 Governments are required to notify IMO of the location of port facilities within their territory that have an approved PFSP. They are required to keep the information up to date for each port facility and resubmit the information on all their facilities at least every five years. The communication of such information should be provided to IMO's GISIS database. The next date for submission of information related to port facilities with a PFSP initially approved on 1 July 2004 – information related to which was required to be resubmitted by 1 July 2009 – is 1 July 2014.

Non-SOLAS port facilities

2.8.14 Designated Authorities should determine what security procedures and measures are appropriate at port facilities occasionally used by SOLAS ships but where a PFSO has not been appointed.

2.8.15 Designated Authorities should appoint a person ashore with responsibility for shore-side security to liaise with SOLAS ships using the port facility. The person can be responsible for a number of non-SOLAS port facilities. The name and contact details of the person responsible for shore-side security should be made available to SOLAS ships intending to use the facility.

2.8.16 In remote areas where ships visit infrequently and there is no person ashore to take responsibility for shore-side security, the Designated Authority should rely on the visiting ships' security measures.

Port security committees

2.8.17 Though not required by the Maritime Security Measures, most port operators have established port security committees to co-ordinate the implementation of the Maritime Security Measures in their port in a consistent manner. Many Designated Authorities have formalized these arrangements and now require port security committees at their ports.

2.8.18 Guidance on the membership and roles of a port security committee is in paragraphs 3.9.3 to 3.9.8.

Port facility security officers

2.8.19 Port facility security officers (PFSOs) appointed by, and reporting to, the management of a port facility play an essential role in establishing and maintaining the security of their port facility. Their responsibilities extend to maintaining effective communication on security matters with the company security officers (CSOs) and ship security officers (SSOs) of ships using, or intending to use, their port facility.

2.8.20 It is the efficiency and effectiveness of individual PFSOs working with their national and local authorities, CSOs and SSOs that underpins the continued successful application of the Maritime Security Measures.

2.8.21 The appointment of PFSOs is essentially a matter for the port facilities that are required by the Designated Authority to have a PFSO.

2.8.22 As PFSOs are likely to be entrusted with security-sensitive information, many Designated Authorities require that they are subjected to security vetting before receiving such information. This requirement should extend to other port facility personnel who perform the responsibilities of a PFSO. It can also extend to senior management at the port facility.

2.8.23 Many Designated Authorities have specified that PFSOs should undertake training courses delivered by training providers approved by the Designated Authority. Guidance on the responsibilities and training of PFSOs is in paragraphs 3.5.1 to 3.5.6.

2.8.24 National authorities should provide guidance to PFSOs on the action to be taken on receipt of a report from a SOLAS ship in their port or port facility on the failure of the ship's security equipment or system or suspension of a security measure which compromises the ship's ability to operate at security levels 1 to 3.

Port facility security assessments

2.8.25 Designated Authorities have to ensure that a port facility security assessment (PFSA) is undertaken for each port facility.

2.8.26 PFSAs can be undertaken by the Designated Authority or by a recognized security organization (RSO) authorized by the Designated Authority.

2.8.27 Guidance on undertaking PFSAs is in subsection 3.6 and in section 5.

2.8.28 When a PFSA has been completed by an RSO, it has to be submitted to, and approved by, the Designated Authority. An RSO cannot approve a PFSA.

2.8.29 Designated Authority personnel should have the experience and training to undertake PFSAs and to assess and approve assessments completed by RSOs.

2.8.30 When completed or approved, PFSAs should be forwarded by the Designated Authority to the PFSO for retention and to allow preparation or amendment of the PFSP.

2.8.31 The Maritime Security Measures specify that PFSAs shall periodically be reviewed and updated, taking account of changing threats and/or minor changes in the port facility, and shall always be reviewed and updated when major changes to the port facility take place.

2.8.32 It is for the Designated Authority to determine the frequency of review of an approved PFSA. Many Designated Authorities review them annually and when there has been a:

.1 significant security incident at the port facility;

.2 change in the shipping operations undertaken at the facility; and/or

.3 change of facility owner or operator.

2.8.33 An essential component of PFSAs undertaken for, or by, Governments and approved by them is an identification of the range of threats and security incidents that could occur. This is addressed in greater detail in section 5.

Port facility security plans

2.8.34 The development and revision of a port facility security plan (PFSP) is the responsibility of the facility's PFSO, having regard to the approved PFSA. Guidance on the preparation and content of PFSPs is provided in subsection 3.7.

2.8.35 Designated Authorities are responsible for establishing the policies and procedures to be included in a PFSP on Declarations of Security and on the security incidents that should be reported to them and the timing of such reports. Further guidance on Declarations of Security is in subsections 2.7 and 3.4.

2.8.36 When completed, the draft PFSP must be submitted to, and assessed and approved by, the Designated Authority.

2.8.37 Designated Authorities should establish the procedures and timescales covering:

.1 preparation and submission of plans;

.2 plan approval;

.3 amendment of approved plans; and

.4 subsequent inspections of port facilities to assess compliance with approved plans.

2.8.38 As part of the approval process, a Designated Authority can propose a modification to a submitted PFSP, or proposed amendment to an approved PFSP, prior to approving the submission. This could occur when the submission does not reflect the conclusions of the PFSA or there is another security issue that the Designated Authority considers to have been inadequately addressed. Such modifications should always follow consultation with the PFSO on the reasons for the modification. The procedures could involve return of the submission to the PFSO for reconsideration and its resubmission incorporating the suggested modification or alternative amendments to meet the Designated Authority's concerns.

2.8.39 Officials of Designated Authorities should have the experience and training to advise on and carry out the above procedures. To assist with approving PFSPs, appendix 2.4 – Sample of a port facility security plan approval form – provides a sample form. Designated Authorities have issued guidance, in varying detail, on the content of PFSPs, including, in some cases, standard templates for their plans. This guidance is described in greater detail in subsection 3.7.

2.8.40 The PFSPs submitted for Government approval are required to include the specific security measures and procedures associated with each of the three security levels. This may also be a requirement for port security plans (PSPs) but, if not, is a recommended practice for an effective PSP.

2.8.41 The PFSP should set out the procedures to be followed when security levels change. While the security level applied to a port, port facility or ship may change from security level 1, through security level 2 to security level 3, it is also possible that the security levels will change directly from security level 1 to security level 3, and security plans should cover this possibility.

2.8.42 The use of firearms in port facilities or on or near ships can pose significant safety risks, in particular in connection with certain dangerous and hazardous substances. If Governments consider it appropriate to allow authorized armed personnel at port facilities or on board ships, steps should be taken to ensure that such personnel are duly authorized and appropriately trained in the use of their weapons and that they are aware of the risks that can arise through the discharge of weapons in port facilities or on board ship. Specific guidelines should be issued by Governments authorizing armed personnel. PFSPs should include specific guidelines on the use of weapons in the vicinity of dangerous goods or hazardous substances.

Security records

2.8.43 Designated Authorities should specify the security records that a port facility is required to keep and be available for inspection, including the period for which they should be kept. The records could cover:

.1 declarations of Security agreed with ships;

- **.2** security threats or incidents;
- **.3** changes in security level;
- **.4** security training undertaken by port facility personnel;
- **.5** security drills and exercises;
- **.6** maintenance of security equipment;
- **.7** internal audits and reviews;
- **.8** reviews of PFSAs;
- **.9** reviews of the PFSP; and
- **.10** any amendments to an approved plan.

Review of an approved PFSP

2.8.44 Designated Authorities should issue guidance on the frequency of reviewing PFSPs. Often, a minimum frequency of once a year is recommended or following:

- **.1** a major security drill or exercise;
- **.2** a security threat or incident involving the port facility;
- **.3** a change in the shipping operations undertaken at the facility;
- **.4** change of the owner or operator of the facility;
- **.5** completion of a review of the PFSA;
- **.6** when an internal audit or inspection by the Designated Authority has identified failings in the facility's security organization and operations to the extent that the approved PFSP may no longer be relevant.

2.8.45 Designated Authorities adopt varying approaches when specifying the amendments that have to be submitted to them. They range from a list of a minimum number of requirements for which the Designated Authority's approval is required to a strict approach whereby any change to an approved PFSP requires their approval.

Amendments to an approved PFSP

2.8.46 Designated Authorities should notify PFSOs of the type of amendments to an approved PFSP that must be approved before they can be implemented. This notification can be provided on approval of the initial PFSP or a subsequent amendment.

2.8.47 If, under exceptional circumstances, the Designated Authority allows a PFSO to amend a PFSP without its prior approval, the amendments must be communicated to the Designated Authority at the earliest opportunity.

Internal audits

2.8.48 PFSPs should establish internal audit procedures to be followed by a port facility operator to ensure the continued effectiveness of the PFSP. To assist PFSOs, Designated Authorities could provide guidance on the following:

- **.1** purpose of the internal audit of the port facility's security (e.g., to identify opportunities for improvement);
- **.2** frequency (e.g., once a year);
- **.3** audit techniques (e.g., site visits and interviews with security personnel);

.4 components of a review;

.5 sample audit report form; and

.6 selection of auditors.

Security measures and procedures

2.8.49 Designated Authorities provide guidance to each of their designated port facilities on the security measures and procedures considered appropriate at each security level. These are based on the facility's PFSA report. Details of the Measures and procedures are provided in subsection 3.3.

2.8.50 Many Governments have classified their ports and port facilities as critical national infrastructure, or use an equivalent designation. In many cases, national standards have been developed covering the installation and maintenance of security equipment, and in such cases Designated Authorities can refer to these national standards when advising port facilities on security equipment and equipment maintenance regimes or when inspecting them. Such standards may address:

.1 fencing, gates, vehicle barriers and lighting;

.2 closed-circuit television (CCTV);

.3 communications and x-ray equipment;

.4 archway and hand-held metal detectors;

.5 perimeter/intruder detection systems;

.6 automated access control equipment (e.g., identification readers or keypads);

.7 information and computer protection systems; and

.8 explosive trace and vapour detection equipment.

Statement of Compliance

2.8.51 Although it is not mandatory under the Maritime Security Measures, Designated Authorities can issue a Statement of Compliance to a port facility. It could indicate:

.1 the name of the port facility;

.2 the types of ship(s) operating at the port facility;

.3 that the port facility complies with the Maritime Security Measures;

.4 the period of validity of the Statement of Compliance (which should not exceed five years); and

.5 the arrangements established by the Designated Authority for subsequent verifications of a Statement of Compliance.

2.8.52 The Maritime Security Measures contain a standard form for use by Designated Authorities (refer to appendix 2.5 – Statement of Compliance of a port facility).

2.8.53 A Statement of Compliance should not be issued unless the Designated Authority has confirmed that:

.1 the port facility has a PFSA undertaken, or approved, by the Designated Authority:

.2 the port facility has a PFSP which has been duly and formally approved by the Designated Authority;

.3 the port facility's security staff have received the necessary training and can implement the security procedures in the approved PFSP; and

.4 any security equipment specified in the PFSP is in place and operating effectively.

2.8.54 A number of Designated Authorities have specified that, for a Statement of Compliance to be valid, it should have at least:

.1 an initial verification before the Statement of Compliance is first issued;

.2 one intermediate verification between the second and third anniversaries of its issuance; and

.3 a renewal verification five years after first being issued.

2.8.55 As the Maritime Security Measures restrict the role of RSOs in approving PFSAs and PFSPs, the authority to undertake Statement of Compliance verifications has generally been retained by Designated Authorities.

2.9 Ship security responsibilities

Appointment and qualifications of security personnel

2.9.1 Shipping companies are responsible for the appointment of CSOs, SSOs and other personnel with security duties.

2.9.2 Presently, the Maritime Security Measures provide guidance on the knowledge and training that these security personnel should have.

2.9.3 From 1 January 2012, IMO's Standards of Training, Certification and Watchkeeping (STCW) Convention and related STCW Code establish mandatory minimum requirements for security-related training and instruction for all SSOs and shipboard personnel serving on SOLAS ships. However, it does not encompass the security-related requirements for CSOs.

2.9.4 The STCW Code further stipulates that SSOs and all shipboard personnel are required to:

.1 meet the appropriate standard of competence; and

.2 be issued with a certificate of proficiency (which can be inspected under the control in the STCW regulations).

2.9.5 The STCW Convention and its related Code were amended at the 2010 Manila Diplomatic Conference and entered into force on 1 January 2012. The approach taken in the amended Convention and its related Code is that of a three-layer certification process. The first layer comprises the issuance of Certificates of Competency, which falls exclusively under the authority of Administrations. The second layer is that of Certificates of Proficiency, which, apart from Certificates of Proficiency under regulations V/1-1 and V/1-2, can be issued on behalf of Administrations by approved training institutions. The third layer is that for Documentary Evidence that can be issued by training institutions approved by Administrations.

2.9.6 Transitional arrangements are specified for SSOs and shipboard personnel who receive security training before January 2012, including the possible need for retraining.

2.9.7 Prior to entry into force of the amended STCW Convention and Code, IMO has advised that, as an interim measure, the ISSC should be accepted as evidence that security-related training of SSOs and shipboard personnel has been conducted in accordance with the Maritime Security Measures.

2.9.8 The STCW Code recognizes that, although shipboard personnel are not security experts, they should receive adequate security-related training so as to acquire the required competencies to perform their assigned duties and to collectively contribute to the enhancement of maritime security.

2.9.9 The STCW Code stipulates that all SSOs and shipboard personnel should receive security-related familiarization training before taking up their duties

2.9.10 Guidance on the responsibilities and qualifications of CSOs, SSOs and shipboard personnel is in subsection 4.5.

2.9.11 Experience to date indicates that Administrations have:

- **.1** required their CSOs and SSOs to attend courses provided by approved training organizations;
- **.2** allowed CSOs to participate in the decision to appoint RSOs engaged by their shipping company in respect of its ships; and
- **.3** taken control measures if inspections detected a lack of security-related training.

Ship security assessments

2.9.12 Ship security assessments (SSAs), including the on-scene survey, are the responsibility of CSOs. Guidance on their undertaking is in subsection 4.7.

2.9.13 Administrations are responsible for providing guidance to CSOs on the security risks that their ships may face on voyages, having regard to the ship type, the sea areas in which the ship operates, and the ports and port facilities that it uses. If a ship changes its trading pattern, the security threats that it faces may significantly change; in such cases, Administrations should be well placed to provide revised guidance on any new threats that the ship may face as a basis for updating the SSA.

2.9.14 The Maritime Security Measures specify that the report of an up-to-date SSA should accompany, or be reflected in, SSPs submitted for approval or when amendments to an approved plan are submitted.

Ship security plans

2.9.15 The development and revision of a ship security plan (SSP) is the responsibility of the shipping company's CSO, having regard to the ship's approved SSA. Guidance on the preparation and content of SSPs is provided in paragraphs 4.8.1 to 4.8.11.

2.9.16 Administrations are responsible for establishing the policies and procedures to be included in an SSP on Declarations of Security and on the security incidents that should be reported to them and the timing of such reports. Further guidance on Declarations of Security is provided in subsection 4.4.

2.9.17 When completed, the SSP has to be submitted to, assessed by and approved by the Administration. This responsibility may be delegated to an RSO provided that the RSO has not assisted in its preparation.

2.9.18 Administrations should establish the procedures and timescales covering:

- **.1** preparation and submission of SSPs;
- **.2** plan approval;
- **.3** amendment of approved plans;
- **.4** subsequent inspection of ships to assess compliance with approved plans.

2.9.19 As part of the approval process, an Administration can propose a modification to a submitted plan, or proposed amendment to an approved SSP, prior to approving the submission. This could occur when the submission does not reflect the conclusions of the SSA or there is another security issue that the Administration considers to have been inadequately addressed.

2.9.20 Such modifications should always follow consultation with the CSO on the reasons for the modification. The procedures could involve return of the submission to the CSO for reconsideration and its resubmission, incorporating the suggested modification or alternative amendments to meet the Designated Authority's concerns.

2.9.21 An Administration's official should have the experience and training to advise on and carry out the above procedures. To assist with approving SSPs, the following website may be useful:
www.dominica-registry.com. Refer to item 3 in appendix 4.8 – Examples of internet sources of guidance material on preparing and validating ship security plans – for further details.

2.9.22 Administrations have issued guidance, in varying detail, on the content of SSPs, including, in some cases, standard templates for their plans. This guidance is described in greater detail in paragraphs 4.8.1 to 4.8.11 and includes:

- **.1** procedures for receiving changes in security level;
- **.2** the time allowed to move between security levels;
- **.3** security-related records that have to be held by the ship;
- **.4** procedures for reporting security system and equipment failures;
- **.5** the circumstances when a master can refuse an inspection, prior to the ship entering port, by Government officials under the Maritime Security Measures' control and compliance measures (subsection 4.10);
- **.6** responses to interdiction at sea;
- **.7** preserving evidence following a security incident;
- **.8** circumstances when a DoS should be requested from a port facility or other ship;
- **.9** procedures for reporting security incidents to the Administration;
- **.10** reports of internal audits and reviews of an approved SSP;
- **.11** amendments to an approved SSP.

2.9.23 Administrations can issue guidance on the frequency of reviewing SSPs. Often, a minimum frequency of once a year is recommended or following:

- **.1** a major security drill or exercise;
- **.2** a security threat or incident involving the ship;
- **.3** a change in shipping operations, including the operator;
- **.4** completion of a review of the SSA;
- **.5** the identification, in an internal audit or inspection by the Administration, of failings in the ship's security operations, to the extent that the approved SSP may no longer be relevant.

2.9.24 A number of Administrations have provided distinct guidance to their CSOs on particular types of ships, based on their assessment of the different security risks that can be faced by the ship operators. The main types are:

- **.1** cruise ships;
- **.2** ro–ro passenger ships;
- **.3** chemical, oil and gas tankers and produce carriers;
- **.4** container ships;
- **.5** ro–ro and general cargo ships;
- **.6** special-purpose ships and mobile offshore drilling units.

2.9.25 Ultimately, an SSP should address all the security threats that the ship may face in service and include appropriate security measures and procedures to mitigate such threats.

2.9.26 Administrations adopt varying approaches when specifying the amendments that have to be submitted to them. They range from a list of a minimum number of requirements for which the Administration's approval is required to a strict approach whereby any change to an approved SSP requires approval.

2.9.27 Administrations should notify CSOs of the type of amendments to an approved SSP that must be approved before they can be implemented. This notification can be provided on approval of the initial SSP or a subsequent amendment.

2.9.28 If the Administration allows a CSO or SSO to amend an SSP without its prior approval, the adopted amendments must be communicated to the Administration at the earliest opportunity.

2.9.29 The SSPs submitted for Government approval are required to include the specific security measures and procedures associated with each of the three security levels. The SSPs should set out the procedures to be followed when security levels change. While the security level applied to a port, port facility or ship may change from security level 1, through security level 2 to security level 3, it is also possible that the security levels will change directly from security level 1 to security level 3, and security plans should cover this possibility.

2.9.30 The use of firearms on or near ships can pose significant safety risks, in particular in connection with certain dangerous and hazardous substances. If Governments consider it appropriate to allow authorized armed personnel on board ships, steps should be taken to ensure that such personnel are duly authorized and appropriately trained in the use of their weapons and that they are aware of the risks that can arise through the discharge of weapons on board ship. Specific regulations should be issued by Governments authorizing armed personnel. SSPs should include specific guidelines on the use of weapons in the vicinity of dangerous goods or hazardous substances. Firearms carried on board ship may have to be reported on arrival in port and may have to be surrendered, or held securely, for the duration of the port visit.

Reporting security system or equipment failures

2.9.31 An SSP should contain details of the procedures to be followed when the ship has to report the failure of its security equipment or system, or suspension of a security measure which compromises the ship's ability to operate at security levels 1 to 3. Such reports, together with any remedial actions that the ship proposes to take and a request for instructions, should be made immediately to the:

.1 Administration;

.2 port facility that the ship is in;

.3 the authorities of the coastal State whose territorial sea the ship is operating in or has indicated that it intends to transit.

Interdiction at sea

2.9.32 Masters have discretion to allow foreign security forces to visit their ship when in international waters. If the master consents and an inspection establishes that an offence may have been committed, jurisdiction remains with the flag State. The flag State can transfer jurisdiction to the inspecting State. Administrations should advise their CSOs on the actions that a master should take in response to such a request to board and inspect, for inclusion in the SSP.

2.9.33 There are an increasing number of circumstances when a ship may be boarded by foreign security forces when in international waters. These can occur under the authority of:

.1 UN Security Council Resolutions relating to the enforcement of sanctions;

.2 bilateral/multilateral agreements relating to the suppression, for example, of nuclear proliferation or drug smuggling. Such agreements are based on prior consent being given by the flag State;

.3 IMO Protocol of 2005 to the Protocol for the Suppression of Unlawful Acts against the Safety of Fixed Platforms Located on the Continental Shelf and the Protocol of 2005 to the Convention for the Suppression of Unlawful Acts against the Safety of Maritime Navigation (the SUA Convention), which entered into force in 2010.

2.9.34 Administrations could consider providing guidance to CSOs on the actions to be taken by the ship when it is boarded under these authorities.

2.9.35 Under the SUA Convention, a request can be made by a Contracting Government to the Convention to board and inspect a ship of another Contracting Government in international waters and take appropriate measures if there are reasonable grounds for believing that a terrorist-related offence has been, is being, or is about to be committed on board. The procedure is based on prior flag State consent, with jurisdiction being retained unless transferred. A ship can be detained if there is evidence that an offence has been committed. The SUA Convention allows the master of a boarded vessel to contact the ship's Administration and the shipping company at the earliest opportunity.

Preserving evidence following a security incident

2.9.36 Administrations, in consultation with their law-enforcement agencies, may wish to provide guidance to CSOs on preserving evidence found on board their ships after an incident.

Reporting security incidents

2.9.37 Administrations are required to specify the types of security incident that have to be reported to them. In such cases, they should provide guidance on their timing, procedures to be followed and their distribution. These procedures are described in greater detail in 4.8.35 to 4.8.37. They should include reporting incidents to local law-enforcement agencies when in a port facility or the adjacent coastal State.

Security records

2.9.38 Administrations should specify the security records that a ship is required to keep and be available for inspection, including the period for which they should be kept (refer to paragraphs 4.8.38 to 4.8.39). The records could cover:

- **.1** Declarations of Security agreed with port facilities and other ships;
- **.2** security threats or incidents;
- **.3** breaches of security;
- **.4** changes in security level;
- **.5** communications relating to the direct security of the ship, such as specific threats to the ship or to port facilities where the ship is or has been;
- **.6** ship security training undertaken by the ship's personnel;
- **.7** security drills and exercises;
- **.8** maintenance of security equipment;
- **.9** internal audits and reviews;
- **.10** reviews of the SSAs;
- **.11** reviews of the SSPs; and
- **.12** any amendments to an approved plan.

Internal audits

2.9.39 SSPs should establish internal audit procedures to be followed by a Company or ship to ensure the continued effectiveness of the SSP. To assist CSOs and SSOs, Administrations could provide guidance on the following:

- **.1** purpose of the internal audit of ship security (e.g., to identify opportunities for improvement);
- **.2** frequency (e.g., once a year);
- **.3** audit techniques (e.g., site visits and interviews with security personnel);
- **.4** components of a review;
- **.5** sample audit report form;
- **.6** selection of auditors.

Security measures and procedures

2.9.40 Administrations should provide guidance to each of their shipping companies and CSOs on the security measures and procedures considered appropriate at each security level for their ships. These are based on the SSAs undertaken for the CSO.

2.9.41 Administrations should require that security equipment receive regular maintenance checks and that these checks be recorded. Security equipment can include:

- **.1** closed-circuit television (CCTV) and lighting;
- **.2** communications and x-ray equipment;
- **.3** archway and hand-held metal detectors;
- **.4** perimeter/intruder detection systems;
- **.5** automated access control equipment;
- **.6** information, including computer, security; and
- **.7** explosive trace and vapour detection equipment.

Continuous Synopsis Records

2.9.42 Administrations must ensure that each ship's Continuous Synopsis Record (CSR) includes the name of the:

- **.1** Administration or RSO that issued the ship's ISSC or Interim ISSC; and
- **.2** if different from above, the organization that carried out the verification leading to the issuance of the Certificate.

Manning levels

2.9.43 Administrations should ensure that, when determining the safe manning level of each national ship, they take into account any additional workload that may result from the implementation of the approved SSP. Consideration should be given to the workload associated with the performance of security responsibilities and the capacity of the shipboard personnel to handle the additional workload; recognizing the need to implement the hours of rest and other measures for addressing and avoiding fatigue among ship personnel.

2.10 International Ship Security Certificates

Introduction

2.10.1 Ships falling under the Maritime Security Measures must carry either the International Ship Security Certificate (ISSC) or, in limited circumstances, the Interim ISSC, both of which are issued by their Administration.

2.10.2 Administrations inspect ships entitled to fly their flag in connection with the issue, intermediate verification and renewal of ISSCs; the issue of Interim ISSCs; and at any other time to assess the ship's compliance with the Maritime Security Measures.

2.10.3 The Maritime Security Measures contain a 'model' International Ship Security Certificate which is referenced in appendix 2.6 – Form of the International Ship Security Certificate. If the Certificate adopted by the Administration is not in English, French or Spanish, the text should include a translation into one of those languages.

Issuance

2.10.4 An ISSC must be issued for a period specified by the Administration which, with one exception, cannot exceed five years. The exception covers the situation when a renewal verification is completed within three months of the expiry date of the existing ISSC. In this situation, the new ISSC becomes valid from the date of completion of the renewal verification to a date not exceeding five years from the expiry date of the existing ISSC.

2.10.5 An ISSC should only be issued or renewed when:

.1 the ship has an approved SSP indicating that it fully addresses all requirements specified in the Maritime Security Measures, as outlined in paragraphs 4.8.1 to 4.8.11, and

.2 the Administration is satisfied, based on objective evidence, that the ship is operating in accordance with the provisions in the approved SSP.

2.10.6 A Certificate should not be issued in cases where there is a minor deviation from the SSP, even when the ship's ability to operate at security levels 1 to 3 is not compromised.

2.10.7 A Certificate can be issued or endorsed by:

.1 the ship's Administration;

.2 an RSO authorized to act on behalf of the ship's Administration; or

.3 another Administration acting on behalf of the ship's Administration.

Verifications

2.10.8 SOLAS ships are subject to verifications of their compliance with the Maritime Security Measures.

2.10.9 Before a ship is put into service and before an Interim ISSC is issued, an interim verification takes place. Although the ISPS Code refers to this type of verification as an 'initial verification', it has become standard industry practice to use the term 'interim verification'.

2.10.10 Subsequent verifications that take place are:

.1 before an ISSC is issued – an *initial verification*;

.2 at least once between the second and third anniversary of the issuance of the ISSC if the validity period is for five years – an *intermediate verification*;

.3 before the ISSC is renewed – a *renewal verification*; and

.4 at other times, at the discretion of the Administration.

2.10.11 The appropriate level of thoroughness in the verification of security systems should be as follows:

.1 100% verification for all technical equipment specified in the SSP; and

.2 a sample audit for all operational (non-technical) security measures, to a level necessary for the auditor to verify the whole operating system.

2.10.12 An initial verification is conducted to ensure that the ship's security system and any security equipment required by the Maritime Security Measures and the approved SSP is in satisfactory condition and fit for the service for which the ship is intended.

2.10.13 An intermediate verification is conducted to ensure that the ship's security system and any security equipment required by the Maritime Security Measures and the SSP remains in satisfactory condition and is fit for the service for which the ship is intended.

2.10.14 A renewal verification is to ensure the ship's security system and any security equipment fully complies with the requirements of the Maritime Security Measures and the approved SSP, is in satisfactory condition and is fit for the service for which the ship is intended.

2.10.15 After verification, the ship's security system and security equipment should be maintained to conform with the provision of the Maritime Security Measures. No changes can be made to the security system or security equipment or to the approved SSP unless agreed by the Administration.

Duration of validity

2.10.16 The duration of a renewed five-year ISSC can vary depending on the date that the renewal verification takes place. If it is completed:

- **.1** within the three months before the expiry of the original ISSC, then the next five-year period starts at the original expiry date;
- **.2** after the expiry of the original ISSC, then the next five-year period starts at the original expiry date;
- **.3** more than three months before the expiry of the original ISSC, then the next five-year period starts at the date of completion of the renewal verification.

2.10.17 If an ISSC has been issued for a period of less than five years, an Administration can extend its validity to a maximum of five years after undertaking the required verification.

2.10.18 If a new ISSC cannot be placed on the ship before the original ISSC expires, the Administration can endorse the original ISSC for an extended period not exceeding five months. The new five-year period starts at the original expiry date.

2.10.19 If a ship is in transit, or its arrival at the port where verification is to take place is delayed, the Administration can endorse the original ISSC to allow the ship to complete its voyage. However, the validity period cannot be extended for longer than three months and the new five-year period starts at the expiry date set for the original ISSC.

2.10.20 If a ship is engaged on short voyages, its ISSC can be extended for a period of up to one month with the new five-year period starting at the expiry date of the original ISSC.

2.10.21 If an intermediate verification is undertaken before the second anniversary of issuance of an ISSC that is valid for five years, its validity period must be reduced to show an expiry date that is no more than three years after the completion date of the verification. However, the original expiry date can be maintained with a further intermediate verification.

Loss of validity

2.10.22 An ISSC can lose its validity when:

- **.1** the required intermediate and renewal verifications have not taken place;
- **.2** it has not been endorsed following an intermediate verification;
- **.3** a new shipping company takes over the operation of the ship; or
- **.4** the ship changes its flag.

2.10.23 On changes of flag, the original Administration should provide the new Administration with copies of all relevant information on the ship's ISSC, including copies of available verification reports.

Remedial actions

2.10.24 The ship's Administration has to be notified immediately when there is a failure of a ship's security equipment or system or suspension of a security measure which compromises the ship's ability to operate at security levels 1 to 3. The notification should be accompanied by any proposed remedial actions.

2.10.25 The ship's Administration has also to be notified when the above circumstances do not compromise the ship's ability to operate at security levels 1 to 3. In such cases, the notification should be accompanied by an action plan specifying the alternative security measure being applied until the failure or suspension is rectified, together with the timing of any repair or replacement.

2.10.26 The Administration may:

.1 approve the alternative security measures being taken and the action plan;

.2 require amendments to such measures;

.3 require additional or alternative measures;

.4 require immediate repair or replacement;

.5 take other appropriate action.

2.10.27 A ship's ISSC may be withdrawn or suspended if the alternative security measures are not applied or the approved action plan is not complied with.

2.10.28 Administrations should provide guidance to their CSOs reminding them of the cumulative effect that individual failures or suspensions of measures could have on the ability of their ships to operate at security levels 1 to 3.

2.10.29 Administrations should also provide guidance to their officials on the action that they should take when receiving a report from a SOLAS ship on the failure of its security equipment or system or suspension of a security measure which compromises the ship's ability to operate at security levels 1 to 3. Administrations should provide clear guidance concerning the impact of the ship's status on the validity of its ISSC.

Ship out of service

2.10.30 Administrations apply widely diverging interpretations of when a SOLAS ship is out of service or laid up; and of the circumstances and passage of time that could lead to consideration of suspension or withdrawal of the ship's ISSC. The Maritime Security Measures do not provide specific direction on these ship out-of-service considerations.

Interim International Ship Security Certificates

2.10.31 Administrations or RSOs may issue an Interim ISSC when:

.1 a ship is on delivery, or prior to its entry or re-entry into service;

.2 a SOLAS ship is changing its flag;

.3 a ship is being transferred from a non-SOLAS State;

.4 the shipping company operating a SOLAS ship changes.

2.10.32 An Interim ISSC can only be issued when the Administration or RSO has conducted interim verification confirming that:

.1 the ship's SSA has been completed;

.2 there is a copy of the SSP on board;

.3 the SSP has been submitted for review and approval and is being implemented;

.4 the ship has a ship security alert system (SSAS);

.5 the CSO has ensured that the necessary arrangements are in place, including drills, exercises and internal audits, for the ship to successfully complete the required verification within six months;

.6 arrangements are in place to carry out the required verification;

.7 the master, SSO and other personnel with specific security duties are familiar with their responsibilities in the Maritime Security Measures and SSP and have been provided with such information in the ship's working language or in a language they understand;

.8 the SSO meets the relevant requirements in the Maritime Security Measures.

2.10.33 Following verification of the items listed above, an Interim ISSC is valid for up to six months.

2.10.34 If a full ISSC is issued to the ship during that six-month period, the Interim ISSC is revoked.

2.10.35 An Interim ISSC cannot be extended.

2.10.36 The Maritime Security Measures contain a 'model' Interim ISSC that is in appendix 2.7 – Form of the Interim International Ship Security Certificate. If the Certificate adopted by the Administration is not in English, French or Spanish, the text should include a translation into one of those languages.

2.10.37 An Administration should not issue a subsequent or consecutive Interim ISSC if it believes that the shipping company intends to avoid full compliance with the Maritime Security Measures for a period beyond the initial six-month validity of the initial Interim ISSC.

2.10.38 ISSCs and Interim ISSCs can be inspected as part of control and compliance measures described in subsection 4.10.

Ship inspections

2.10.39 Administrations undertake inspections of their SOLAS ships as initial, intermediate and renewal verifications of the ship's ISSC. At their discretion, Administrations may also conduct:

- **.1** additional inspections on ships flying their flag to assess compliance with the Maritime Security Measures;
- **.2** covert tests of the Maritime Security Measures and procedures of a ship flying their flag.

2.10.40 A sample of a ship security inspection checklist which can be used for verifications and other inspections is attached as appendix 2.8 – Sample of a ship security inspection checklist.

2.10.41 To assist shipping companies, Administrations and their authorized RSOs have sought to link the timing of verifications required under the Maritime Security Measures with other verifications or inspections, including, particularly, those required under the International Safety Management (ISM) Code. Combining inspections in this way can be of significant benefit to the shipping industry. However, in ports where ISM auditors are not always available, this joint approach may not be practicable and could unduly delay shipping schedules.

2.10.42 The training and experience required for those undertaking verifications and inspections under the Maritime Security Measures can differ for those undertaking other forms of verification or inspection.

2.11 Control and compliance measures

Introduction

2.11.1 Governments can apply specific control and compliance measures to foreign-flagged SOLAS ships using, or intending to use, their ports when assessing their compliance with the Maritime Security Measures. Elements of these control and compliance measures are unique, including:

- **.1** the authority to require ships to provide security-related information prior to entering port;
- **.2** the authority to inspect ships intending to enter into port when there are clear grounds for doing so once the ship is within the territorial sea, and the right of a master to refuse such an inspection; and
- **.3** the authority to refuse to allow a ship to enter port or to expel a ship from port.

2.11.2 Those authorized to undertake control and compliance measures under the Maritime Security Measures may also carry out control functions in respect of foreign-flagged vessels under other SOLAS Convention provisions as well as under other Conventions adopted by IMO and under International Labour Organization (ILO) Conventions. The exercise of such control measures is traditionally described as "port state control". Governments often co-operate through regional Memoranda of Understanding (MOUs) on port state control.

2.11.3 Governments should not give more favourable treatment to ships flying the flag of a State which is not a signatory to the SOLAS Convention and should apply the same control and compliance measures. However, ships that fall outside the Maritime Security Measures because of their size are subject to the Measures established by the Government.

2.11.4 Under the Maritime Security Measures, RSOs cannot apply control and compliance measures on behalf of Government.

Duly authorized officers

2.11.5 Governments can authorize duly authorized officers to apply the control and compliance measures under the Maritime Security Measures. Their authorization is usually through the Administration and the officers may also undertake other control functions.

2.11.6 Duly authorized officers applying control and compliance measures under the Maritime Security Measures should:

- **.1** be knowledgeable of the Maritime Security Measures and shipboard operations;
- **.2** be able to communicate with the ship's master, SSO and other officers in English;
- **.3** receive the training necessary to fully undertake the control functions that they are authorized to carry out. They may be assisted by persons with specialized search expertise;
- **.4** receive the training necessary to ensure their proficiency in safety procedures when boarding a ship, particularly if boarding is to take place at sea. This training should specifically cover emergency evacuation procedures and procedures for entering enclosed spaces on ships;
- **.5** when boarding a ship, carry and present identification documentation which includes their authorization to impose control measures. Procedures should be in place to allow a ship's master or SSO to verify the identity of duly authorized officers; and
- **.6** when on board, comply with the security measures and procedures that are in place on the ship unless they are incompatible with the control activities being undertaken.

Pre-arrival information procedures

2.11.7 If requested to do so, a ship has to provide security-related information prior to entering into a port. The port State should specify the information required and provide the names and contact details of who should receive the information. This information can be assessed to establish the security risk that a particular ship may pose and to determine whether control and compliance measures should be taken in respect of the ship.

2.11.8 Most Governments have specified the minimum time before arrival in port that a ship should notify its intention to arrive and provide the necessary security-related information. The time can vary between 24 and 96 hours prior to arrival. Special arrangements may apply when ships are on a short international voyage or undertake intensive short sea scheduled services on a daily basis, such as passenger ro–ro ferries.

2.11.9 IMO has developed a standard data set of the security-related information that a ship might be expected to provide (refer to appendix 4.6 – Standard data set of security-related pre-arrival information). The standard data set does not preclude a Government from requesting further security-related information on a regular basis or in specified circumstances. When Governments require additional information, the shipping industry should be appropriately advised.

Clear grounds

2.11.10 A duly authorized officer, on analysing the security-related information provided by a foreign-flagged ship and any other relevant information available relating to the ship intending to enter port, may consider that there are clear grounds that the ship may not be in compliance with the Maritime Security Measures. Examples of such clear grounds could include:

- **.1** evidence or reliable information that the ship has serious security deficiencies;
- **.2** receipt of a reliable report or complaint that the ship does not comply with the requirements in the Maritime Security Measures;

.3 evidence or reliable information that the ship had:

- a ship/port interface which did not comply with the Maritime Security Measures and did not take either appropriate additional security measures or complete a DoS with the port facility; or
- a ship-to-ship activity with another ship which did not comply with the Maritime Security Measures and did not either take appropriate security measures or complete a DoS with the other ship.

.4 evidence or reliable information that the ship had:

- a ship/port interface which did not have to comply with the Maritime Security Measures and did not either take appropriate additional security measures or complete a DoS with the port facility; or
- a ship-to-ship activity which did not have to comply with the Maritime Security Measures and did not either take appropriate additional security measures or complete a DoS with the non-SOLAS vessel.

.5 evidence that the ship holds a sequentially issued Interim ISSC, contrary to the Maritime Security Measures;

.6 failure of the ship to provide all of the requested security-related information identified in paragraph 2.12.20.

2.11.11 At the end of an inspection, the duly authorized officer should ensure that the ship's master or SSO is provided with a report giving the results of the inspection, details of any action taken by the officer and a list of any non-compliances to be rectified by the master, SSO or shipping company. The report should be made in accordance with the format set out in appendix 2.14 – Report of the imposition of a control and compliance measure.

2.11.12 The duly authorized officer could at this time:

.1 request the ship to rectify the non-compliance;

.2 require the ship to proceed to a specified location within the territorial sea or internal waters of the port State;

.3 undertake a detailed inspection of the ship, if the ship is within the territorial sea of the port State; and/or

.4 deny entry into port.

2.11.13 If a deficiency is identified as a result of a detailed inspection, this could lead to the imposition of further control measures.

2.11.14 If a ship that has been advised of the intention to take control measures under the Maritime Security Measures decides to withdraw its intention to enter port, the control measures proposed by the duly authorized officer no longer apply. Any other steps that are taken in respect of the ship must be based on, and consistent with, international law.

Ship inspection in port

2.11.15 Under the Maritime Security Measures, a ship can also be inspected to assess its compliance when in port. Normally, an inspection starts with verifying the presence and validity of the ship's ISSC or Interim ISSC. A copy of a Certificate is not accepted as being a valid ISSC or Interim ISSC.

2.11.16 On the basis of observation, a duly authorized officer can establish that there are clear grounds for believing that the ship is not in compliance with the requirements of the Maritime Security Measures.

2.11.17 A duly authorized officer may not have been challenged on boarding the ship or may find that restricted areas on the ship are not secured.

2.11.18 A duly authorized officer could check:

.1 that the ship is operating at the security level applying to the port facility, or at a higher security level set by the ship's Administration;

.2 that security drills have been carried out at the required interval; and

.3 the records of the last 10 ports of call and any ship-to-ship activity undertaken during the period of the last 10 ports of call.

2.11.19 Examples of clear grounds warranting an inspection could include:

.1 evidence that the ship's ISSC is not valid or has expired;

.2 evidence or observation that the ship's crew are not familiar with essential shipboard security procedures or cannot carry out ship security drills;

.3 evidence or observation that key members of the ship's crew are unable to communicate with crew members with security responsibilities.

2.11.20 The clear grounds that could apply to a ship intending to enter port could also apply to a ship in port.

2.11.21 If there are clear grounds, or no valid ISSC or Interim ISSC is on board, control measures could be applied to the ship. Any control measures must be proportionate with the identified security deficiencies. In deciding the control measures that should be applied, the duly authorized officer may consider if the ship can:

.1 maintain communication with the port facility;

.2 prevent unauthorized access to the ship and to restricted areas on the ship; and

.3 prevent the introduction of unauthorized weapons, incendiary devices or explosives to the ship.

2.11.22 If a duly authorized officer considers that the ship is not in conformity with the requirements of the Maritime Security Measures, parts of the ship's SSP may be inspected.

2.11.23 Parts of an SSP are confidential and can only be inspected by a duly authorized officer with the consent of the ship's Administration. The confidential parts of an SSP are the:

.1 identification of restricted areas and measures to prevent access to them;

.2 procedures for responding to security threats;

.3 procedures for responding to the instructions received from the ship's Administration at security level 3;

.4 details of the duties of ship personnel with security responsibilities;

.5 procedures for inspecting, testing and calibrating security equipment on the ship;

.6 locations of the SSAS activation points; and

.7 guidance on the use of the SSAS.

2.11.24 The control measures that could be applied to a ship in port include:

.1 more detailed inspection of the ship, including searches – which could lead to the imposition of more stringent control measures;

.2 delaying the ship;

.3 detention of the ship;

.4 restrictions on operation – including unloading or loading and its movement within the port;

.5 expulsion from the port; and

.6 lesser administrative or corrective measures.

Notifications

2.11.25 When control measures are taken with respect to a ship, the ship's Administration and the RSO that issued the ship's ISSC or Interim ISSC should be notified without delay.

2.11.26 Under the Maritime Security Measures, Administrations are required to establish a contact point that can be available at any time to receive and act upon reports from Governments exercising control and compliance measures.

2.11.27 Refusing a ship the right to enter port, the detention of a ship or the expulsion of a ship from port has to be reported to the Consular representative of its flag State.

2.11.28 The control measures taken in respect of a ship under the Maritime Security Measures must also be reported to IMO.

2.11.29 Control measures should only be imposed until the non-compliance which gave rise to them is rectified.

2.11.30 Every effort should be taken to avoid undue detention or delay. The Maritime Security Measures provide for compensation to be claimed for loss or damage if a ship is unduly delayed.

Immediate security threat

2.11.31 Denial of entry into port or expulsion from port should only be imposed if the duly authorized officer believes that the ship poses an immediate security threat and that there is no other appropriate means of removing the threat.

2.11.32 In such instances, other States whose ports the ship is known to be intending to visit, and any relevant coastal States, should be informed, in confidence, of the circumstances which led to the denial or expulsion.

2.11.33 The same procedure could be followed if a ship intending to enter port refuses to permit an inspection when notified by a duly authorized officer of the intention to take control measures.

Experience to date

2.11.34 The reports of Port State Control Memoranda of Understanding indicate that security-related deficiencies represent some 3% to 5% of the deficiencies found on SOLAS ships. Ships with security-related deficiencies are almost invariably found to have safety or other deficiencies.

2.11.35 Further guidance on aspects of control and compliance measures from the perspective of ship operators is in subsection 4.10. A more detailed description of implementing control and compliance measures can be found in the Procedures for Port State Control booklet which is referenced on IMO publications webpage at: www.imo.org

2.12 Ship security communications

Requirement for alert and identification systems

2.12.1 Under the Maritime Security Measures, all SOLAS ships have to have a ship security alert system (SSAS).

2.12.2 Under provisions elsewhere in the SOLAS Convention, the following SOLAS ships are required to be fitted with an automatic identification system (AIS):

- **.1** passenger ships irrespective of size;
- **.2** cargo ships of 300 gross tonnage and upwards engaged on international voyages;
- **.3** cargo ships of 500 gross tons and upwards not engaged on international voyages.

2.12.3 Also under provision elsewhere in the SOLAS Convention, the following SOLAS ships engaged on international voyages have to be fitted with a long-range identification and tracking (LRIT) system:

- **.1** passenger ships, including high-speed craft;
- **.2** cargo ships, including high-speed craft, of 300 gross tonnage and upwards; and
- **.3** mobile offshore drilling units.

Ship security alert systems

2.12.4 A ship security alert system (SSAS) transmits a covert alarm to one or more competent authorities ashore indicating that the security of the ship is under threat or has been compromised. Ship security alerts can be activated in the event of any serious security incident, including acts of piracy and armed robbery against the ship.

2.12.5 Guidance on installation and operation of SSASs on ships is in paragraphs 4.6.1 to 4.6.11. These details need not be included in SSPs but can be included in a separate document known to the master, SSO or other senior shipboard personnel selected by the Company.

2.12.6 Administrations designate one or more competent authorities ashore to receive ship security alerts from their SOLAS ships. Any designated competent authority should be able to obtain a covert verification from the ship and alert the country's security forces responsible for initiating the security response to acts of violence against ships.

2.12.7 Administrations have to establish an effective means of communication between their competent authorities and the security force responsible for the response.

2.12.8 Many Administrations have designated CSOs and a selected Maritime Rescue Co-ordination Centre (MRCC), or equivalent agency, as their competent authorities. Protocols have to be in place to ensure immediate communication between CSOs receiving a ship security alert and the selected competent authority (which is the point of contact with the responding security force). CSOs are often in the best position to seek verification of alerts from their ships. Covert verification can be achieved by pre-arranged exchanges of messages.

2.12.9 Other Administrations have designated a MRCC as their sole competent authority for the receipt of ship security alerts. In such cases, the MRCC should establish procedures for verifying individual ship security alerts.

2.12.10 Unless directed by the Administration or security force, a competent authority who receives a ship security alert should not overtly acknowledge its receipt to the ship.

2.12.11 Administrations should provide guidance to competent authorities on the procedures to be followed on the:

- **.1** prioritization of ship security alerts;
- **.2** distinction between covert and overt alarms;
- **.3** receipt of false security alerts and distress/security double alerts; and
- **.4** testing ship security alert systems and associated communication procedures.

2.12.12 IMO has requested that information be provided on the receipt of false security alerts and distress/security double alerts.

2.12.13 Administrations should ensure that ships flying their flag test ship security systems and associated communication procedures on a regular basis. When doing so, it should be made clear that it is a TEST alert.

2.12.14 In consultation with their responding security forces, Administrations should develop protocols on notifying MRCCs in the vicinity of the ship, their Governments, and the Administrations or response organizations in adjacent countries of the receipt of an alert.

2.12.15 Upon receiving notification of a security alert from a ship entitled to fly its flag, the Administration must immediately notify the State(s) in the vicinity of which the ship is presently operating. If a security alert is received from a ship that is not entitled to fly its flag, that Contracting Government must immediately notify the relevant Administration and, if appropriate, the State(s) in the vicinity of which the ship is presently operating.

Automatic identification systems

2.12.16 Regulation 19 of SOLAS chapter V requires AIS to be fitted aboard all ships of 300 gross tonnage and upwards engaged on international voyages, cargo ships of 500 gross tonnage and upwards not engaged on international voyages and all passenger ships irrespective of size. The requirement became effective for all ships by 31 December 2004.

2.12.17 Ships fitted with AIS are expected to maintain AIS in operation at all times except where international agreements, rules or standards provide for the protection of navigational information. Performance standards for AIS were adopted in 1998.

2.12.18 The regulation requires that AIS shall:

- **.1** provide information – including the ship's identity, type, position, course, speed, navigational status and other safety-related information – automatically to appropriately equipped shore stations, other ships and aircraft;
- **.2** receive automatically such information from similarly fitted ships;
- **.3** monitor and track ships; and
- **.4** exchange data with shore-based facilities.

2.12.19 AIS-generated ship data should not be available on open-source internet sites as it is considered to be detrimental to the safety and security of ships and port facilities, and undermines the efforts of IMO and its Member States to enhance the safety of navigation and security in the international maritime transport sector.

Pre-arrival notification

2.12.20 As explained in subsection 2.11 on control and compliance measures, a ship intending to enter a port of another Contracting Government may be required to provide the following information to responsible officials:

- **.1** confirmation of a valid ISSC and the name of its issuing authority;
- **.2** the security level at which it is currently operating;
- **.3** the security level at which it operated in the last 10 ports of call where it conducted a ship/port interface;
- **.4** any special or additional security measures that were taken in the last 10 ports of call where it conducted a ship/port interface, e.g., Declarations of Security;
- **.5** confirmation that the appropriate ship security procedures were maintained during any ship-to-ship activity during the last 10 ports of call, e.g., with ships that are not required to comply with the Maritime Security Measures or persons and goods rescued at sea;
- **.6** other practical security-related information, but not details of the SSP. Examples include:
  - information contained in the Continuous Synopsis Record;
  - the location of the ship at the time of reporting;
  - the expected time of arrival;
  - crew and passenger lists;
  - general description of cargo being carried;
  - person(s) responsible for appointing crew and other shipboard personnel;
  - information on charter parties.

2.12.21 The Contracting Government may seek supplementary information as a condition of entry or, subsequent to entry, additional information to validate the data set provided. The request for supplementary information may not include details of the SSP.

2.12.22 Details of a ship's responsibilities in providing the above information are documented in paragraphs 4.6.13 to 4.6.15.

2.12.23 Experience to date indicates that Administrations have established standing requirements on:

- **.1** the information to be provided;
- **.2** how the information is to be provided; and
- **.3** the time period required for submission of pre-arrival information.

2.12.24 IMO has developed the standard data set in this respect (refer to appendix 4.6 – Standard data set of security-related pre-arrival information). Administrations are expected to advise shipping companies of these requirements (as their ships will not be requested for the information by duly authorized officers).

Long-range identification and tracking of ships

2.12.25 Long-range identification and tracking (LRIT) was developed at IMO as a means of enhancing maritime security by providing information on ship identity and its current location in sufficient time for a Contracting Government to evaluate the security risk posed by a ship off its coast and to respond, if necessary. A robust international scheme for long-range identification and tracking of ships is an important and integral element of maritime security. Contracting Governments can request LRIT information from foreign-flag ships transiting within 1,000 nm of their coasts, or intending to enter their ports. A number of countries therefore link pre-arrival notification to a request for LRIT data.

2.12.26 The LRIT regulation in the SOLAS Convention (refer to chapter V, regulation 19-1) entered into force on 1 January 2008, with all ships now required to be compliant with the exception of ships that are operating exclusively in coastal areas defined by its Administration and are fitted with an AIS.

2.12.27 LRIT is a satellite-based tracking system designed to utilize existing shipboard equipment such as the Global Maritime Distress and Safety System (GMDSS) to track SOLAS-class vessels over 300 gross tonnage on international voyages. Four times daily and at six-hour intervals, ships are required to transmit LRIT information, which comprises:

- **.1** the ship's identity;
- **.2** the ship's location (latitude and longitude); and
- **.3** the date and time of the position.

2.12.28 Unlike AIS, LRIT communication is addressed (i.e., it is a secure point-to-point transmission of information) rather than a broadcast.

2.12.29 While routine tracking is every six hours, the performance standards stipulate that onboard terminals must be capable of being remotely reconfigured to transmit LRIT information as frequently as every 15 minutes. Once communication has been established, the satellite terminal automatically responds to subsequent polling requests.

2.12.30 Each Administration must have a data centre (DC) to which its ships transmit. The DC is the repository of all of the flag State's LRIT information and is connected to the wider International LRIT system via the International Data Exchange (IDE), through which all information is routed to other DCs. A Government not wishing to establish its own DC may utilize the services of another DC. Each Administration can associate itself with only one DC. The majority of Administrations contract their DC services to third-party service providers.

2.12.31 LRIT information is collected by each Administration by means of its DC and shared with requesting Contracting Governments based on strict entitlements defined in the SOLAS regulation. In addition to establishing or joining a DC, each Government that has flag vessels must formally appoint an application service provider to:

.1 conduct conformance tests on those ships;

.2 manage the associated communications between the ship, the communications service provider and the DC; and

.3 issue ships with a conformance test report.

2.12.32 A Contracting Government is entitled to request and receive LRIT data about ships:

.1 entitled to fly its own flag irrespective of where the ships are located;

.2 flying the flag of another Contracting Government that have indicated their intention to enter a port facility under the jurisdiction of the requesting Contracting Government; and

.3 flying the flag of another Contracting Government that are navigating within 1000 nautical miles of the coast of the requesting Contracting Government.

2.12.33 An Administration may at any time, in order to meet security or other concerns, decide not to provide LRIT information about its ships to another Contracting Government. In such a case, the Administration concerned must communicate its decision to IMO, which, in turn, is required to inform all Contracting Governments of the action. To date, no Administration has done so.

2.12.34 International agreement restricts the use of data to Contracting Governments and Search and Rescue (SAR) authorities. Contracting Governments can share within their own Government the data that they receive in response to a request from another Contracting Government. However, data requested or received by a SAR authority within a Contracting Government must only be used for SAR purposes.

2.12.35 As LRIT is a user-pay system, all requesting data centres must pay DCs supplying information for the information that is received. Experience to date indicates that a regular LRIT position report typically costs the equivalent of US$0.25 while a poll costs US$0.50 and a terminal reconfiguration US$3.00.

2.12.36 All requests for and receipts of LRIT information are logged in a journal maintained by the IDE. This journal is used for costing and billing, as well as for auditing purposes.

2.12.37 The International Mobile Satellite Organization (IMSO) provides oversight of the international LRIT system and conducts annual audits of each LRIT data centre.

2.13 Alternative security agreements

Introduction

2.13.1 Governments can conclude bilateral or multilateral alternative security agreements (ASAs) for short international voyages on fixed routes between dedicated port facilities. These agreements allow the security measures and procedures applied to the port facilities and ships to differ from those required under the Maritime Security Measures.

2.13.2 Elsewhere in the SOLAS Convention, a short international voyage is defined in the context of life-saving appliances and arrangements as: "... an international voyage in the course of which a ship is not at any time more than 200 miles from a port or place in which the passengers and crew could be placed in safety. Neither distance between the last port of call in the country in which the voyage begins and the final port of destination nor the return voyage shall exceed 600 miles. The final port of destination is the last port of call in the scheduled voyage at which the ship commences its return voyage to the country in which the voyage began."

Application

2.13.3 The port facilities included in an ASA can only interface with ships operating on the fixed routes to which the ASA applies.

2.13.4 All ships operating on the fixed route between the port facilities covered by an ASA have to be covered by that ASA.

2.13.5 Third-party flagged vessels can be covered by an ASA if their Administration ensures that their ships fully comply with the provisions in the ASA.

2.13.6 The ships covered by an ASA cannot conduct any ship-to-ship activity with a ship not covered by the ASA or ship/port interfaces at any other port facility.

Procedure

2.13.7 A combined PFSA and SSA should be undertaken by the national authorities and other relevant Government organizations (e.g., customs and immigration services) of the States involved.

2.13.8 The combined security assessment should be based on a shared understanding of the security risks likely to be associated with the port facilities, the ships and the voyages to be covered by the proposed agreement. It should cover all ship/port interfaces at the port facilities and any ship-to-ship activities to be undertaken between the ships.

2.13.9 When undertaking such a combined security assessment, national authorities should consult the relevant authorities in any country likely to be affected by the operation of the proposed agreement.

2.13.10 The combined assessment should identify the security measures and procedures appropriate at the port facilities and to the ships involved. All parties to the agreement should approve the combined security assessment and the resulting security measures and procedures.

2.13.11 The respective national authorities should then take the necessary actions to ensure that the required security measures and procedures are applied and maintained at the port facilities and on the ships for the duration of the agreement.

2.13.12 The security procedures should ensure that the required control measures applying to embarking passengers and vehicles are carried out at the port facility prior to the loading of a ship when the ship, such as a ro–ro ferry, has a short turn-round time.

2.13.13 National authorities concluding such agreements are required to notify IMO by accessing the alternative security agreement screen in GISIS (refer to paragraph 2.19.2) at http://gisis.imo.org/ and providing the following information:

- **.1** ships and port facilities covered by the agreement;
- **.2** name of agreement;
- **.3** fixed route covered by the agreement;
- **.4** information on consultation with other Governments;
- **.5** date of entry into force of the agreement;
- **.6** periodicity of review of the agreement; and
- **.7** has a security agreement been withdrawn?

2.13.14 Further guidance on ASAs is in subsections 3.2 and 4.2.

Review

2.13.15 The operation of an ASA should be continually monitored and reviewed in the light of experience. A review should take place if there is any significant security threat or incident involving the port facilities or ships covered by the agreement.

2.13.16 Under the Maritime Security Measures, ASAs have to be reviewed every five years.

Experience to date

2.13.17 ASAs have covered such aspects of international ferry services as:

- **.1** ship security alerts;
- **.2** identification and screening of security personnel;
- **.3** reciprocal recognition of SSP approvals;
- **.4** acceptance of minor differences in regulatory requirements; and
- **.5** alternative security requirements to those in the Maritime Security Measures.

2.14 Equivalent security arrangements

2.14.1 National authorities can allow port facilities, groups of port facilities and ships to implement other security measures equivalent to those in the Maritime Security Measures. Such measures have to be at least as effective as those prescribed in the Maritime Security Measures. Few national authorities have allowed equivalent security arrangements.

2.14.2 Designated Authorities can allow a port facility or a group of port facilities to implement security measures or procedures equivalent to those in the Maritime Security Measures without having to appoint a PFSO or submit a PFSP. However, these equivalent security arrangements (ESAs) are allowed only under limited circumstances, applying to port facilities with more than occasional use by SOLAS ships but without frequent services, or involving special operations (e.g., berths used by SOLAS ships at naval facilities with military security measures and procedures).

2.14.3 As with port facilities used only occasionally by SOLAS ships, the equivalent security arrangements allowed by Designated Authorities should identify a person ashore with responsibility for shore-side security, including the completion of a DoS.

2.14.4 ESAs should not be used as a stop gap, allowing port facilities frequently used by SOLAS ships to delay or avoid full implementation of the Maritime Security Measures.

2.14.5 Similarly, ESAs should not allow SOLAS ships to avoid full compliance with the requirements of the Maritime Security Measures.

2.14.6 National authorities concluding such arrangements are required to notify IMO by accessing the equivalent security arrangement for ships or for port facilities screen in GISIS at http://gisis.imo.org/ and providing the following information:

- **.1** name of ships or port facilities;
- **.2** name of the arrangement; and
- **.3** description of the arrangement.

2.14.7 A limited number of ESAs have been reported to IMO. A number apply to port facilities occasionally used by SOLAS ships. For ships, some apply to ships operating regional shipping services while others appear to apply to ships trading internationally.

2.14.8 The limited particulars of individual ESAs do not allow any useful assessment of experiences to date.

2.15 Enforcement actions

Introduction

2.15.1 Governments are ultimately responsible for ensuring that their port facilities and SOLAS ships fully comply with the Maritime Security Measures.

2.15.2 For SOLAS ships, failure to comply with the Maritime Security Measures could lead to Governments applying control measures against non-compliant ships. The application of control measures in this way can ultimately have significant implications for all ships flying the State's flag, and is best avoided by ensuring compliance with the Maritime Security Measures.

2.15.3 Security inspections of their port facilities and SOLAS ships by national authorities can result in enforcement action to ensure correction of any identified security deficiencies and prevent such deficiencies recurring in future.

2.15.4 The enforcement actions following the identification of security deficiencies will depend on:

.1 whether the deficiencies prevent the port facility or SOLAS ship from continuing to operate at security levels 1 to 3;

.2 whether the deficiencies compromise the ability of the port facility or SOLAS ship from continuing to operate at security levels 1 to 3;

.3 the extent of the sanctions available to the national authorities under their legislation.

2.15.5 Whatever the ultimate sanctions available to a national authority are, Governments should take a stepped approach when seeking to ensure that the port facility or ship corrects an identified deficiency which does not prevent the port facility or ship from continuing to operate at security levels 1 to 3. A more robust approach may have to be taken if a port facility or ship has a security deficiency which compromises its ability to operate at security levels 1 to 3.

Stepped approach

2.15.6 A stepped approach follows distinct steps:

.1 advice to the port facility or ship on correcting the deficiency;

.2 further persuasion of the port facility or ship on the need to correct the deficiency;

.3 formal notification of the requirement to correct the deficiency;

.4 commencement of proceedings to impose sanctions for the failure to correct the deficiency;

.5 the imposition of sanctions for failing to correct the deficiency.

2.15.7 An example of a stepped approach is shown below:

Type of enforcement action	Seriousness of contravention	Impact on operator	Legal basis for action
Counselling	Minor	Low	None required
Notice of non-compliance	Minor	Low	None required
Compliance agreement (in lieu of penalty)	Moderate	Low to medium	Required
Fine	Moderate	Medium	Required
Suspension or restriction of activities	Significant	Medium to high	Required
Withdrawal of Certificate or Statement of Compliance	Significant	Medium to high	Required
Imposition of penalties	Significant	High	Required

2.15.8 The procedures followed at each step should be taken in the knowledge that ultimately sanctions may have to be imposed. The maintenance of evidence of the deficiency and of records of the actions taken at each stage could be essential if proceedings are taken imposing sanctions and if they are to be upheld in any subsequent appeal proceedings

Counselling

2.15.9 Once a deficiency has been identified, details should be recorded and evidence collected and protected. The deficiency should immediately be discussed with the PFSO or SSO to establish what action is needed to correct the deficiency. Advice could be offered on the appropriate actions to take.

2.15.10 Temporary alternative security procedures or measures could be agreed with the port facility or ship which should be applied until the original deficiency has been corrected. Records should be kept of all discussions with the PFSO or SSO. A period should be agreed in which the deficiency should be corrected and a further inspection undertaken.

2.15.11 If it is established that the deficiency has not been corrected within the agreed time, efforts should be made to persuade the PFSO or SSO of the need to correct the deficiency and to maintain the agreed temporary alternative security procedures or measures. At this stage, the national authority may seek to involve the port facility operator or, in the case of ships, the CSO. Records should be kept of any discussion and the PFSO or SSO should be advised in writing of the deficiency and the action required to correct the deficiency.

Formal notification

2.15.12 If informal advice and persuasion has not secured correction of the deficiency, or if the deficiency is serious or recurring, the PFSO, master or SSO should receive a formal notification, in writing, describing the deficiency, the action needed to correct it, and the PFSO's, master's or SSO's responsibility to remedy the deficiency. Emphasis could be placed on the possible security and safety implications of the continued deficiency for the ship and those using the facility. A sample of such a notice for a SOLAS ship is shown in appendix 2.9 – Sample of a notice of non-compliance; the notice for a port facility would be similar.

2.15.13 The formal notification should set a period of time within which the deficiency should be corrected. Also, it should advise that failure to correct the discrepancy within that period could lead to the commencement of formal proceedings to achieve compliance, which, in turn, could lead to sanctions being imposed on the port facility or ship. The formal notification should be issued to the senior management of the port facility or shipping company rather than to the PFSO/SSO/CSO.

2.15.14 Once again, it is important to record all contacts and to retain and protect correspondence and evidence relating to the deficiency.

Serious security deficiencies

2.15.15 Serious security deficiencies are those which compromise the ability of the port facility or SOLAS ship to continue to operate at security levels 1 to 3.

2.15.16 Immediate action may need to be taken to secure correction of such deficiencies and, initially, the inspector should discuss with the PFSO, master or SSO alternative security measures and procedures of equal effect which could be put in place to allow the facility or ship to operate at security levels 1 to 3. If such alternatives are identified and there is no immediate security risk, the port facility or ship should be given reasonable time to introduce them.

Restriction or suspension of activities

2.15.17 If alternatives cannot be found or applied within a reasonable time-frame, or if agreed alternative security procedures or measures have not been put in place, the national authority could, in the most serious cases, have the authority to be able to restrict or suspend specified activities at a port facility or on a ship.

2.15.18 A restriction notice could limit the activities that could be undertaken at the port facility or on the ship until action has been taken to correct the serious security deficiency.

2.15.19 When an immediate security risk linked to a specific activity has been identified, a suspension notice could stop the activity being undertaken by the port facility or ship pending correction of the serious security deficiency.

2.15.20 A restriction or suspension notice could be lifted when the national authorities consider that:

.1 the serious deficiency has been corrected, or

.2 agreed security measures or procedures of equal effect are in place and operating effectively.

Suspension or withdrawal of an approved PFSP or SSP

2.15.21 There could be circumstances when cumulative security failings at a port facility or on a ship could lead to the:

.1 suspension or withdrawal of the approved PFSP, and the Statement of Compliance if one has been issued; or

.2 suspension or withdrawal of the approved SSP and ISSC.

2.15.22 National authorities may require completion of a PFSA and submission of an amended PFSP before reinstating a PSFP which has been suspended or withdrawn.

2.15.23 Similarly, a new SSA may have to be undertaken and an amended SSP submitted before a suspended or withdrawn SSP can be reinstated and the initiation of the procedures leading to the re-issue of an ISSC.

Imposition of penalties

2.15.24 In the occasional situation that none of the preceding steps has resulted in correction of the deficiency, the national authority may commence proceedings to seek sanctions against the port facility or ship operator. The procedures should be clearly stated in national legislation and are likely to include the right to appeal against the imposition of sanctions.

2.15.25 The proceedings could involve hearings before an administrative or judicial tribunal where the national authority is required to explain and, if necessary, defend the actions that it has taken to seek correction of the deficiency. The documentary evidence of the actions taken and of the deficiency could be essential to the success of the national authority's case.

2.15.26 The sanctions that can be imposed on a port facility or ship for failure to correct an identified deficiency should, again, be specified in national legislation. The authority to impose sanctions may rest with a senior official within the national authority or judicial body. Sanctions could include administrative, civil and criminal penalties. The national authority may be required to sustain its case through any appeal procedures that might follow the imposition of sanctions. The sanctions should be effective, proportional and persuasive.

2.16 Training of Government officials with security responsibilities

Introduction

2.16.1 Government officials undertake an extensive range of responsibilities under the Maritime Security Measures relating to all aspects of the security of port facilities and ships. Ensuring individual officials have the knowledge and competencies needed to undertake their responsibilities can make a significant contribution to the effective implementation of the Maritime Security Measures.

2.16.2 The following paragraphs provide guidance on the competencies that Government officials could have to allow them to successfully undertake their responsibilities relating to the implementation or oversight of the Maritime Security Measures.

Duties of officials

2.16.3 The duties of officials working in Designated Authorities could include:

.1 advising on, and overseeing, the implementation of the Maritime Security Measures to port facilities;

.2 drafting and implementing national legislation and regulations implementing the Maritime Security Measures for port facilities;

.3 consulting ports and related industries on security issues;

.4 communicating the applicable security level;

.5 determining which port facilities used by SOLAS ships have to appoint a PFSO and prepare a PFSP;

.6 appointing a person ashore, who has responsibility for shore-side security at port facilities occasionally used by SOLAS ships, to liaise with SOLAS ships using that facility;

.7 authorizing RSOs to undertake port facility related tasks for the Designated Authority and subsequently monitoring their activities and outputs;

.8 advising on security threats;

.9 undertaking, reviewing and approving PFSAs, including those undertaken by RSOs;

.10 determining policy on Declarations of Security;

.11 determining the requirements for port facilities to report security incidents;

.12 determining the security records to be kept by port facilities and the duration of retention;

.13 advising on the preparation and content of PFSPs;

.14 reviewing and approving PFSPs and determining the types of amendments to an approved plan that have to be submitted for approval;

.15 undertaking inspections and verification relating to the issue and endorsement of Statements of Compliance, and

.16 undertaking inspections of port facilities to assess their compliance with the Maritime Security Measures.

2.16.4 The duties of officials working in Administrations could include:

.1 advising on and overseeing the implementation of the Maritime Security Measures for ships;

.2 drafting and implementing national legislation and regulations implementing the Maritime Security Measures for ships;

.3 consulting the shipping and related industries on security issues;

.4 communicating the applicable security level;

.5 authorizing RSOs to undertake delegated responsibilities for the Administration and subsequently monitoring their activities and outputs;

.6 advising on security threats;

.7 advising on the preparation of SSAs;

.8 determining policy on Declarations of Security;

.9 determining the requirements for reports of security incidents from ships;

.10 determining the security records to be kept by ships and the duration of retention;

.11 advising on the preparation and content of SSPs;

.12 assessing and approving SSPs and determining the types of amendments to an approved plan that have to be submitted for approval;

.13 undertaking inspections and verification relating to the issue and endorsement of International Ship Security Certificates;

.14 issuing Interim International Ship Security Certificates;

.15 exercising control and compliance measures under the Maritime Security Measures to foreign-flagged vessels using, or intending to use, their ports;

.16 undertaking inspections of their SOLAS ships to assess their compliance with the Maritime Security Measures;

.17 advising on the security procedures and measures appropriate for non-SOLAS vessels;

.18 issuing Certificates of Proficiency to shipboard personnel under the STCW Convention and Code (refer to paragraphs 2.9.1 to 2.9.11 and subsection 4.5).

Training requirements

2.16.5 Given the range of duties that Government officials can exercise under the Maritime Security Measures, their training should impart an appropriate level of knowledge of:

.1 the drafting and implementing of national legislation, including regulations;

.2 the requirements of the Maritime Security Measures relating to port facilities and ships;

.3 the supervision of RSOs authorized to undertake duties for national authorities;

.4 the security threats that could be experienced at port facilities and on ships;

.5 risk assessments of security incidents;

.6 the security measures and procedures appropriate to mitigate security threats that could occur at port facilities and ships;

.7 the completion and assessment of PFSAs;

.8 the preparation and content of SSAs;

.9 the preparation, content, submission and approval of PFSPs and SSPs;

.10 the content, submission and approval of amendments to approved PFSPs and SSPs;

.11 the undertaking of inspections or verifications associated with the issue and endorsement of Statements of Compliance of a port facility;

.12 the undertaking of inspections or verifications associated with the issue and endorsement of International Ship Security Certificates;

.13 the undertaking of inspections and assessments relating to the issue of Interim International Ship Security Certificates;

.14 the exercise of control and compliance measures in respect of foreign-flagged vessels to assess their compliance with the requirements of the Maritime Security Measures;

.15 the undertaking of security inspections of port facilities and SOLAS ships to assess their compliance with national security requirements – including the collection and protection of evidence relating to identified security deficiencies where enforcement action may be required.

2.16.6 For a national authority to be confident that its inspectors are adequately qualified to carry out their delegated responsibilities, it is recommended that the authority have an approved training curriculum. Under such a scenario, the training may be delivered by external training organizations according to specifications determined by the national authority. The basic or core training elements, the details of which are shown as a sample curriculum in appendix 2.10 – Sample of a core training curriculum for officials in national authorities – could include:

.1 knowledge of the national authority's legislative framework;

.2 knowledge of the international maritime security framework;

.3 knowledge of the maritime industry over which the authority has jurisdiction;

.4 the responsibilities of the national authority specified in the Maritime Security Measures;

.5 the responsibilities delegated to inspectors (attending the course);

.6 a code of conduct;

.7 description of the authority's regulatory oversight programme;

.8 procedures for preparing, conducting and reporting the results of verifications;

.9 procedures for handling cases of non-compliance;

.10 procedures for observing or participating in exercises;

.11 procedures for issuing, renewing, suspending and withdrawing certificates and other forms of authorization; and

.12 procedures for conducting awareness and education activities with industry and labour associations, port security committees and the public.

2.16.7 Experience to date indicates that:

.1 for maximum effect and to facilitate practical sessions and involvement of participants in discussion, the course size should be in the range of 6–12 participants;

.2 complementary to the above and as an integral part of the basic training requirements, workshops could be held on report writing, presentations, interviews and consultations; and

.3 in the spirit of continuous learning, as the qualified personnel become more experienced, they should have access to more advanced training in specialized areas (such as methodologies for conducting threat and risk assessments, techniques for investigating serious contraventions of regulatory requirements) and participation in emergency response and preparedness exercises.

Code of conduct

2.16.8 IMO issued a Code of Good Practice for Port State Control Officers in 2007 and invited its member Governments and regional port state control regimes to bring the Code to the attention of officials exercising port and coastal State actions. The 28-point Code is based on the following three main principles:

.1 integrity – the state of moral soundness, honesty and freedom from corrupting influences or motives;

.2 professionalism – applying accepted professional standards of conduct and technical knowledge; and

.3 transparency – implying openness and accountability.

2.16.9 The Code may be accessed at IMO's internet site for its circulars:
http://docs.imo.org/Category.aspx?cid=538 or at the internet sites of regional port State control regimes.

2.16.10 Experience to date indicates that some national authorities have adapted the Code for their Government officials and incorporated it into their training curriculum and oversight manuals as a code of conduct.

Identification documents

2.16.11 Government officials are entitled, as part of their duties, to enter port facilities or to board ships and as such should carry appropriate identification documents issued by the Government. Identification documents should include a photograph of the holder of the document. They should also include the name of the holder or have a unique identification number. If the identity document is in a language other than English, French or Spanish, a translation into one of those languages should be provided.

2.16.12 Government officials should present their identification document when requested to do so at access points to port facilities and when boarding a ship.

2.16.13 Port facility and ship security personnel should be able to verify the authenticity of identity documents issued to Government officials and Governments should establish procedures, and provide contact details, to facilitate such validation.

2.16.14 Emergency response services and pilots should also carry appropriate identification documents and present them when boarding a vessel. The authenticity of such identification documents should be capable of being verified.

2.16.15 Only the person in charge of an emergency response team need present an identification document when accessing a port facility or boarding a ship and should inform the relevant security personnel of the number of emergency response personnel entering or boarding.

2.16.16 Government officials who are properly identified and acting in the course of their duties, emergency response personnel and pilots should not be required to surrender their identity documents when entering a port facility or boarding a ship. The issue of visitor identification documents by a port facility or a ship may not be appropriate when Government officials, emergency response personnel or pilots have produced an identity document which can be verified.

2.16.17 Government officials should not be subject to search by port facility or ship security personnel. Any search requirement in an approved security plan could be waived for emergency response personnel responding to an emergency or for a pilot boarding a ship once their identity has been verified.

2.16.18 Port facility security officers should assist ship security officers to verify the identification of Government officials, emergency response personnel or pilots intending to board a ship.

2.17 National oversight

Introduction

2.17.1 Under the Maritime Security Measures, Governments have the responsibility to assess the continuing effectiveness of the security measures and procedures required of their port facilities, shipping companies and ships and the RSOs that are authorized to act on their behalf. Through control and compliance measures, Governments can also assess the compliance of foreign-flagged ships using, or intending to use, their ports.

2.17.2 An effective oversight programme should include continuous monitoring and evaluation of the Government's own performance in the implementation, application and operation of its specific security responsibilities under the Maritime Security Measures.

2.17.3 A national oversight programme should allow Governments to determine the extent to which:

.1 it has met all its obligations under the Maritime Security Measures;

.2 appropriate advice and guidance has been offered to their port facility operators, shipping companies, ships and RSOs;

.3 their port facility operators, shipping companies, ships and RSOs understand and meet their obligations under the Maritime Security Measures;

.4 their port facilities implement the security measures and procedures in their PFSPs;

.5 their SOLAS ships implement the security measures and procedures in their SSPs;

.6 foreign-flagged vessels using their ports comply with the Maritime Security Measures;

.7 inspections, verifications, audits, reviews and control measures promptly identify non-conformities;

.8 immediate action is taken to correct non-conformities; and

.9 their officials undertaking inspections, verifications, reviews and control measures to assess compliance with the Maritime Security Measures have the required training and conduct themselves in a professional manner.

2.17.4 Although not mandatory, a set of governing principles, such as the one shown below, may influence how a national authority intends to implement its oversight programme:

- **.1** *transparency*, by officials being as open as legislation and confidentiality requirements permit;
- **.2** *fairness*, by dealing with non-compliance through actions that are authorized, impartial and appropriate to the risk imposed by the non-compliance while ensuring that there is access to appeal procedures;
- **.3** *timeliness*, by making decisions in a timely manner;
- **.4** *consistency*, by interpreting, administering and enforcing legislation in a consistent manner; and
- **.5** *confidentiality*, by applying all appropriate measures to protect confidentiality or sensitive information.

Seafarer access considerations

2.17.5 The 2002 SOLAS Conference that adopted the Maritime Security Measures and associated Conference resolutions was aware of the potential for the fundamental human rights of seafarers to be adversely affected by the imposition of a security regime on international shipping. It was recognized that seafarers would have the primary duties and responsibilities for implementing the security regime for ships. At the same time, there was concern that the emphasis on port facility security could result in ships and seafarers being viewed as a potential threat to security rather than as partners in the effective implementation of the security regime.

2.17.6 One of the resolutions adopted at the 2002 Diplomatic Conference urged Contracting Governments to take the need to afford special protection to seafarers and the critical importance of shore leave into account when implementing the provisions of the Maritime Security Measures.

2.17.7 In this regard, it was recognized that:

- **.1** there may be conflicts between security and human rights as well as between security and the efficient movement of ships and cargoes in international trade (that is essential to the global economy);
- **.2** there should be a proper balance between the needs of security, the protection of the human rights of seafarers and port workers, and the requirement to maintain the safety, security and working efficiency of ships by allowing access to ship support services (e.g., loading stores, repair and maintenance of essential equipment, and other vital activities that are appropriately undertaken while moored at port facilities);
- **.3** the ISPS Code must not be interpreted in a manner that is inconsistent with existing international instruments protecting the rights and freedoms of maritime and port workers; and
- **.4** in approving PFSPs, Contracting Governments should be aware of the need for seafarers' shore leave and access to shore-based welfare facilities and medical care.

2.17.8 IMO considers that an essential part of national oversight activities is to verify that PFSPs contain provisions to facilitate:

- **.1** shore leave by seafarers;
- **.2** shore access by ships' crews for operational and safety reasons; and
- **.3** the access of legitimate visitors – including those undertaking maintenance or repairs on the ship and representatives of welfare organizations – to and from ships.

2.17.9 National authorities should ensure that:

- **.1** arrangements have been put in place to monitor the effective implementation of such provisions;
- **.2** there are unbiased and non-discriminatory practices in allowing access to shore, i.e., they are irrespective of ships' flags and the nationalities of individual crew members;

.3 neither seafarers nor their legitimate visitors should have to pay for the implementation of such provisions; and

.4 all port facility security personnel are fully aware of the necessity to provide an adequate protection of seafarers' rights and of the humanitarian significance of shore leave.

2.17.10 Contracting Governments and representative organizations of seafarers and shipowners are encouraged to report to IMO any instances where the human element has been adversely impacted by the implementation of the provisions of the Maritime Security Measures. They are requested to bring instances of unfair and selective practices in providing shore leave and access to the shore-based facilities in foreign ports to the attention of IMO's Maritime Safety and Facilitation Committees.

Port facility inspections

2.17.11 The frequency of inspections of port facilities should be determined by the Designated Authority. Inspections can be programmed and arranged in advance or they can be unannounced. Inspections can be undertaken in connection with:

.1 initial, intermediate and renewal verification of the port facility's Statement of Compliance;

.2 following up a report of a security incident; and

.3 assessments of the port facility's compliance with the Maritime Security Measures.

2.17.12 The Designated Authority can undertake covert testing of the security measures and procedures at their port facilities.

2.17.13 Those undertaking inspections for the Designated Authority should have the power to enter port facilities and inspect all or, if appropriate, a sample of the facility's security measures, procedures, documentation and records. Areas for inspection could include:

.1 access control, including to restricted areas;

.2 handling of cargo;

.3 delivery of ship's stores and bunkers;

.4 monitoring the port facility;

.5 handling threats, breaches of security and security incidents;

.6 security communications;

.7 audits and amendments;

.8 procedures for shore leave and visitors to the ship;

.9 procedures for ship-to-shore interface activities;

.10 evacuation procedures; and

.11 protection of sensitive security information, e.g., the security plan.

2.17.14 Appendix 2.11 – Sample of a port facility security inspection report form provides a template for reporting on the results of inspections. It provides examples of questions that could be asked, or issues pursued, when undertaking an inspection as well as including questions that could be asked on the qualifications of:

.1 port facility security officers;

.2 personnel with security responsibilities; and

.3 personnel without security responsibilities.

2.17.15 Those undertaking inspections should record:

- **.1** the security procedures and measures inspected;
- **.2** their observations on the security procedures and measures;
- **.3** the identification of any deficiencies;
- **.4** the action(s) required of the port facility to correct any identified deficiencies, and
- **.5** the action to be taken by the inspector or the Designated Authority.

2.17.16 The identification of deficiencies may lead to enforcement action by the Designated Authority (refer to subsection 2.15).

2.17.17 Experience to date indicates that some Governments exercise oversight to ensure the sustained implementation of PFSPs through frequent and continuous spot checks of the security conditions at port facilities. These checks are sometimes done when officials are conducting other business at the facility.

2.18 Additional security-related instruments and guidance issued by IMO

Introduction

2.18.1 The following paragraphs refer to the security instruments and guidance issued by IMO on:

- **.1** non-SOLAS vessels;
- **.2** port security;
- **.3** the Suppression of Unlawful Acts (SUA) Convention;
- **.4** offshore activities; and
- **.5** specific security issues, including:
  - piracy and armed robbery;
  - drug smuggling;
  - stowaways;
  - illegal migration; and
  - the security of dangerous goods.

2.18.2 This guidance does not relate specifically to the Maritime Security Measures, and Governments retain complete discretion as to the extent they consider the guidance should be reflected, if at all, in PFSAs, PFSPs, SSAs and SSPs prepared under the Maritime Security Measures.

Non-SOLAS vessels

2.18.3 The Maritime Security Measures do not apply to non-SOLAS vessels. However, Governments were specifically encouraged by IMO to establish appropriate measures to enhance the security of ships and port facilities not covered by the Maritime Security Measures, including mobile offshore drilling units on location, and fixed and floating platforms not covered by the Maritime Security Measures.

2.18.4 Governments have complete discretion as to the action they take in respect of their ships and port facilities that are not covered by the Maritime Security Measures. As a result, several Governments have extended the Maritime Security Measures, in whole or part, to domestic passenger shipping services and the port facilities they use; some of these envisage extension to domestic cargo services if a risk assessment establishes the need.

2.18.5 Some Governments have applied security requirements to all their ships and port facilities, including fishing vessels and recreational craft and also covering fishing ports and marinas. Others have focused on harbour craft or other craft that engage in ship-to-ship activities with ships that are covered by the Maritime Security Measures.

2.18.6 The action taken by Governments in respect of non-SOLAS vessels should rest on an objective assessment of the security risk that such vessels can pose for themselves or through their interaction with ships covered by the Maritime Security Measures.

2.18.7 IMO has developed a risk assessment and management tool (refer to section 5) to allow Government officials responsible for administering non-SOLAS vessels and their operators to consider:

.1 the security risks associated with each category of vessel;

.2 the security measures and procedures that operators of non-SOLAS vessels could take to mitigate the identified risks.

2.18.8 Some national authorities offer guidance to non-SOLAS vessel operators aimed at:

.1 enhancing security awareness;

.2 fostering links between the operators of such vessels and the Government's maritime security services;

.3 establishing procedures to facilitate reporting of suspicious activities and other security concerns to the Government's maritime security services.

2.18.9 As part of enhancing security awareness, national authorities may wish to develop security policies and procedures to ensure that all operators and crew of non-SOLAS vessels are aware of the basic security measures applying to their vessel. In appropriate circumstances, passengers could also be advised on the basic security measures applying to the vessel on which they are travelling.

2.18.10 National authorities may recommend basic security familiarization training for all crew members of non-SOLAS vessels to enable them to respond to security threats. In high-risk areas, such training should allow for assessment of their response capability. The proficiency training provided for owners and operators of pleasure craft could also encompass security awareness familiarization.

2.18.11 Any guidance to operators of non-SOLAS vessels should cover the likely need to agree a Declaration of Security when undertaking ship-to-ship activities with a SOLAS ship or when entering a port facility where the Maritime Security Measures apply.

2.18.12 To enhance the control exercised by national authorities controlling port arrivals and departures, non-SOLAS vessels engaged on international voyages could be required to provide arrival and departure information including:

.1 particulars of the vessel;

.2 date/time of arrival or departure;

.3 position of the vessel off, or in, the port;

.4 particulars of master/owner/shipping line/agent;

.5 purpose of call;

.6 cargo on board;

.7 crew and passenger lists;

.8 emergency contact information.

2.18.13 Similarly, pleasure craft and other non-SOLAS vessels could be requested to provide voyage information, including time of departure, destination and planned route. This information can assist the relevant authorities with their traffic monitoring activities and facilitate search and rescue operations if the vessel is in distress.

2.18.14 Difficulties can arise and delays occur when Government organizations are unable to establish the identity of non-SOLAS vessels engaged on international voyages. The requirement for a unique IMO identification number does not apply to:

.1 vessels solely engaged in fishing;

.2 vessels without mechanical means of propulsion;

.3 pleasure yachts;

.4 vessels on special service, e.g., light vessels and SAR vessels;

.5 hopper barges;

.6 hydrofoils and air-cushion vehicles;

.7 floating docks and similar structures;

.8 wooden vessels;

.9 seagoing merchant vessels of less than 100 gt; and

.10 all cargo vessels of less than 300 gt.

2.18.15 In such cases, Administrations could consider establishing procedures to allow the identity of their non-SOLAS vessels to be confirmed by other Governments without significant delay. This form of liaison is well developed in many regional counter-narcotics agreements. Governments could also consider recommending to operators of non-SOLAS vessels that the fitting of automated tracking equipment on their vessels would result in several benefits, including:

.1 enhanced safety and security;

.2 rapid emergency response to accidents and casualties;

.3 enhanced SAR capabilities; and

.4 enhanced control of smuggling and illegal migration.

Port security

2.18.16 IMO approved the guidance on wider port security provided in the ILO/IMO Code of practice on security in ports. This guidance relates to port areas which include port facilities as defined in, and designated under, Maritime Security Measures.

2.18.17 The guidance in the Code of practice suggests that Governments should:

.1 develop a port security strategy;

.2 identify port areas required to appoint a port security committee and appoint a port security officer (PSO);

.3 prepare and approve port security assessments (PSAs); and

.4 prepare and approve port security plans (PSPs).

2.18.18 PSAs and PSPs should be approved by the Designated Authority responsible for port facility security under the Maritime Security Measures.

2.18.19 The provisions in a PSP should not conflict with or override any security measures and procedures contained in the approved PFSPs of the port facilities located within the port area.

2.18.20 Several European Governments have enacted legislation establishing a port security authority for some of their port areas. The port security authority appoints a PSO and is required to submit a port security risk assessment and port security plan to the Designated Authority for approval.

SUA Convention

2.18.21 Most Contracting Governments that have adopted the SOLAS Convention have also ratified the Organization's Convention for the Suppression of Unlawful Acts against the Safety of Maritime Navigation (SUA Convention) and the related 2005 Protocols (see 2.9.33). The original 1988 SUA treaties provided the legal basis for action to be taken against persons committing unlawful acts against ships, including the seizure of ships by force; acts of violence against persons on board ships; and the placing of devices on board which are likely to destroy or damage the ship. Contracting Governments are obliged either to extradite or to prosecute alleged offenders.

2.18.22 Two new Protocols to the Convention for the Suppression of Unlawful Acts against the Safety of Maritime Navigation, 1988 and its Protocol for the Suppression of Unlawful Acts against the Safety of Fixed Platforms located on the Continental Shelf, 1988 (the SUA Treaties) were adopted on 14 October 2005. The two new Protocols expand the scope of the original Convention and Protocol to address terrorism by including a substantial broadening of the range of offences and introducing provisions for boarding suspect vessels.

2.18.23 The revision took into account developments in the UN system relating to countering terrorism. The relevant UN Security Council Resolutions and other instruments, including the International Convention for the Suppression of Terrorist Bombings (1997) and the International Convention for the Suppression of the Financing of Terrorism (1999), are directly linked to the new SUA Protocol.

2.18.24 Drafted to criminalize the use of a ship "when the purpose of the act, by its nature or context, is to intimidate a population, or to compel a Government or an international organization to do or to abstain from doing any act", these new instruments represent another significant contribution to the international framework to combat terrorism.

2.18.25 The 2005 amendments to the SUA Convention and the related Protocol entered into force on 28 July 2010.

Offshore activities

2.18.26 Although the Maritime Security Measures do not extend to offshore activities or installations located on a State's Continental Shelf, Governments with significant offshore activities, particularly those linked to exploiting oil or gas reserves, have developed specific security requirements applying to ships engaged in offshore activities, to mobile offshore drilling units on location and to fixed and floating platforms. When foreign-flagged ships are engaged in offshore supply or support activities on a State's Continental Shelf, they can be covered by both the requirements of the Maritime Security Measures and any additional security requirements set by the coastal State.

2.18.27 Under their national law, a limited number of Governments have defined fixed platforms located on their Continental Shelf as port facilities, requiring the appointment of a PFSO and the preparation of a PFSP. Such provisions can extend to include floating production storage and offloading (FPSO) vessels associated with oil and gas exploitation.

Specific security issues

2.18.28 It is for Governments to determine the extent to which the guidance issued by IMO on the following is reflected when undertaking PFSAs and SSAs and in PFSPs and SSPs:

- **.1** piracy and armed robbery;
- **.2** drug smuggling;
- **.3** stowaways;
- **.4** illegal migration; and
- **.5** the security of dangerous goods.

2.19 Information to IMO

Introduction

2.19.1 Through their national authorities, Governments are required to provide IMO with information on their national contact points and details on other aspects of their responsibilities, including legislation, RSOs, security agreements and arrangements, designated port facilities and PFSP approvals.

Global Integrated Shipping Information System

2.19.2 IMO Secretariat launched the Global Integrated Shipping Information System (GISIS) in 2005 to allow:

- **.1** direct reporting by Member States in compliance with existing requirements; and
- **.2** access to data compiled by the Secretariat.

2.19.3 The GISIS website may be accessed at: http://gisis.imo.org

2.19.4 GISIS has two login options: a member login and a public user login. The former is limited to IMO Member States and organizations with consultative or observer status at IMO whereas the public user login has read-only access to a limited amount of the information provided in GISIS.

National contact points

2.19.5 Effective international application of the Maritime Security Measures is dependent on the maintenance of strong communication links and liaison between port facility and ship operators on the one hand, and the national contact points to which they can express security concerns and from which they can seek security advice on the other.

2.19.6 To this end, Governments are required to provide IMO with up-to-date information on the points of contact for their national authorities. The template designed for this purpose is included as appendix 2.12 – Details of national authority contact points. A separate form is to be completed for each of the following national contact points:

- **.1** national authority responsible for ship security;
- **.2** national authority responsible for port facility security;
- **.3** recipient of ship security alerts;
- **.4** recipient of security-related communications from other Governments;
- **.5** recipient of security concerns from ships and requests for advice and assistance on ship-related security incidents and issues; and
- **.6** those who have been designated to be available at all times to receive and act upon reports from Governments exercising control and compliance measures.

2.19.7 Unless this information is regularly updated, the ability of CSOs and SSOs to communicate with PFSOs and national contact points is adversely affected, particularly when updating SSAs or seeking advice on security issues.

2.19.8 To facilitate the exchange of the information specified in the Maritime Security Measures between Governments and IMO, Governments have been asked to designate a single national contact point with responsibility for the exchange of the required information. The name and contact details must be kept updated.

Port facilities

2.19.9 Governments are also required to provide IMO with up-to-date information on their designated port facilities. The template designed for this purpose is included as appendix 2.13 – Details of port facilities – and it includes such details as:

- **.1** location;
- **.2** name of security point of contact (typically the PFSO);
- **.3** date of the original approval of the PFSP;
- **.4** date of most recent review of the PFSP;
- **.5** date of most recently issued Statement of Compliance, if applicable; and
- **.6** date of any withdrawal or amendment of a PFSP.

2.19.10 Any changes, including newly-listed port facilities, location covered by an approved PFSP and withdrawal of an approved PFSP, should be provided at the earliest opportunity along with the date when the change took effect.

2.19.11 Governments are required to provide an updated list of their ISPS Code-compliant port facilities at five-yearly intervals. The next updated list has to be submitted by 1 July 2014.

2.19.12 SOLAS chapter XI-2, regulation 13 requires that Contracting Governments communicate to IMO information on:

- **.1** the names and contact details of their national authority or authorities responsible for ship and port facility security;
- **.2** the locations within their territory covered by the approved PFSPs;
- **.3** the names and contact details of those who have been designated to be available at all times to receive and act upon the ship-to-shore security alerts;
- **.4** the names and contact details of those who have been designated to be available at all times to receive and act upon any communications from Contracting Governments exercising control and compliance measures; and
- **.5** the names and contact details of those who have been designated to be available at all times to provide advice or assistance to ships and to whom ships can report any security concerns.

2.19.13 The information listed under paragraph 2.19.12 can be submitted to IMO by making appropriate entries into its GISIS database. GISIS comprises a number of modules; the relevant module for communicating information in accordance with SOLAS chapter XI-2, regulation 13 is the maritime security Module. IMO Secretariat has recently amended the information to be provided for Contracting States' port facilities, and this is reflected in appendix 2.13 – Details of port facilities.

National legislation

2.19.14 Under the SOLAS Convention, Governments are required to transmit to IMO "....the text of laws, decrees, orders and regulations which have been promulgated on the various matters within the scope of the present Convention."

2.19.15 Experience to date indicates that few Governments have provided copies of the required texts.

Additional information

2.19.16 Governments are required to provide the name and contact details of any RSO authorized to act on their behalf together with details of its delegated responsibilities and any conditions attached to the exercise of such authority.

2.19.17 As described in subsection 2.13, Governments that have concluded an alternative security agreement are required to provide IMO with the information listed in paragraph 2.13.13.

2.19.18 As described in subsection 2.14, Governments that have allowed any equivalent security arrangements at port facilities or on ships are required to provide IMO with the information listed in paragraph 2.14.6.

Appendix 2.1

Implementation questionnaire for Designated Authorities

Source: MSC.1/Circ.1192, May 2006

This questionnaire may be used by Designated Authorities to examine the status of implementation of the Government's responsibilities for port facility security as specified in the Maritime Security Measures. When completing the questionnaire, the answers should be sufficiently detailed in order to gain a full understanding of the approach taken by the Contracting Government in implementing the Maritime Security Measures and prevent the drawing of erroneous conclusions.

Implementation process

1 Who is the Designated Authority? (SOLAS regulation XI-2/1.11)

2 What is the national legislative basis for the implementation of the ISPS Code? (SOLAS regulations XI-2/2 and XI-2/10)

3 What guidance to industry was released to implement the ISPS Code? (SOLAS regulations XI-2/2 and XI-2/10)

4 What are the means of communication with port facilities regarding ISPS Code implementation? (SOLAS regulations XI-2/3 and XI-2/10)

5 What processes are in place to document initial and subsequent compliance with the ISPS Code? (SOLAS regulation XI-2/10.2)

6 What is the Contracting Government's definition of a port facility? (SOLAS regulation XI-2/1.1)

7 What are the procedures used to determine the extent to which port facilities are required to comply with the ISPS Code, with particular reference to those port facilities that occasionally serve ships on international voyages? (SOLAS regulations XI-2/1, XI-2/2.2)

8 Has the Contracting Government concluded in writing bilateral or multilateral agreements with other Contracting Governments on alternative security agreements? (SOLAS regulation XI-2/11.1)

9 Has the Contracting Government allowed a port facility or group of port facilities to implement equivalent security arrangements? (SOLAS regulation XI-2/12.1)

10 Who has the responsibility for notifying and updating IMO with information in accordance with SOLAS regulation XI-2/13? (SOLAS regulation XI-2/13)

Port facility security assessment (PFSA)

11 Who conducts PFSAs? (SOLAS regulation XI-2/10.2.1, ISPS Code, sections A/15.2 and 15.2.1)

12 How are PFSAs conducted and approved? (ISPS Code, sections A/15.2 and 15.2.1)

13 What minimum skills are required for persons conducting PSFAs? (ISPS Code, section A/15.3)

14 Are PFSAs used for each port facility security plan? (ISPS Code, section A/15.1)

15 Do single PFSAs cover more than one port facility? (ISPS Code, section A/15.6)

16 Who is responsible for informing IMO if the single PFSA covers more than one port facility? (ISPS Code, section A/15.6)

17 What national guidance has been developed to assist with the completion of PFSAs? (SOLAS regulation XI-2/10.2.1)

18 What procedures are in place for determining when re-assessment takes place? (ISPS Code, section A/15.4)

19 What procedures are in place for protecting the PFSAs from unauthorized access or disclosure? (ISPS Code, section A/15.7)

Port facility security plans (PFSPs)

20 How are port facility security officers designated? (ISPS Code, section A/17.1)

21 What are the minimum training requirements that have been set by the Contracting Government for PFSOs? (ISPS Code, section A/18.1)

22 Are procedures used to determine the individuals/organizations responsible for the preparation of the PFSP? If yes, please describe.

23 Are procedures in place to protect PFSPs from unauthorized access? (ISPS Code, sections A/16.7 and A/16.8)

24 What procedures are in place for approval and subsequent amendments of the PFSPs? (ISPS Code, section A/16.6)

Security levels

25 Who is the authority responsible for setting the security level for port facilities? (SOLAS regulation XI-2/3.2)

26 What are the procedures for communicating security levels to port facilities by the responsible authority? (SOLAS regulation XI-2/3.2)

27 What are the procedures for communicating port facilities' security levels to ships? (SOLAS regulations XI-2/4.3 and XI-2/7.1)

28 What are the contact points and procedures for receiving ships' security level information in the Contracting Government and for notifying ships of contact details? (SOLAS regulation XI-2/7.2)

Declaration of Security

29 What procedures are used to determine when a Declaration of Security is required? (SOLAS regulation XI-2/10.3, ISPS Code, section A/5.1)

30 What is the minimum timeframe that a Declaration of Security is required to be retained? (ISPS Code, section A/5.6)

Delegation of tasks and duties

31 What tasks and duties have the Contracting Government delegated to recognized security organizations (RSOs) or others? (ISPS Code, section A/4.3)

32 To whom have these tasks and duties been delegated? What oversight procedures are in place? (SOLAS regulation XI-2/13.2)

Appendix 2.2

Implementation questionnaire for Administrations

Source: MSC.1/Circ.1193, May 2006

This questionnaire may be used by Designated Authorities to examine the status of implementation of the Government's responsibilities for ship security as specified in the Maritime Security Measures. When completing the questionnaire, the answers should be sufficiently detailed in order to gain a full understanding of the approach taken by the Contracting Government in implementing the Maritime Security Measures and prevent the drawing of erroneous conclusions.

Implementation process

1 What is the national legislative basis for the implementation of the ISPS Code? (SOLAS regulations XI-2/2 and XI-2/4)

2 What guidance to industry was released to implement the ISPS Code? (SOLAS regulations XI-2/2, XI-2/4, XI-2/5 and XI-2/6)

3 What are the means of communication developed by the Administration with (a) ships, and (b) companies, regarding ISPS Code implementation? (SOLAS regulations XI-2/3 and XI-2/4)

4 What processes are in place to document verification and certification of initial and subsequent compliance with the ISPS Code? (SOLAS regulation XI-2/4.2)

5 Has the Contracting Government nominated a point of contact for ships to request assistance or report security concerns? If yes, provide the name and contact details. (SOLAS regulation XI-2/7.2)

6 Have officers been duly authorized to exercise control and compliance measures on security grounds and has guidance been issued to them? (SOLAS regulation XI-2/9)

7 Has guidance been issued to companies and ships on the provision of information to other Contracting Governments when applying control and compliance measures, including the records to be retained by the ship in respect of the last ten calls at port facilities? (SOLAS regulation XI-2/9)

8 Has the Contracting Government concluded, in writing, bilateral or multilateral agreements with other Contracting Governments on alternative security agreements? (SOLAS regulation XI-2/11.1)

9 Has the Administration allowed a ship or group of ships to implement equivalent security arrangements? (SOLAS regulation XI-2/12.1)

10 Who has the responsibility for notifying and updating IMO with information in accordance with SOLAS regulation XI-2/13? (SOLAS regulation XI-2/13)

Ship security assessments (SSAs)

11 Who conducts SSAs? (ISPS Code, sections A/8.2 and 8.3)

12 Has national guidance been developed to assist with the completion of the on-scene security survey? (ISPS Code, section A/8.4)

Ship security plans (SSPs)

13 Who approves SSPs? (ISPS Code, sections A/9.1 and 9.2)

14 How are company and ship security officers designated? (ISPS Code, sections A/11.1 and A/12.1)

15 What are the minimum training requirements that have been set by the Administration for CSOs and SSOs? (ISPS Code, sections A/13.1 and A/13.2)

16 Has guidance been issued on the development and approval of SSPs? (ISPS Code, sections A/9.2 and 9.4)

17 Are procedures in place to protect SSPs from unauthorized access? (ISPS Code, section A/9.7)

18 What procedures are in place for approval and subsequent amendments of the SSPs? (ISPS Code, sections A/9.5 and 9.5.1)

19 Do SSPs contain a clear statement emphasizing the master's authority? (ISPS Code, section A/6.1)

20 Is the original or a translation of the SSP available in English, French or Spanish? (ISPS Code, section A/9.4)

21 Who verifies the ship security system including SSPs? (ISPS Code, section A/19.1.2)

22 Has the Administration specified the periods when renewal, intermediate and additional verifications shall be carried out? (ISPS Code, section A/19.1.1)

23 Who issues the International Ship Security Certificate (ISSC)? (ISPS Code, section A/19.2.2)

24 Has the Administration specified the period of validity of ISSCs? (ISPS Code, section A/19.3.1)

25 Does the Administration have procedures in place for the issue of Interim ISSCs? (ISPS Code, section A/19.4)

26 Has the Administration specified the minimum period for which records of activities addressed in the SSP shall be kept on board? (ISPS Code, section A/10.1)

Security levels

27 Who is the authority responsible for setting the security level for ships? (SOLAS regulation XI-2/3.1)

28 What are the procedures for communicating security levels to ships by the responsible authority? (SOLAS regulation XI-2/3.1)

29 Have procedures been notified for a ship to comply with the security level set by the Contracting Government for a port facility whose security level is higher than that set for the ship by the Administration? (SOLAS regulations XI-2/4.3 and XI-2/4.4)

30 Are procedures in place to provide advice to ships in cases where a risk of attack has been identified? (SOLAS regulation XI-2/7.3)

Declaration of Security

31 What procedures are used to determine when a Declaration of Security is required? (ISPS Code, section A/5.1)

32 What is the minimum time frame that a Declaration of Security is required to be retained? (ISPS Code, section A/5.7)

Delegation of tasks and duties

33 What tasks and duties, if any, have the Administration delegated to recognized security organizations (RSOs)? (ISPS Code, section A/4.3)

34 To whom have these tasks and duties been delegated? Based on what criteria and under what conditions has the status of RSO been granted by the Administration to those organizations? What oversight procedures are in place? (SOLAS regulation XI-2/13.2)

35 What procedures are in place to ensure that the RSO undertaking the review and approval process for an SSP was not involved in the preparation of the SSA or SSP? (ISPS Code, section A/9.2.1)

Appendix 2.3

Criteria for selecting recognized security organizations

Related document: MSC/Circ.1074, June 2003

Demonstrating organizational effectiveness

- Clear lines of managerial oversight for the proposed delegation of authority.
- Relevant qualifications and experience of key personnel proposed for the delegation of authority, including security clearances – these should be matched with their proposed work assignments.
- Planned training of key personnel during the duration of the delegation to ensure that qualifications are maintained and upgraded as necessary.
- Replacement strategy for key personnel.
- Company code of ethics or code of conduct.
- Successful testing of procedures established to avoid unauthorized disclosure of, or access to, security-sensitive material.
- Successful completion of similar activities to those identified in the proposed delegation of authority – this may require the RSO to identify recent examples of other national authorities which awarded similar delegations of authority.
- Adequate management of records and internal quality-control systems.

Demonstrating technical capabilities for ship-related delegations

- Appropriate knowledge of ship operations, including design and construction considerations.
- Appropriate knowledge of the requirements and guidance specified in the special measures and relevant national legislation, regulations, policies and operating procedures.
- Appropriate knowledge of current security threats and patterns and their relevance to ship operations.
- Experience in the application and maintenance of security and surveillance equipment and systems installed on board ships.
- Appropriate knowledge of their operational limitations, including techniques used to circumvent them.
- Experience in assessing the likely security risks that could occur during ship operations, including the ship/port interface, and identifying options to minimize such risks.

Demonstrating technical capabilities for port-related delegations

- Appropriate knowledge of port operations, including design and construction considerations.
- Appropriate knowledge of the requirements and guidance specified in the special measures and relevant national legislation, regulations, policies and operating procedures.
- Experience in assessing the likely security risks that could occur during port facility operations, including at the ship/port interface, and identifying options to minimize such risks.
- Appropriate knowledge of current security threats and patterns and their relevance to port operations.
- Experience in the application and maintenance of security and surveillance equipment and systems installed in port areas.
- Appropriate knowledge of their operational limitations, including techniques used to circumvent them.

Appendix 2.4
Sample of a port facility security plan approval form

PORT FACILITY SECURITY PLAN APPROVAL FORM		File number:	
Type of port facility:			
Name of port facility:			
Location			
Port ID number:			
UN locator			
Statement of Compliance date of issue (dd/mm/yyyy):		Date of expiry (dd/mm/yyyy):	
Name of operator:		Address of operator:	
Telephone:	Fax:	E-mail:	
Name of PFSO:		24 h contact number	
Telephone:	Fax:	E-mail:	
Designated Authority Security Office:	Address		
Telephone:	Fax:	E-mail:	
Approved Date:	Follow-up action required:		Date reviewed:

Reviewed by:

....

Print name *Signature*

APPROVAL DOCUMENT SECTIONS

(Check the box when section completed)

Section 1 – Organizational structure of the port facility. . . . . ☐

Section 2 – Security and communication equipment . . . . . ☐

Section 3 – Drills and exercises . . . . . ☐

Section 4 – Records and documentation . . . . . ☐

Section 5 – Communications . . . . . ☐

Section 6 – Security procedures during Interfacing . . . . . ☐

Section 7 – Declarations of Security . . . . . ☐

Section 8 – Response to a change in the security level. . . . . ☐

Section 9 – Security procedures for access control . . . . . ☐

Section 10 – Security procedures for restricted areas . . . . . ☐

Section 11 – Security procedures for handling cargo . . . . . ☐

Section 12 – Security procedures for delivery of ship's stores and bunkers . . . . . ☐

Section 13 – Security procedures for monitoring. . . . . ☐

Section 14 – Response to security threats, breaches of security and security incidents. . . . . ☐

Section 15 – Audits and amendments . . . . . ☐

Section 1 – Organizational structure of the port facility	
Requirement – the plan identifies:	**Plan reference**
Name of security organization	
Name of operator	
Name and position of PFSO and 24 h contact information	
Duties and responsibilities of the PFSO	
Duties and responsibilities of personnel with security responsibilities	
Training requirements of the PFSO and port facility personnel with designated security responsibilities	
The security organization's links with other national or local authorities with security responsibilities	
Comments:	

Section 2 – Security and communication equipment	
Requirement – the plan includes:	**Plan reference**
Procedures for maintaining security and communication systems and equipment	
Procedures for identifying and correcting failures or malfunctions of security equipment or systems	
A description of security equipment for access control	
A description of security equipment for monitoring the port facility and surrounding area	
A description of how monitoring is achieved by any combination of lighting, security guards on foot or in vehicles, waterborne patrols, automatic intrusion-detection devices and surveillance equipment	
If an automatic intrusion-detection device is used, it activates an audible or visual alarm, or both, at a location that is continuously attended or monitored	
Monitoring is able to function continuously, including during periods of adverse weather or power disruption	
Monitoring equipment covers access and movements adjacent to ships interfacing with the port facility	
Comments:	

Section 3 – Drills and exercises	
Requirement – the plan includes provision for:	**Plan reference**
Security drills to be conducted every three months	
Security drills to test individual elements of the PFSP, including the response to security threats, breaches of security and security incidents, taking into account the types of operations, personnel changes, the types of ships interfacing with the facility and other relevant circumstances	
Security exercises to fully test the PFSP, including the active participation of facility personnel who have security responsibilities, relevant Government officials, the CSO and any available Ship Security Officers	
Security exercises to check communication and notification procedures, elements of co-ordination, resource availability and response	
Security exercises to be conducted at least once every calendar year, with no more than 18 months between them.	
Comments:	

Section 4 – Records and documentation	
Requirement – the plan includes provision for the PFSO to keep the following records:	**Plan reference**
Rate of inspections specified in the plan	
Security training, including dates, duration, description and names of participants	
Security drills & exercises, including dates, description, names of participants and any best practices or lessons learned	
Security threats, breaches of security and security incidents, including date, time, location, the response to them and the person to whom they were reported	
Changes in the security level, including the date, time that notification was received and the time of compliance with the requirement of the new level	
Maintenance, calibration and testing of equipment used for security, including the date and time of the activity and the equipment involved	
Declarations of security in respect of the port facility	
Internal audits and reviews of security activities	
Security assessment information, including the PFSA, each periodic review, the dates conducted and their findings	
The plan, including each periodic review date conducted, the findings, and any recommended amendments	
Amendments to the plan, including the dates of their approval and implementation	
Records of inspections and patrols	
A list, by name or position, of the persons who have security responsibilities	
An up-to-date list containing the names of screening officers (if applicable)	
For at least two years and to be available to Government officials on request. In the case of the plan and its related PFSA, the retention time is for at least two years after the plan's expiry date	
Protected from unauthorized access or disclosure, including the plan	
If in electronic format, protected from deletion, destruction and revision	
Comments:	

Section 5 – Communications	
Requirement – the plan addresses:	**Plan reference**
Procedures that allow for effective communications between personnel with security responsibilities with respect to the ship's interfacing with the facility and with port operators, if applicable, the Designated Authority and local law-enforcement agencies	
The means of alerting and obtaining the services of waterside patrols and specialist search teams, including bomb searches and underwater searches	
Back-up communications to ensure internal and external communications	
Comments:	

Section 6 – Security procedures during interfacing	
Requirement – the plan includes procedures for:	**Plan reference**
Co-ordinating with ships interfacing with the port facility and the port operator, if applicable.	
Assisting SSOs in confirming the identity of those seeking to board the ship, when requested	
Facilitating shore leave for ship's personnel or personnel changes, as well as access of visitors to the ship, including representatives of seafarers' welfare and labour organizations	
Comments:	

Section 7 – Declarations of Security	
Requirement – the plan makes provision for:	**Plan reference**
The requirements and procedures for completing Declarations of Security	
A DoS to be completed before an interface starts between a port facility and a ship if they are operating at different security levels	
A DoS to be completed before an interface starts between a port facility and a ship if one of them does not have an approved security plan	
A DoS to be completed before an interface starts between a port facility and a ship if the interface involves a cruise ship, a ship carrying dangerous goods or the loading or transfer of dangerous goods	
A DoS to be completed before an interface starts between a port facility and a ship if the security officer of either of them identifies security concerns about the interface	
Comments:	

Section 8 – Response to a change in the security level	
Requirement – the plan contains procedures for ensuring that, when the operator of the port facility is notified of an increase in the security level:	**Plan reference**
The port facility complies with the required additional security procedures within the specified time period after the notification	
The Designated Authority receives a report indicating compliance or non-compliance with the security level	
If the increase is to security level 3, the port facility evaluates the need for additional security procedures	
Comments:	

Section 9 – Security procedures for access control	
Requirement – the plan includes procedures for:	**Plan reference**
At all security levels:	
Preventing unauthorized access to the port facility by persons, weapons, incendiaries, explosives, dangerous substances and devices	
At security level 1:	
Establishing control points for restricted access that should be bounded by fencing or other barriers	
Verifying the identity of every person seeking to enter a controlled access area and the reasons for which they seek entry	
Screening of persons, goods and vehicles for weapons, explosives or incendiaries at the rate specified in the plan	
Checking vehicles used by those seeking entry to the port facility	
Verifying the identity of port facility personnel and those employed within the port facility, and their vehicles	
Restricting access to exclude those not employed by the port facility or working within it, if they are unable to establish their identity	
Searches of persons, personal effects, vehicles and their contents at the rate specified in the plan	
Denying or revoking of a person's authorization to enter or remain on a port facility if they are not authorized or fail to identify themselves	
Determining the appropriate access controls for deterring unauthorized access to the port facility, including its restricted areas	
Identifying access points that must be secured or attended to deter unauthorized access	
Screening or searching unaccompanied baggage at the rate(s) specified in the plan	
At security level 2:	
Increasing the frequency of screening persons and goods	
Authorized screening of all unaccompanied baggage by means of x-ray equipment	
Additional personnel to guard access points and for perimeter patrols	
Limiting the number of access points to the port facility	
Impeding movement through the remaining access points, e.g., security barriers	
Increasing the frequency of searches of persons, personal effects and vehicles	
Denying or revoking access to persons who are unable to provide a verifiable justification for seeking access	
Co-ordinating with the Designated Authority, appropriate law-enforcement agencies, and port operator, if applicable, to deter waterside access to the facility	
At security level 3:	
Additional screening of unaccompanied baggage	
Co-ordinating with emergency response personnel and other port facilities	
Granting access to those responding to the security incident or security threat	
Suspending all other access to the port facility	
Suspending cargo operations within all, or part, of the port facility	
Evacuating the port facility or part thereof	
Restricting pedestrian and vehicular movements	
Increasing monitoring of the security patrols within the port facility, if appropriate	
Directing all movements relating to all, or part, of the port facility	
Comments:	

Section 10 – Security procedures for restricted areas	
Requirement – The plan makes provision for designating restricted areas, including those listed below, and specifying measures and procedures, as appropriate to the facility's operations at each security level:	**Plan reference**
At all security levels:	
Land areas adjacent to ships interfacing with the port facility	
Embarkation and disembarkation areas, holding and processing areas for passengers and ship's personnel, including search points	
Areas designated for loading, unloading or storage of cargo and ship's stores	
Areas in which security-sensitive information is kept, including cargo documentation	
Areas where dangerous goods and hazardous substances are held	
Vessel traffic management system control rooms, aids to navigation and port control buildings, including security and surveillance control rooms	
Areas where security and surveillance equipment is stored or located	
Essential electrical, radio and telecommunication, water and other utility installations	
Locations in the port facility where it is reasonable to restrict access by vehicles and persons	
At security level 1:	
Providing permanent or temporary barriers to surround the restricted area	
Procedures for securing all access points not actively used and providing physical barriers or security guards to impede movement through the remaining access points	
Procedures for controlling access to restricted areas, such as a pass system that identifies an individual's entitlement to be within the restricted area	
Procedures for examining the identification and authorization of persons and vehicles seeking entry, and clearly marking vehicles allowed access to restricted areas	
Procedures for patrolling or monitoring the perimeter of restricted areas	
Procedures for using security personnel, automatic intrusion-detection devices or surveillance equipment/systems to detect unauthorized entry or movement in the restricted areas	
Procedures for controlling the movement of vessels in the vicinity of ships using the port facility	
Procedures for designating temporary restricted areas, if applicable, to accommodate port facility operations, including restricted areas for segregating unaccompanied baggage that has undergone authorized screening by a ship operator	
Procedures for conducting a security sweep (both before and after) if a temporary restricted area is designated	
At security level 2:	
Procedures for enhancing physical barriers, use of patrols or intrusion-detection devices	
Procedures for reducing the number of access points and enhancing controls applied at the remaining access points	
Procedures for restricting parking of vehicles adjacent to ships	
Procedures for reducing access to restricted areas and movements and storage in them	
Procedures for using surveillance equipment that records and monitors continuously	
Procedures for increasing the number and frequency of patrols, including the use of waterside patrols	
Procedures for establishing and restricting access to areas adjacent to restricted areas	
Enforcing restrictions on access by unauthorized craft to the waters adjacent to ships using the port facility	

Section 10 – Security procedures for restricted areas *(continued)*	
At security level 3:	
Procedures for designating additional restricted areas adjacent to the security incident or threat to which access is denied	
Procedures for searching restricted areas as part of a security sweep of all or part of the port facility	
Comments:	

Section 11 – Security procedures for handling cargo	
Requirement – the plan includes procedures for:	**Plan reference**
At all security levels:	
Identifying cargo that is accepted for loading onto ships interfacing with the port facility	
Identifying cargo that is accepted for temporary storage in a restricted area while awaiting loading or pick up	
At security level 1:	
Verifying that cargo, containers and cargo transport units entering the port facility match the invoice or other cargo documentation	
Routine inspection of cargo, containers, transport units and cargo storage areas before and during handling operations to detect evidence of tampering, unless unsafe to do so	
Verifying that the cargo entering the facility matches the delivery documentation	
Searching vehicles entering the port facility	
Examining seals and other methods used to detect evidence of tampering when cargo, containers or cargo transport units enter the port facility or are stored there	
At security level 2:	
Detailed checking of cargo, containers, and cargo transport units in or about to enter the port facility or cargo storage areas, for weapons, explosives and incendiaries	
Intensified inspections to ensure that only documented cargo enters the port facility, is temporarily stored there and then loaded onto the ship	
Detailed search of vehicles for weapons, explosives and incendiaries	
Increasing the frequency and detail of examinations of seals and other methods used to prevent tampering	
Increasing the frequency and intensity of visual and physical inspections	
Increasing the frequency of the use of scanning/detection equipment, mechanical devices or dogs	
Co-ordinating enhanced security measures with shippers or those acting on their behalf in accordance with an established agreement and procedures	
At security level 3:	
Restricting or suspending cargo movements or operations in all or part of the port facility	
Confirming the inventory and location of certain dangerous cargoes in the port facility	
Comments:	

Section 12 – Security procedures for delivery of ship's stores and bunkers	
Requirement – the plan include procedures for:	**Plan reference**
At security level 1:	
Checking ship's stores	
Requiring advanced notification of the delivery of ship's stores or bunkers, including a list of stores, and driver and vehicle registration information in respect of delivery vehicles	
Inspecting delivery vehicles at the rate specified in the plan	
At security level 2:	
Detailed checking of ship's stores	
Detailed searches of delivery vehicles	
Co-ordinating with ship personnel to check the order against the delivery note prior to entry to the port facility	
Escorting delivery vehicles in the port facility	
At security level 3:	
Restricting or suspending the delivery of ship's stores and bunkers	
Refusing to accept ship's stores in the port facility	
Comments:	

Section 13 – Security procedures for monitoring	
Requirement – the plan establishes the procedures and equipment needed at each security level and the means of ensuring that monitoring equipment will be able to perform continually, including consideration of the possible effects of weather or of power disruptions, including:	**Plan reference**
At security level 1:	
The security measures to be applied, which may be a combination of lighting, security guards or use of security and surveillance equipment, to allow port facility security personnel to:	
– Observe the general port facility area, including shore- and water-side accesses;	
– Observe access points, barriers and restricted areas; and	
– Allow port facility security personnel to monitor areas and movements adjacent to ships, including augmentation of lighting provided by the ship itself.	
At security level 2:	
Additional procedures to increase the coverage and intensity of lighting and surveillance equipment, including the provision of additional lighting and surveillance	
Procedures for increasing the frequency of foot, vehicle or waterborne patrols	
Procedures for assigning additional security personnel to monitor and patrol	
At security level 3:	
Procedures for switching on all lighting in, or illuminating the vicinity of, the port facility	
Procedures for switching on all surveillance equipment capable of recording activities in or adjacent to the port facility	
Procedures to maximize the length of time that surveillance equipment can continue to record	
Comments:	

Section 14 – Response to security threats, breaches of security and security incidents	
Requirement – the plan addresses procedures at all security levels for:	**Plan reference**
Responding to security threats, breaches of security and security incidents, including provisions to maintain critical port facility and interface operations	
Evacuating the port facility in case of security threats and security incidents	
Reporting security threats, breaches of security, and security incidents to the Designated Authority	
Briefing port facility personnel on potential threats to security and the need for vigilance	
Securing non-critical operations in order to focus response on critical operations	
Reporting security threats, breaches of security and security incidents to the appropriate law-enforcement agencies, the Designated Authority and, if applicable, the port operator	
Comments:	

Section 15 – Audits and amendments	
Requirement – the plan addresses when an audit is required and the timing for submitting audit-based amendments, as follows:	**Plan reference**
The PFSA relating to the facility is altered	
An independent audit or the Designated Authority's testing of the port facility security organization identifies failings in the organization or questions the continuing relevance of significant elements of the approved plan	
Security incidents or threats involving the port facility have occurred	
There is a new operator of the port facility, a change in operations or location, or modifications to the port facility that could affect its security	
If the audit results require an amendment to be made to the PFSA or plan, the PFSO submits an amendment to the Designated Authority for approval within 30 days after completion of the audit	
If the operator of a port facility submits other amendments to the approved plan, they are to be submitted at least 30 days before they take effect	
Comments:	

PFSP REVIEW

APPROVED ....

DISAPPROVED ....

COMMENTS:

....

....

....

....

Appendix 2.5
Form of a Statement of Compliance of a port facility
Source: Part B of the ISPS Code

Statement number . . . .

Issued under the provisions of

Part B of the International Ship and Port Facility Security (ISPS) Code
by the Government of [*insert name and official seal, if appropriate*]

Name of the port facility . . . .

Address of the port facility . . . .

. . . .

THIS IS TO CERTIFY that:

- the compliance of this port facility with the provisions of chapter XI-2 of the SOLAS Convention and part A of the ISPS Code has been verified; and
- this port facility operates in accordance with its approved port facility security plan (PFSP). This plan has been approved for the types of operations, types of ship or activities or other relevant information listed below (*delete non-applicable categories*):
  - Passenger ship
  - Passenger high-speed craft
  - Cargo high-speed craft
  - Bulk carrier
  - Oil tanker
  - Chemical tanker
  - Gas carrier
  - Mobile offshore drilling unit
  - Cargo ships other than those referred to above

This Statement of Compliance is valid until . . . ., subject to verifications (as indicated overleaf)

Issued at . . . .
(*place of issue*)

Date of issue . . . .

. . . .
(*Signature of the duly authorized official issuing the document*)

(*Seal or stamp of the issuing authority, as appropriate*)

Endorsement for verifications

The Government of [*insert name*] has established that the validity of this Statement of Compliance is subject to [*insert relevant details of the verifications e.g., mandatory annual or unscheduled*].

THIS IS TO CERTIFY that, during a verification carried out in accordance with paragraph 16.62.4 of part B of the ISPS Code, the port facility was found to comply with the relevant provisions of chapter XI-2 of the SOLAS Convention and part A of the ISPS Code.

First verification

Signed: ....
(Signature of authorized official)

Place: ....

Date: ....

(Seal or stamp of authority, as appropriate)

Second verification

Signed: ....
(Signature of authorized official)

Place: ....

Date: ....

(Seal or stamp of authority, as appropriate)

Third verification

Signed: ....
(Signature of authorized official)

Place: ....

Date: ....

(Seal or stamp of authority, as appropriate)

Fourth verification

Signed: ....
(Signature of authorized official)

Place: ....

Date: ....

(Seal or stamp of authority, as appropriate)

Appendix 2.6

Form of the International Ship Security Certificate

Source: Part A of the ISPS Code

INTERNATIONAL SHIP SECURITY CERTIFICATE

(Official seal) *(State)*

Certificate number: . . . .

Issued under the provisions of the

INTERNATIONAL CODE FOR THE SECURITY OF SHIPS AND OF PORT FACILITIES
(ISPS CODE)

Under the authority of the Government of

. . . .
(name of State)

by

. . . .
(person(s) or organization authorized)

Name of ship: . . . .

Distinctive number or letters: . . . .

Port of registry: . . . .

Type of ship: . . . .

Gross tonnage: . . . .

IMO Number:. . . . .

Name and address of the Company: . . . .

Company identification number: . . . .

THIS IS TO CERTIFY:

1. that the security system and any associated security equipment of the ship has been verified in accordance with section 19.1 of part A of the ISPS Code;
2. that the verification showed that the security system and any associated security equipment of the ship is in all respects satisfactory and that the ship complies with the applicable requirements of chapter XI-2 of the Convention and part A of the ISPS Code;
3. that the ship is provided with an approved ship security plan.

Date of initial/renewal verification on which this Certificate is based . . . .

This Certificate is valid until . . . ., subject to verifications in accordance with section 19.1.1 of part A of the ISPS Code.

Issued at . . . .

(Place of issue of the Certificate)

Date of issue . . . .

(Signature of the duly authorized official issuing the Certificate)

(Seal or stamp of issuing authority, as appropriate)

Endorsement for intermediate verification

THIS IS TO CERTIFY that at an intermediate verification required by section 19.1.1 of part A of the ISPS Code the ship was found to comply with the relevant provisions of chapter XI-2 of the Convention and part A of the ISPS Code.

Intermediate verification

Signed: ....
(Signature of authorized official)

Place: ....

Date: ....

(Seal or stamp of authority, as appropriate)

Endorsement for additional verifications*

Additional verification

Signed ....
(Signature of authorized official)

Place ....

Date ....

(Seal or stamp of the authority, as appropriate)

Additional verification

Signed ....
(Signature of authorized official)

Place ....

Date ....

(Seal or stamp of the authority, as appropriate)

Additional verification

Signed ....
(Signature of authorized official)

Place ....

Date ....

(Seal or stamp of the authority, as appropriate)

---

* This part of the certificate shall be adapted by the Administration to indicate whether it has established additional verifications as provided for in section 19.1.1.4. of part A of the ISPS Code.

Additional verification in accordance with section A/19.3.7.2 of the ISPS Code

THIS IS TO CERTIFY that at an additional verification required by section 19.3.7.2 of part A of the ISPS Code the ship was found to comply with the relevant provisions of chapter XI-2 of the Convention and part A of the ISPS Code.

Signed: . . . .
(Signature of authorized official)

Place: . . . .

Date: . . . .

(Seal or stamp of authority, as appropriate)

Endorsement to extend the certificate if valid for less than 5 years where section A/19.3.3 of the ISPS Code applies

The ship complies with the relevant provisions of part A of the ISPS Code, and the Certificate shall, in accordance with section 19.3.3 of part A of the ISPS Code, be accepted as valid until . . . .

Signed: . . . .
(Signature of authorized official)

Place: . . . .

Date: . . . .

(Seal or stamp of authority, as appropriate)

Endorsement where the renewal verification has been completed and section A/19.3.4 of the ISPS Code applies

The ship complies with the relevant provisions of part A of the ISPS Code, and the Certificate shall, in accordance with section 19.3.4 of part A of the ISPS Code, be accepted as valid until . . . .

Signed: . . . .
(Signature of authorized official)

Place: . . . .

Date: . . . .

(Seal or stamp of authority, as appropriate)

Endorsement to extend the validity of the certificate until reaching the port of verification where section A/19.3.5 of the ISPS Code applies or for a period of grace where section A/19.3.6 of the ISPS Code applies

This Certificate shall, in accordance with section 19.3.5/19.3.6* of part A of the ISPS Code, be accepted as valid until . . . .

Signed: . . . .
(Signature of authorized official)

Place: . . . .

Date: . . . .

(Seal or stamp of authority, as appropriate)

Endorsement for advancement of expiry date where section A/19.3.7.1 of the ISPS Code applies

In accordance with section 19.3.7.1 of part A of the ISPS Code, the new expiry date† is . . . .

Signed: . . . .
(Signature of authorized official)

Place: . . . .

Date: . . . .

(Seal or stamp of authority, as appropriate)

* Delete as appropriate.

† In case of completion of this part of the Certificate, the expiry date shown on the front of the Certificate shall also be amended accordingly.

Appendix 2.7
Form of the Interim International Ship Security Certificate
Source: Part A of the ISPS Code

INTERIM INTERNATIONAL SHIP SECURITY CERTIFICATE

(Official seal) (State)

Certificate Number: ....

Issued under the provisions of the

INTERNATIONAL CODE FOR THE SECURITY OF SHIPS AND OF PORT FACILITIES
(ISPS CODE)

Under the authority of the Government of

....
(name of State)

by

....
(person(s) or organization authorized)

Name of ship: ....

Distinctive number or letters: ....

Port of registry: ....

Type of ship: ....

Gross tonnage: ....

IMO Number:. ....

Name and address of the Company: ....

Company identification number: ....

Is this a subsequent, consecutive Interim Certificate? Yes/No*

If Yes, date of issue of initial Interim Certificate: ....

* Delete as appropriate.

THIS IS TO CERTIFY THAT the requirements of section A/19.4.2 of the ISPS Code have been complied with.

This Certificate is issued pursuant to section A/19.4 of the ISPS Code.

This Certificate is valid until . . . .

Issued at . . . .

(Place of issue of the Certificate)

Date of issue . . . .

(Signature of the duly authorized official issuing the Certificate)

(Seal or stamp of issuing authority, as appropriate)

Appendix 2.8
Sample of a ship security inspection checklist

Administrations may wish to include any specific requirements (e.g., security records have to be maintained for 12 months) they have regarding ship security into the checklist's generic questions. Administrations may also wish to provide guidance on which questions are applicable to initial, intermediate, renewal and additional inspections.

As an alternative to the format provided, Administrations may wish to amend the inspection details, change the order of questions to suit their inspection process, or add separate comments sheets and use the question number as the cross reference, as opposed to having space for comments against each question.

SHIP SECURITY INSPECTION DETAILS	
Ship name	
IMO Number	
Type of vessel	
Date of inspection	
Place of inspection	
Type of inspection	Initial/Intermediate/Renewal/Additional
Company name	
Company ID number	
Company security officer name	
Master name	
Ship security officer name	

Questions	Yes/No	Comments
APPROACHING THE SHIP		
1 Is the ship identification number marked as defined in SOLAS XI-1/3?		
Access control (preventing unauthorized access and articles)		
2 Are procedures implemented for identification of ship's personnel, passengers and visitors, as specified in SSP?		
3 If a pass system is defined in the SSP, is it implemented?		
4 If a pass system is defined in the SSP, are crew passes withdrawn when that person leaves ship permanently?		
5 If a pass system is defined in the SSP, are visitor or contractor passes withdrawn on exiting the ship?		
6 Are the ship's deck and access points illuminated?		
7 Is there controlled access to the ship?		
8 Is there effective control of the embarkation of persons and their effects?		
9 Is checking of identification of all persons seeking to board the ship carried out?		
10 Are inspections and searching of persons carried out in designated areas (as applicable)?		
11 Are access points that should be secured or attended secured or attended?		
12 Is 'search' signage, as defined in SSP, in place?		
13 For ro–ro ships, is the frequency for searching vehicles for the designated ship, as described in the SSP, maintained (as applicable)?		

Questions		Yes/No	Comments
14	For passenger ships, are checked and unchecked persons segregated as well as embarking and disembarking passengers?		
15	For passenger ships, are the unattended spaces adjoining areas to which passengers and visitors have access secured or locked?		
16	For passenger ships, is there at least one male and one female searcher at each access point?		
17	For passenger ships, do passengers have passes and return them when leaving the ship at the end of the voyage?		
18	If the ship has operated at security level 2, have the additional protective access–control measures defined in the SSP been implemented?		
19	If the ship has operated at security level 3, have the further specific protective access–control measures defined in the SSP been implemented?		
20	At security level 3, are the provisions in the SSP for limited access by authorized officers understood?		
IN MASTER'S CABIN / SHIP'S OFFICE			
Certification and ship security plan			
21	Is the ISSC an original version?		
22	Do the entries on the ISSC agree with those on CSR, SMC and DOC?		
23	Is there an up-to-date Continuous Synopsis Record on board?		
24	Do the entries on the CSR agree with those on Certificate of Registry and Class Certificate?		
25	Is the version of the SSP held on board the same as the Administration's approved version?		
26	Is the SSP based on the current ship security assessment (SSA)?		
27	Are the SSP, SSA, and other associated material, protected from unauthorized access or disclosure?		
28	If the SSP is in electronic format, is it protected against unauthorized deletion, destruction or amendment?		
29	If the SSP is in electronic format, is the password changed regularly?		
30	Is the SSP written in the working language of the ship (if not English, French or Spanish, a translation into one of these languages has to be provided)?		
31	Is the organizational structure, as described in the SSP, for security known, understood and implemented?		
32	Are the relationships with company, port facilities, other ships and relevant authorities known and understood by key personnel?		
33	Are the evacuation procedures in case of security threats or breaches of security known and implemented?		
34	Are amendments to the SSP or changes in the security equipment approved, if required, by the Administration?		
35	Are these changes clearly identified in the SSP?		
36	Have any amendments to the SSP been implemented?		
37	Does the SSO understand the procedure to follow for changes to the SSP?		
38	Are the company policy and objectives on security understood?		
39	Does the master understand the discretion available for safety and security?		

Questions	Yes/No	Comments
40 Has the company ensured that the necessary support is given to the CSO to enable the master and SSO to fulfil their duties and responsibilities?		
41 Are contact points available for flag State and other relevant Contracting Governments?		
Security records		
42 Have security records (training, drills and exercises; threats and incidents; breaches of security; change in security level; internal audits and reviews; implementation of amendments to the SSP; maintenance, calibration and testing of security equipment) been kept up to date?		
43 Are security records kept for the length of time specified in the SSP?		
44 Are security records in the ship's working language?		
45 Are records protected from unauthorized access or disclosure?		
46 When kept in electronic format, are records protected to prevent unauthorized deletion, destruction or amendment?		
Ship/port and ship/ship interface		
47 Has the ship acted upon the security levels set by the Administration?		
48 Are procedures understood regarding instructions from Contracting Governments when at security level 3?		
49 Is receipt of instructions by the Administration, including change of the security levels, acknowledged?		
50 Has the ship informed nearby coastal States regarding its operating at security level 2 or security level 3, if applicable?		
51 Has the ship acted upon the security levels set by the Contracting Government of ports?		
52 Is provision made to ensure that the ship does not have a lower security level than the port facility?		
53 Have the Contracting Government and the PFSO been informed if the ship has been at a higher security level than the port facility?		
54 Have the pre-arrival notifications been handled correctly?		
55 Have the procedures for interfacing with the port facility specified in the SSP been followed?		
56 Are Declarations of Security completed as defined in the SSP?		
57 Is the procedure understood regarding failure by the port facility or other ship to acknowledge a DoS?		
58 Are appropriate ship security procedures maintained during any ship-to-ship interface?		
Security training, drills and exercises		
59 Does the ship security officer have a certificate of proficiency as SSO?		
60 Has familiarization training for all personnel been carried out?		
61 Do all crew have certificates of proficiency in security awareness?		
62 Do all crew with designated security duties have certificates of proficiency in security duties?		
63 Are security briefings for the crew carried out?		
64 Have security drills been carried out as detailed in the SSP?		
65 Have security exercises been carried out at the frequency shown in the SSP?		

Questions	Yes/No	Comments
Responding to security threats and breaches of security		
66 Has the CSO obtained information on the assessments of threat for the ports of call and their protective measures, and is this available on board?		
67 Is information available from Contracting Governments with respect to security threats?		
68 Are procedures for raising the alarm for security threats or security incidents known and implemented?		
69 Are procedures known, and can they be implemented, for responding to security threats or breaches of security, including maintenance of critical operations of the ship or ship/port interface?		
70 Have any breaches of security been reported to the Administration, port facility, coastal State or CSO?		
71 Have breaches of security been investigated and mitigation measures implemented?		
72 Have potential on-board weapons that could be used by those committing a breach of security been identified?		
73 Are such potential weapons controlled to prevent their unauthorized use?		
Audit and review		
74 Have internal audits been undertaken as required by the SSP?		
75 Have reviews of security been undertaken as required by the SSP?		
76 Have personnel been conducting internal audits of the security activities independent of the activities being audited (unless this is impracticable, due to the size and the nature of the company or of the ship)?		
77 Have the non-conformities of previous internal audits and reviews been properly dealt with?		
78 Have the non-conformities of previous external audits been properly dealt with?		
TOUR OF SHIP		
Restricted areas		
79 Are the restricted areas identified and known?		
80 Are the procedures in the SSP implemented to prevent unauthorized access to restricted areas?		
81 If coded key pads are used, is the code changed regularly, as required by the SSP?		
82 Are there any other ship-critical areas that should be restricted areas?		
83 Are there other areas that require extra vigilance at heightened security levels (e.g., galley or crew mess)?		
84 Is the functionality of emergency escapes maintained while preventing unauthorized access to restricted areas?		
85 If the ship has operated at security level 2, has the additional monitoring, as defined in the SSP, been implemented?		
86 If the ship has operated at security level 3, have measures been taken in co-operation with those responding to the incident or threat?		
Monitoring security		
87 Is there an effective deck watch while the ship is in port or anchored off?		
88 Are areas surrounding the ship, particularly to seaward, monitored?		

Questions	Yes/No	Comments
89 Is there lighting on all decks and access points whilst berthed?		
90 Is CCTV used when available?		
91 Have patrols, as detailed in the SSP, been implemented?		
92 For ro–ro ships, are car decks monitored when loading and unloading?		
93 If the ship has operated at security level 2, have the additional measures defined in the SSP been implemented?		
94 If the ship has operated at security level 3, have measures been taken in co-operation with those responding to the incident or threat?		
Handling of cargo, stores and unaccompanied baggage		
95 Is handling of cargo supervised as defined in the SSP?		
96 For ro–ro ships, are vehicles searched prior to loading according to the frequency in the SSP?		
97 Is handling of deck, engine-room and catering stores supervised as defined in the SSP?		
98 Is handling of unaccompanied baggage supervised as defined in the SSP?		
99 If the ship has operated at security level 2, have the additional measures defined in the SSP been implemented?		
100 If the ship has operated at security level 3, have measures been taken in co-operation with those responding to the incident or threat?		
Security equipment		
101 Can the sending of a test SSAS message be demonstrated?		
102 Are procedures (testing, activation, de-activation and resetting, limiting false alarms) understood for the use of the SSAS, including the location of activation buttons?		
103 Are tests of the SSAS conducted at the frequency specified in the SSP?		
104 Is any other security equipment inspected, tested and calibrated as applicable?		
105 Is all security equipment in working order?		
Communication		
106 Are the crew able to communicate on security issues?		
107 Is the CSO 24-hour contact known to key personnel?		

Appendix 2.9
Sample of a notice of non-compliance

Name of ship	IMO Number	Type of ship	Flag State	Date of inspection	Place of inspection

Item number		Deficiency	Regulatory reference	Due date	Date rectified	Date checked
1.						
2.						
X						

The information on this form is collected under the authority of ....

Date ....

Signature of inspector (for the national authority)

Date ....

Signature of authorized representative acknowledging receipt

Appendix 2.10
Sample of a core training curriculum for officials in national authorities

Core training element	Main topics
Overview of international maritime security framework	• IMO's role and structure, decision-making process, member States, Conventions, Codes of Practice (including ILO/IMO Code of practice on security in ports) • Port State control MOUs • History of SOLAS amendments 2002 • Role of regional organizations
Overview of national authority's legislative, policy and organizational framework	• Legislation, regulations and other legal instruments (including planned amendments) • Approval process • National policy statements • Inter-departmental/agency roles and co-ordination mechanisms • Bilateral & multilateral agreements • Organizational structure of national authority and link to responsible Minister
Overview of maritime industry under national authority's jurisdiction	• Key statistics on maritime trade, port activity and ship movements • Current and planned industry initiatives • Industry associations • Major security incidents
National authority responsibilities under SOLAS amendments 2002 and ISPS Code	• List of responsibilities, comparison with port and shipping industry responsibilities and link to legislative framework (which may prescribe a broader set of responsibilities)
Responsibilities delegated to officials	• Delegation of authority or equivalent document empowering officials • Official identification cards • Delegations to RSOs
Code of conduct	• National authority's code of conduct
National authority's regulatory oversight programme	• Programme structure & elements • Ships, port facilities & other entities under the programme's jurisdiction • Operational policies
Verification procedures	• Pre-approval verification process, including techniques and checklists • Post-approval/monitoring process, including techniques and checklists • Report writing
Procedures for handling non-compliance	• Enforcement principles and continuum of enforcement actions • Techniques for handling non-compliance and promoting voluntary compliance • Forms and reports
Procedures for observing or participating in exercises	• Types of exercises • Planning considerations and evaluating exercise results • Role of inspectors
Procedures for administering authorizations	• Certificate issuance process • Certificate renewal process
Procedures for conducting awareness and education activities	• Identifying target audiences • Types of delivery mechanisms • Promotional items

Appendix 2.11
Sample of a port facility security inspection report form

<table>
<tr><td colspan="3">PORT FACILITY SECURITY INSPECTION REPORT FORM</td><td>File number:</td></tr>
<tr><td colspan="4">Type of port facility:</td></tr>
<tr><td colspan="4">Name of port facility:</td></tr>
<tr><td colspan="4">Location:</td></tr>
<tr><td colspan="4">Port ID number</td></tr>
<tr><td colspan="4">UN locator</td></tr>
<tr><td colspan="2">Statement of Compliance number:</td><td colspan="2">Security level:</td></tr>
<tr><td colspan="2">Statement of compliance date of issue (dd/mm/yyyy):</td><td colspan="2">Date of expiry (dd/mm/yyyy):</td></tr>
<tr><td colspan="2">Name of operator:</td><td colspan="2">Address of operator:</td></tr>
<tr><td colspan="2">Telephone:</td><td>Fax:</td><td>E-mail</td></tr>
<tr><td colspan="2">Name of PFSO:</td><td colspan="2">24 h contact number:</td></tr>
<tr><td colspan="2">Telephone:</td><td>Fax:</td><td>E-mail:</td></tr>
<tr><td colspan="4">Type of inspection: ☐ Initial ☐ Intermediate ☐ Renewal
☐ Additional (includes monitoring and follow-up)</td></tr>
<tr><td colspan="4">Areas of Inspection (tick the appropriate boxes below):</td></tr>
<tr><td colspan="4">☐ 1 Documents and records
☐ 2 Access control
☐ 3 Restricted area access control
☐ 4 Handling of cargo
☐ 5 Delivery of ship's stores and bunkers
☐ 6 Security procedures for monitoring
☐ 7 Procedures for threats and security incidents
☐ 8 Security communications
☐ 9 Audits and amendments
☐ 10 Procedures for shore leave and visitors to the ship
☐ 11 Procedures for interfacing with ship security activities
☐ 12 Evacuation procedures
☐ 13 Security procedures to protect the security plan</td></tr>
<tr><td colspan="2">Deficiencies found
☐ Yes ☐ No</td><td colspan="2">Supporting documentation
☐ Yes ☐ No</td></tr>
<tr><td>Date of last inspection (dd/mm/yyyy)</td><td colspan="3">Have there been any changes to the port facility since the plan was approved?</td></tr>
<tr><td>Security office:</td><td colspan="3">Address:</td></tr>
<tr><td colspan="2">Telephone:</td><td>Fax:</td><td>E-mail:</td></tr>
<tr><td colspan="2">Date of inspection (dd/mm/yyyy)</td><td colspan="2">Name and signature of inspector
Name:
Signature:</td></tr>
</table>

Note: This inspection report must be retained at the port facility's office for a period specified by the Designated Authority and be available for consultation by a Government official at all times.

<table>
<tr><th>1 Documents and records</th><th>☐ Compliant</th><th>☐ Action required</th></tr>
<tr><td colspan="3">Plan references:
Possible questions for PFSO/security personnel and follow-up actions:
• Does the PFSO keep the records or are they kept elsewhere? If so, has the PFSO documented their existence, location and the name/position of the person responsible? Verify.
• Are they complete as required and kept for at least 1 year? Verify.
• Are the PFSP and related PFSA kept for at least 1 year after the day on which the PFSP expires? Verify.
• How are records protected from unauthorized access or disclosure? Verify.
• Are records kept electronically? If so, how are they protected from deletion, destruction and revision? Verify.
• Are computers password-protected and how often are the passwords changed? Verify.
• Interview port facility personnel to verify information recorded.</td></tr>
<tr><td colspan="3">Observations:</td></tr>
<tr><td colspan="3">Action required by operator (if necessary):</td></tr>
<tr><td colspan="3">Action by inspector (if necessary):</td></tr>
</table>

<table>
<tr><th>2 Access control</th><th>☐ Compliant</th><th>☐ Action required</th></tr>
<tr><td colspan="3">Plan references:
Possible questions for PFSO/security personnel and follow-up actions:
• Gates/barriers
– Are gates secured (manned/locked)? Verify.
– Do gates have card accesses? Test card access with a number of cards.
– Do gates have keys? Test the keys of those authorized.
– Inspect gates/barriers to ensure they are in good condition.
• Fencing
– Inspect fences to ensure they are in good condition.
– Verify that fences are clear of equipment/vehicles and debris against them.
– Who patrols/checks the fences? Verify this with the person(s) named.
– Who do personnel report breaches or damage to fencing to?
– Are logs maintained for patrols of fence line or maintenance? Verify.
• Rail security
– Are access controls established where rail lines enter the facility? Verify.
– Who monitors the activity at rail access points? Verify that they are monitored.
• Identification
– Observe what types of ID are valid to access the port facility and verify that these types of ID are detailed in the PFSP.
When would a person be denied access to the facility or a restricted area? Is a log kept? Verify.</td></tr>
<tr><td colspan="3">Observations:</td></tr>
<tr><td colspan="3">Action required by operator (if necessary):</td></tr>
<tr><td colspan="3">Action by inspector (if necessary):</td></tr>
</table>

3 Restricted area access control	☐ Compliant ☐ Action required
Plan references:	
Possible questions for PFSO/security personnel and follow-up actions: • What ID is valid to access or remain in a restricted area? Verify. What procedures are in place to issue passes, record their issuance, and record their loss? Verify. • What procedures are in place for verifying the identity of Government officials? Verify by interviewing a Government official at the facility/vessel or through observation. • What procedures are in place for verifying the identity of emergency responders? • What procedures are in place for verifying visitors? Truck drivers? Crew? Verify through interview or observation. • How are keys and passes controlled for restricted areas? Verify records. • What is the process for reporting lost keys, passes or access cards? Is a record kept of those that are lost? Inspect it. • Are persons subject to additional security measures when working in restricted areas? Verify through observation. • Are persons entering the facility or restricted area recorded in a log? If so, verify the log. • What is the procedure for crew access? What procedures are in place to ensure that only authorized crew are allowed back on the vessel? Verify this information with the SSO. • What procedures are in place for visitors to access restricted areas or the vessel? Verify through observation where possible.	
Observations	
Actions required by operator (if necessary):	
Actions taken by inspector (if necessary):	

4 Handling of cargo	☐ Compliant ☐ Action required
Plan references:	
Possible questions for PFSO/security personnel and follow-up actions: • What procedures are followed to deter cargo tampering? Verify by observation. • How is cargo identified and accepted for loading onto vessels? Verify through observation. • How long is cargo stored at the facility prior to loading? Are there temporary storage areas? How is this cargo inspected prior to loading? • Is there an inventory of dangerous cargoes? Are these cargos segregated from the remainder of the cargo at the port facility? Are they subject to additional security procedures? If so, are they detailed in the PFSP? Verify • Are vehicles carrying cargo inspected? If so, how? Are these procedures detailed in the PFSP? Observe.	
Observations	
Action required by port (if necessary):	
Action by inspector (if necessary):	

5 Delivery of ship's stores	☐ Compliant ☐ Action required
Plan references:	
Possible questions for PFSO/security personnel and follow-up actions: • How are security guards advised of deliveries of ship's stores? Verify. • Are all ship's stores deliveries scheduled in advance?	
Observations:	
Actions required by operator (if necessary):	
Actions taken by inspector (if necessary):	

6 Security procedures for monitoring	☐ Compliant	☐ Action required
Plan references:		
Possible questions for PFSO/security personnel and follow-up actions: • Alarms, motion detectors and lights – Who responds to alarm activations? Is the alarm company local? Do the alarms call the police? – Are they silent or audible? Inspect alarms by testing. – Where are the motion-detection devices located? Inspect them by testing. – Who is responsible to ensure that facility lighting is in good working order? Verify. – What are the maintenance procedures for alarms, motion detectors and lights? Verify. • Control/surveillance rooms – Is this area restricted? – Is it signed? Verify signage. – Who has access? How is access controlled/secured? Verify. – How many persons are on duty throughout the day? Do they have other security responsibilities that may take them away from monitoring camera activity? Is the control room ever unattended? – How is the surveillance equipment maintained? Are records of maintenance and occurrences kept in the control room? Inspect records if kept in the control room. – Are images recorded when cameras are motion-activated, continuously recorded or not capable of recording? – What is the length of recording time? How long are the recordings kept before re-recording? • Security rounds – Who conducts security rounds? What do the security rounds entail? As part of security rounds, are passes verified and unfamiliar persons questioned? – Are the times and results recorded? Verify these records. – Are security sweeps conducted before (and/or after) a vessel interfaces with the dock? What is the procedure for security sweeps? Verify with facility personnel. – Are all restricted areas patrolled? If so, what is the frequency? • Waterside security – Who patrols the waterside of the port facility? – How does the PFSO contact the police or service provider for assistance? Verify the contact number in the PFSP. – Do security rounds include a patrol of the waterside and lands adjacent to the water? Who conducts patrols of the lands adjacent to the waterside? What is their frequency? – Are there surveillance cameras directed at the waterside of the port facility? Do they record activity? If so, how?		
Observations:		
Actions required by operator (if necessary):		
Actions taken by inspector (if necessary):		

7 Procedures for responding to security threats, breaches of security and security incidents	☐ Compliant ☐ Action required

Plan references:

Possible questions for PFSO/security personnel and follow-up actions :

- Reporting security incident and threats
  - What are the procedures for reporting suspicious activities?
  - Do facility personnel use security incident reports at the facility? Are these logged? Inspect the records. Are they submitted to the Designated Authority?
  - What procedures do facility personnel follow if they receive a bomb threat on their phone? Or discover a suspicious package on the dock? Or discover a suspicious person or activity occurring in the facility? Verify by asking facility personnel.
- Response procedures
  - What is the responsibility of the PFSO when notified of an increase in security level?
  - How does the PFSO respond to a specific security threat or breach? Verify with the PFSP's response procedures.
  - How do personnel with security responsibilities respond to a specific security threat or breach? Verify that their response coincides with the procedure described in the PFSP.

Observations:

Actions required by operator (if necessary):

Actions taken by inspector (if necessary):

8 Security communications	☐ Compliant ☐ Action required

Plan references:

Possible questions for PFSO/security personnel and follow-up actions:

- Are personnel equipped with radios for security communication purposes?
- What channel is used for security communications? Verify with other personnel.
- Test the communication system and backup system (radio, telephone, etc.) by requesting the PFSO to contact someone on the facility or on board the vessel.
- Are additional communication procedures put into effect when security levels increase? Verify.
- Where signs are used to advise facility personnel of a change in security level, verify by inspecting the signs and asking personnel if they are aware of their usage.
- Ask personnel to identify the PFSO.
- Ask the PFSO if the vessel (or ship's agent) advises the facility of its security level prior to arrival? How is this information communicated? Verify.
- What are the maintenance procedures for communications equipment? Verify.
- Ask the PFSO under what circumstances a DoS should be completed.
- Who has the authority to complete a DoS at the facility? Verify that this coincides with the PFSP and/or the persons listed.
- How are communication procedures established when a vessel interfaces? Verify this information with the SSO.
- If a radio or cellular phone is used, test this by requesting the PFSO to contact the vessel.
- How is the delivery and inspection of ship's stores co-ordinated? Verify this information with the SSO.
- How is information concerning the contact of reciprocal security officers, SSAS activation, security threats, breaches and incidents conveyed? Verify with the SSO.
- How is crew access controlled? Verify with the SSO.

Observations:

Actions required by operator (if necessary):

Actions taken by inspector (if necessary):

9 Audits and amendments	☐ Compliant	☐ Action required
Plan references: Possible questions for PFSO/security personnel and follow-up actions: • Are annual audits of the PFSP based on the date of the original plan's approval? • Do audits take place whenever there is a new operator, a change in operations or location, or modification to the port facility that could affect its security? • Is there evidence (in the form of audit plans, audit reports, meeting minutes, records of follow-up or remedial actions) of audits being undertaken ? • Does the person who conducted the audit have the relevant formal qualifications and work experience? Verify by requesting examples of previous audit reports.		
Observations:		
Actions required by operator (if necessary):		
Actions taken by inspector (if necessary):		

10 Procedures for shore leave and visitors to the ship	☐ Compliant	☐ Action required
Plan references: Possible questions for PFSO/security personnel and follow-up actions: • What are the procedures in place for facilitating shore leave? Verify these with the PFSP. • Who is responsible for escorting or continuously monitoring seafarers transiting through restricted areas of the port? Verify with the security personnel who are identified as being responsible.		
Observations:		
Actions required by operator (if necessary):		
Actions taken by inspector (if necessary):		

11 Procedures for interfacing with ship security activities	☐ Compliant	☐ Action required
Plan references: What are the procedures in place? Verify these with the PFSP.		
Actions required by operator (if necessary):		
Actions taken by inspector (if necessary):		

12 Evacuation procedures	☐ Compliant	☐ Action required
Plan references: Possible questions for PFSO/security personnel and follow-up actions:		
Observations:		
Actions required by operator (if necessary):		
Actions taken by inspector (if necessary):		

13 Security procedures to protect the security plan	☐ Compliant	☐ Action required
Plan references: • Is the PFSP made available only to people who have a legitimate need to know how to fulfil their official duties or contractual obligations? • Is the PFSP handled with due care and only in accordance with authorized procedures? • Is the PFSP kept in a safe place and accessed only in accordance with authorized procedures?		
Observations:		
Actions required by operator (if necessary):		
Actions taken by inspector (if necessary):		

Additional questions on qualifications of port facility personnel

For port facility security officer:

Q: What training have you received to become a PFSO?

For personnel with security responsibilities:

Q: What are your role and duties concerning security?

Q: What training have you received to perform these duties?

Q: How do your security duties change at security levels 2 and 3?

For personnel without security responsibilities:

Q: What security orientation or training have you received?

Q: Do you have a facility identification card?

Q: What is a security level and what is the meaning of each level?

Q: What procedures are required of you at each security level?

Additional comments

Date	Comments

Appendix 2.12

Details of national authority contact points

Source: IMO Circular letter No. 2514, December 2003

1	Contact type*	
2	Organization/authority/ department	
3	First name	
4	Surname	
5	Title	
6	Post	
7	Specific responsibilities	
8	Condition of authority†	
9	Address	
10	Phone	
11	Fax	
12	Mobile	
13	E-mail	
14	Telex	

* One copy of the form is to be used for each organization according to its contact type:
- National authorities responsible for ship security;
- National authorities responsible for port facility security;
- Proper recipients of SSAS alerts;
- Proper recipients of maritime security-related communications from other Contracting Governments;
- Proper recipients of requests for assistance with security incidents;
- Names of recognized security organizations (RSOs) approved by the State.

† Condition of authority is only applicable in the case of recognized security organizations.

Appendix 2.13

Details of port facilities

Source: IMO Circular letter No. 2514, December 2003

1	Detail of the port	Name of port	
		Status*	
		Port ID number	
		UN .ocator	
2	Port facility name		
3	Assigned port facility number†		
4	Alternative names for port (if applicable)		
5	Port facility description		
6	Location	Longitude	
		Latitude	
7	Port facility security point of contact		
8	Port facility has taken part in alternative arrangement		
9	Port facility has approved port facility security plan		
10	Date of the original approval of the port facility security plan		
11	Date of most recent review of the port facility security plan		
12	Date of most recently issued Statement of Compliance, if applicable		
13	Has this port facility security plan been withdrawn?		
14	Port facility security plan withdrawal date		

* Whether the port is open or closed.

† Port facility number should not be duplicated.

Appendix 2.14
Report of the imposition of a control and compliance measure

Source: MSC/Circ.1111, June 2004

(Reporting authority)

(Address)

(Telephone and fax)

Copy to: Master
Duly authorized officer administrative office

If control measures, other than lesser administrative measures, are taken, additional copies of this report shall be provided to:

- Administration ☐
- Recognized security organization ☐
- IMO ☐
- Port State of ship next port call (if denied entry or expelled) ☐

1 Name of reporting authority:. . . . . 2 Date of inspection: . . . .

3 Place of inspection:. . . . .

4 Name of ship: . . . . 5 Flag of ship: . . . .

6 Type of ship: . . . . 7 Call sign: . . . .

8 IMO Number: . . . . 9 Gross tonnage: . . . .

10 Year of build: . . . .

11 Recognized security organization: . . . .

12 Registered owner (from Continuous Synopsis Record (CSR)): . . . .

13 Registered bareboat charterer, if applicable (from CSR): . . . .

14 Company (from CSR): . . . .

15 IMO Company identification number: . . . .

16 ISSC issuing authority: . . . . 17 Dates of issue/expiry: . . . .

18 Ship security level: . . . .

19 Reason(s) for non-compliance: . . . .

. . . .

. . . .

. . . .

. . . .

20 Action taken by duly authorized officer:. . . . .

. . . .

. . . .

21 Specific control measures taken (marks as follow: "x" actions taken, "-" no actions taken)

None	☐
Lesser administrative measures	☐
More detailed inspection	☐
Ship departure delayed	☐
Restricted ship operation	
Cargo operation modified or stopped	☐
Ship directed to other location in port	☐
Ship detained	☐
Ship denied entry into port	☐
Ship expelled from port	☐

22 Corrective action taken by ship or Company: . . . .

. . . .

. . . .

Issuing office: . . . . Duly authorized officer

Name: . . . .

Telephone/fax: . . . . Signature: . . . .

Section 3 – Security responsibilities of port facility and port operators

3.1 Introduction

3.1.1 This section provides guidance on the security responsibilities of port facility and port operators under the Maritime Security Measures. Following a general description of the security framework, guidance is offered on:

- **.1** security levels;
- **.2** security personnel;
- **.3** port facility security assessments (pfsas);
- **.4** port facility security plans (PFSPs);
- **.5** PFSP implementation;
- **.6** Statements of Compliance;
- **.7** port security; and
- **.8** guidelines for non-SOLAS marinas, ports and harbours.

3.1.2 Primarily addressed to those responsible for port facility security, the guidance is also relevant to those exercising security responsibilities for the port facility and the Government officials that regulate them.

3.1.3 To facilitate comparisons of the responsibilities of port facility operators with those of Governments and their Designated Authorities, the chart below references the equivalent subsections and paragraphs in section 2.

Port facility operator responsibilities	Maritime Security Measures	Cross-reference to responsibilities for Designated Authorities
3.2.1–3.2.3	Defining the port facility	2.8.1–2.8.12
3.2.5–3.2.8	Recognized security organizations	2.5
3.2.9–3.2.10	Alternative security agreements	2.13
3.2.11	Equivalent security arrangements	2.14
3.3	Changing security levels	2.6
3.4	Declarations of Security	2.7
3.5.1–3.5.6	Port facility security officers	2.8.19–2.8.24
3.6	Port facility security assessments	2.8.25–2.8.33
3.7	Port facility security plans	2.8.34–2.8.42
3.8.8	Reporting security incidents	2.9.38
3.9	Port security	2.18.16–2.18.20
3.9.3–3.9.8	Port security committees	2.8.17–2.8.18
3.10	Guidelines for non-SOLAS marinas, ports and harbours	2.18.3–2.18.15

3.2 Security framework

Defining the port facility

3.2.1 In the Maritime Security Measures, a *port facility* is defined as the location where the ship/port interface occurs. Governments are responsible for identifying which port facilities fall under the Maritime Security Measures and the extent to which they apply to facilities which occasionally serve ships on international voyages. However, port facility operators can assist this process by complying with requests to provide information on the types and frequency of ships using the port facility, their trading patterns, and the cargoes handled, passenger numbers and origins and other security-related information.

3.2.2 Once a port facility has been identified as falling under the Maritime Security Measures, the next step is to establish its geographic boundary. Experience to date indicates that this can be a challenging process due to the need to carefully consider a range of factors, including:

- **.1** where passengers embark and disembark;
- **.2** where dangerous goods or high-value cargoes are handled;
- **.3** where containers are loaded, unloaded and stored (both in the short and long term);
- **.4** the economic significance of the port facility;
- **.5** the proximity of the port facility to populated areas;
- **.6** the areas of risk or vulnerability identified by the PFSA;
- **.7** the location of pipelines and related valves (including on the water side);
- **.8** the location of natural barriers (e.g., tree lines, drainage channels and inlets); and
- **.9** the location of existing man-made barriers (e.g., fences, walls, roads, access gates).

3.2.3 Experience has also shown that the preparation of a map delineating the area of each port facility should be considered as it can:

- **.1** present the boundary in a way that is clear and easily understood;
- **.2** show all natural and man-made features which form the boundary or are adjacent to it;
- **.3** be inserted into the PFSA and PFSP;
- **.4** include distances, directions and co-ordinates; and
- **.5** be easily amended to reflect future changes to the boundary or existing features.

3.2.4 Guidance on preparing a map may be downloaded from the following internet site: www.infrastructure.gov.au/transport/security/maritime/pdf/GuidancePaperMappingStandardsforPorts.pdf

Recognized security organizations

3.2.5 As indicated in subsection 2.5, recognized security organizations (RSOs) may advise or provide assistance to port facilities on PFSAs and PFSPs, including their completion.

3.2.6 Port authorities and port facility operators may be appointed as RSOs provided that they have the appropriate security-related expertise (refer to appendix 2.3 – Criteria for selecting recognized security organizations).

3.2.7 RSOs may not approve, verify or certify work products that they have either developed or used sub-contractors to develop.

3.2.8 Experience to date indicates that, when a port or port facility operator intends to enter into a contract for the services of an RSO, sound business practice encourages preparation of a formal written agreement signed by both parties. As a minimum, this agreement should:

- **.1** specify the scope and duration of the work;
- **.2** identify the main points of contact within the port/port facility and the RSO;
- **.3** detail the data to be provided to the port Administration/port facility operator;

- **.4** identify the legislation, policies, procedures and other work instruments to be provided to the RSO;
- **.5** specify the records to be maintained by the RSO and made available as necessary;
- **.6** specify any reports to be provided regularly, including changes in capability (e.g., loss of key personnel); and
- **.7** specify a process for resolving performance-related issues.

Alternative security agreements

3.2.9 Alternative Security Agreements are agreements between national Governments on how to implement the Maritime Security Measures for short international voyages using fixed routes between port facilities within their jurisdiction (subsection 2.13). To date, such Agreements usually cover international ferry services and address such topics as acceptance of minor differences in regulatory requirements and security arrangements.

3.2.10 Operators of ports and port facilities covered by such Agreements should ensure that they are fully aware of the implications for their operations.

Equivalent security arrangements

3.2.11 For port facilities with limited or special operations (e.g., terminals attached to factories or quaysides with occasional operations (subsection 2.14), it may be appropriate for operators to implement security measures equivalent to those prescribed in the Maritime Security Measures. Details of ESAs could be included in the PFSP.

3.3 Changing security levels

3.3.1 Governments are responsible for setting security levels and communicating changes rapidly to those who need to be informed, including port and port facility operators (subsection 2.6). This requires Governments, usually through their Designated Authorities, to compile and maintain an accurate set of contact details. In turn, this requires operators of ports and port facilities to promptly communicate changes in contact details.

3.3.2 In addition to having security plans specifying the security measures and procedures in place at each security level, port and port facility operators should ensure that their plans identify the Measures and procedures to be implemented when a ship is operating at a higher security level, set by its Administration, than that applying at their port or port facility.

3.3.3 Experience to date provides examples of:

- **.1** for port operators, the PSO being identified as the point of contact;
- **.2** for port facility operators, the PFSO being identified as the point of contact;
- **.3** in each case, the manager of the PSO/PFSO being identified as an alternate;
- **.4** for ports with a PSO, the line of notification of changes can be a three-step 'fan-out' process:
  - Designated Authority to PSO
  - PSO to PFSOs and other port stakeholders
  - PFSOs to key facility personnel and ship security officers (SSOs)
- **.5** for ports without a PSO, the line of change notification can be a two-step process:
  - Designated Authority to PFSO(s) and other port stakeholders
  - PFSOs to key port facility personnel and SSOs
- **.6** PSOs/PFSOs regularly testing lines of communication;
- **.7** multiple means of communicating with contacts, i.e., by telephone, e-mail and fax.

3.4 Declarations of Security

3.4.1 A Declaration of Security (DoS) is a written agreement between a port facility and a ship visiting that facility on their respective security responsibilities during the visit (subsection 2.7). The requirement for a port facility to initiate, complete and retain a DoS is determined by the Designated Authority and includes the conditions under which ships and port facilities may request a DoS.

3.4.2 The Maritime Security Measures contain a model form for a Declaration of Security between a port facility and a ship (refer to appendix 3.1 – Declaration of Security Form). As well as including information on the identity of the port facility and ship, the form specifies the type of activity to be covered, its duration and the security level applying to the particular ship/port interface. If a ship is operating at a higher security level than the port facility, the ship/port interface should take place at its higher security level.

3.4.3 Normally, the DoS is completed by the PFSO. However, if the Designated Authority determines otherwise, it may be handled by another person responsible for shore-side security, on behalf of the port facility. When completed, it must be signed and dated both by the PFSO (or their alternate) designated by the Designated Authority and by the ship's master or ship security officer (SSO). Unless there are exceptional circumstances, the DoS only takes effect after it has been signed by both parties in a language common to both parties.

3.4.4 When a ship initiates a DoS, the request shall be acknowledged by the port facility; however, the port facility does not have to comply with the request.

3.4.5 When a port facility initiates a DoS, the request shall be acknowledged by the ship's master or SSO; in this instance, the ship must comply with the request if the ship intends to interface with the port facility.

3.4.6 The conditions under which a DoS may be requested are referenced in paragraph 2.7.3 and should be documented in the PFSP.

3.4.7 The PFSP should detail the procedures to be followed and the security measures and procedures to be implemented when responding to a request for a DoS or initiating a DoS. For a ship/port interface, these could include the respective responsibilities accepted by the port facility and the ship in accordance with their security plans to:

- **.1** ensure the performance of all security duties;
- **.2** monitor restricted areas to ensure that only authorized personnel have access;
- **.3** control access to the port facility and ship;
- **.4** monitor the port facility, including berthing areas and areas surrounding the ship;
- **.5** monitor the ship, including berthing areas and areas surrounding the ship;
- **.6** handle cargo and unaccompanied baggage;
- **.7** monitor the delivery of ship's stores;
- **.8** control the embarkation of persons and their effects;
- **.9** ensure that security communication is readily available between the ship and the port facility.

3.4.8 Experience to date provides examples of:

- **.1** when the port facility's security measures documented in the DoS are extracted from the PFSP, care being taken to omit sensitive security information such as security standards;
- **.2** the PFSO notifying the Designated Authority if a ship:
  - for any reason, refuses a request for a DoS (in addition to denying it entry to the facility);
  - requesting a DoS is at security level 3.
- **.3** the DoS being kept on file for three years (which may be longer than the minimum specified by the Designated Authority), so as to be aware of any trends in DoS requests;

.4 at port facilities occasionally used by SOLAS ships, the person ashore responsible for shore-side security (in place of a PFSO) having clear authority to agree a DoS with a SOLAS ship intending to engage in a ship/port interface at the facility.

3.5 Security personnel

Port facility security officers

3.5.1 Each designated port facility operator is required to appoint a port facility security officer (PFSO). Refer to paragraphs 2.8.19 to 2.8.24 for more information. Individual PFSOs establish and maintain the security of their port facility and are responsible for maintaining effective contacts with the CSOs and SSOs of ships using their port facility on which efficient operation of the Maritime Security Measures depends.

3.5.2 A PFSO may be responsible for one or more port facilities. Also, it is required to give the PFSO the necessary support to perform the duties listed below, including access to required training.

3.5.3 The duties of a PFSO include:

.1 conducting a comprehensive security survey of the port facility, taking into account the approved PFSA;

.2 ensuring the development and maintenance of the PFSP;

.3 implementing and testing the PFSP;

.4 undertaking regular security inspections of the port facility to ensure that appropriate security measures are in place;

.5 recommending and incorporating, as appropriate, modifications to the PFSP in order to correct deficiencies and take into account relevant changes to the port facility;

.6 enhancing security awareness and vigilance of port facility personnel;

.7 ensuring that adequate training has been provided to personnel responsible for the security of the port facility;

.8 reporting to relevant authorities and maintaining records of incidents which threaten the security of the port facility;

.9 co-ordinating the implementation of the PFSP with appropriate CSOs and SSOs;

.10 co-ordinating with security services, as appropriate;

.11 ensuring standards for personnel responsible for security of the port facility are met;

.12 ensuring security equipment is properly operated, tested, calibrated and maintained;

.13 liaising and co-ordinating appropriate actions with an SSO if advised that:

- a ship is at a higher security level than that of the port facility;
- a ship is encountering difficulty in complying with the applicable Maritime Security Measures, including instructions issued by the Contracting Government if the port facility is at security level 3; or in implementing the relevant measures and procedures detailed in the SSP;

.14 reporting a ship that is at a higher security level than that of the port facility to the competent authority; and

.15 assisting SSOs in confirming the identity of those seeking to board ships when requested;

3.5.4 In connection with the last duty identified above, PFSOs should actively seek to facilitate shore leave for ships' crews, crew changes and access of visitors to ships, including representatives of seafarers' welfare and labour organizations.

3.5.5 Each person performing the duties of a PFSO should be able to satisfactorily demonstrate the competencies listed in appendix 3.2 – Competency matrix for port facility security officers. Persons who have satisfactorily completed a training course for PFSOs which is recognized by the Designated Authority should be considered to have met this requirement.

3.5.6 Experience to date includes examples of PFSOs and those appointed to undertake their duties:

.1 being required to have documentary evidence of their appointment and training;

.2 being required to have security clearances, particularly if they have access to sensitive security information provided by the Contracting Government (e.g., information on national threats);

.3 being restricted to port facility employees, that is, PFSOs are not contracted in from an external company (e.g., security company or consultancy);

.4 having an approved documented list of security and non-security duties (non-security duties should not interfere with the ability to carry out security duties);

.5 being active members of port security committees; and

.6 reporting to a senior member of the port facility operator's management team.

Other port facility personnel with security-related duties

3.5.7 Other port facility personnel with security-related duties (e.g., guards, access control officers, training officers and relevant port facility managers) are also required to have the knowledge and training required to carry out their assigned duties. They should be able to satisfactorily demonstrate the competencies listed in appendix 3.3 – Competency matrix for port facility personnel with security duties. Persons who have satisfactorily completed a recognized training course should be considered to have met this requirement.

3.5.8 Experience to date includes examples of other port facility personnel with security-related duties:

.1 being required to meet the same or similar requirements as PFSOs (see paragraph 3.5.5 above), with personnel being allowed to demonstrate competency by the following alternative means:

- having evidence of equivalent service for a period of at least six months in total during the preceding three years; or
- passing an approved test.

.2 before being assigned to their duties, receiving security-related familiarization training, provided by the PFSO or equally qualified person, in their assigned duties in accordance with the provisions specified in the PFSP;

.3 being required to have documentary evidence of their training; and

.4 being listed in the PFSP.

All other port facility personnel

3.5.9 All other port facility personnel should receive adequate security-related training so as to contribute collectively to the enhancement of maritime security at the port facility. They should be able to satisfactorily demonstrate the competencies listed in appendix 3.4 – Competency matrix for port facility personnel without security duties.

3.5.10 Experience to date provides examples of personnel without security-related duties being expected to:

.1 receive familiarization training sufficient to enable them to:

- report a security incident;
- know the procedures to follow when they recognize a security threat;
- take part in security-related emergency and contingency procedures;

.2 receive security-related training, provided by the PFSO or equally qualified person, at least once in their career at the port facility; and

.3 have documentary evidence of their training.

Security clearances

3.5.11 Port facility operators can be required to comply with any instructions issued by their Government regarding the application of any security clearance procedures for port facility personnel.

3.5.12 Security clearances are the means of verifying that personnel whose duties require access to restricted areas or security-sensitive information do not pose a risk to maritime security. The vetting associated with these clearances is more stringent than the pre-employment background checks conducted by port facility operators.

3.5.13 Experience to date includes examples of Governments requiring security clearance for:

.1 the senior managers of a port facility;

.2 the PFSO and those appointed to undertake the duties of the PFSOs; and

.3 in some cases, all those working in any capacity within port areas.

3.6 Port facility security assessments

Introduction

3.6.1 Governments, normally through their Designated Authority, are responsible for carrying out port facility security assessments (PFSAs) or authorizing RSOs to do so on their behalf (subsection 2.5). In practice, the undertaking of PFSAs requires the involvement of port facility operators due to their in-depth knowledge of the port facility's assets, infrastructure, vulnerabilities and past security incidents.

3.6.2 The PFSA may be considered to be a risk analysis of all aspects of a port facility's operations in order to determine which parts of it are more susceptible, and/or more likely, to be vulnerable. It is an essential and integral part of developing or updating the PFSP.

Conducting PFSAs

3.6.3 The PFSA is required to include the following four elements:

.1 identification and evaluation of important assets and infrastructure;

.2 identification of possible threats to them and the likelihood of their occurrence;

.3 identification, selection and prioritization of countermeasures and procedural changes and their level of effectiveness in reducing vulnerabilities; and

.4 identification of weaknesses, including human factors, in the infrastructure, policies and procedures.

3.6.4 A risk-assessment and management tool that encompasses these four elements is described in section 5 along with a list of port security assessment techniques accessible on the internet.

Preparing PFSA reports

3.6.5 A report shall be prepared upon completion of the PFSA. It provides the means by which a PFSA is approved and is required to:

.1 summarize how the assessment was conducted;

.2 describe each vulnerability found during the assessment;

.3 describe the countermeasures that could address each vulnerability; and

.4 be protected from unauthorized access or disclosure.

3.6.6 As indicated above, the report must be protected from unauthorized access or disclosure. Upon approval, some Member States provide a numbered copy to an approved list of individuals within the Designated Authority and port facility and establish procedures for how the report is to be retained and accessed.

PFSA coverage of multiple facilities

3.6.7 Contracting Governments may allow a PFSA to cover more than one port facility if the operator, location, equipment and design of these port facilities are similar. If such an arrangement is allowed, details must be communicated to IMO.

3.6.8 Experience to date indicates that no such arrangements have been submitted to IMO.

Updating PFSAs

3.6.9 PFSAs are to be reviewed and updated periodically or when major changes to the port facility take place (paragraphs 2.8.25 to 2.8.33).

3.6.10 Experience to date includes examples of Designated Authorities producing guidance material recommending that PFSAs should be updated within a short time (e.g., 45 days) after a major change or security incident at the port facility, the term 'major' being used to describe changes to physical structures or operations, or incidents that are sufficient to have an impact on port/port facility operations.

3.6.11 In the absence of any major changes or incidents, PFSAs are reviewed at least every five years, with a shorter period (two to three years) for larger port facilities.

3.7 Port facility security plans

Introduction

3.7.1 Port facility security plans (PFSPs) shall be developed and maintained based on the results of approved PFSAs conducted at each port facility. The close inter-relationship between PFSAs and PFSPs is shown by the example of a PFSA/PFSP approval process illustrated in appendix 3.5 – Example of a port facility security assessment and plan approval process.

3.7.2 PFSPs must be approved by the Designated Authority. RSOs cannot approve them but may assist in their preparation.

3.7.3 PFSPs may be developed by PFSOs or by RSOs acting on their behalf. When RSOs are acting on their behalf, PFSOs continue to be responsible for ensuring that they are properly prepared.

Preparing and maintaining PFSPs

3.7.4 All PFSPs must provide details of:

- **.1** measures designed to prevent weapons or any other dangerous substances and devices intended for use against persons, ships or ports, and the carriage of which is not authorized, from being introduced into the port facility or on board a ship;
- **.2** measures designed to prevent unauthorized access to the port facility, to ships moored at the facility, and to restricted areas of the facility;
- **.3** procedures for responding to security threats or breaches of security, including provisions for maintaining critical operations of the port facility or ship/port interface;
- **.4** procedures for responding to any security instructions the Contracting Government in whose territory the port facility is located may give at security level 3;
- **.5** procedures for evacuation in case of security threats or breaches of security;

.6 duties of port facility personnel assigned security responsibilities and of other facility personnel on security aspects;

.7 procedures for interfacing with ship security activities;

.8 procedures for the periodic review of the plan and updating;

.9 procedures for reporting security incidents;

.10 identification of the PFSO, including 24-hour contact details;

.11 measures to ensure the security of the information contained in the plan;

.12 measures designed to ensure effective security of cargo and the cargo handling equipment at the port facility;

.13 procedures for auditing the plan;

.14 procedures for responding in case the ship security alert system (SSAS) of a ship at the port facility has been activated; and

.15 procedures for facilitating shore leave for ship's personnel or personnel changes, as well as access of visitors to the ship, including representatives of seafarers' welfare and labour organizations.

3.7.5 In addition, PFSPs should detail:

.1 the port facility's security organization;

.2 the security organization's links with other relevant authorities, the communication systems necessary to allow its effective continuous operation and ships within or approaching the port facility;

.3 the basic security level 1 measures, both operational and physical, that will be in place;

.4 the additional security measures that will allow the port facility to progress without delay to security level 2 and, when necessary, to security level 3;

.5 the procedures for the regular review or audit of the PFSP and for its amendment in response to experience or changing circumstances;

.6 the procedures for reporting incidents to the appropriate Contracting Government's contact points; and

.7 the procedures for interacting with ships which are operating at a higher security level, set by the ship's Administration, than that applying at the port or port facility.

3.7.6 Internet sites which have been developed by Member States to illustrate how PFSPs may be prepared and updated are shown in appendix 3.6 – Examples of internet sources of guidance material on preparing, updating and implementing port facility security plans. The appendix also identifies sources of information on best practices.

3.7.7 Due to considerations of conflict of interest, personnel conducting internal audits of the security measures specified in PFSPs or evaluating their implementation are required to be independent of the Measures being audited unless this is impracticable, due to the size and nature of the port facility.

3.7.8 PFSPs are required to be protected from unauthorized access or disclosure. If PFSPs are kept in electronic format, procedures must be put in place to prevent their unauthorized deletion, destruction or amendment.

3.7.9 Subject to approval by the Designated Authority, a PFSP may cover multiple facilities if their operators, location, type of operation, equipment and design are similar.

3.8 PFSP implementation

Introduction

3.8.1 Proposed measures in amended PFSPs may not be implemented until authorized by the Designated Authority.

3.8.2 The security measures in PFSPs should be implemented within a reasonable period of their approval. Some Member States require PFSPs to specify when proposed measures will be in place and for PFSOs to contact the Designated Authority and discuss contingency plans if there is likely to be any delay.

Planning and conducting drills and exercises

3.8.3 To ensure the effective implementation of PFSPs, drills are required to be carried out on each element at a recommended minimum interval of three months. These are usually organized by PFSOs, who are responsible for testing the effective implementation of PFSPs.

3.8.4 To ensure the effective implementation and co-ordination of PFSPs, PFSOs are required to participate in exercises at a recommended minimum interval of once each calendar year, with no more than 18 months between the exercises. These exercises are usually planned and co-ordinated by port authorities and conducted on a port-wide basis; they may be:

.1 full-scale or live;

.2 tabletop simulation or seminar;

.3 combined with other exercises organized by Government agencies or port authorities to test emergency response or commerce resumption plans.

3.8.5 Drills and exercises take up organizational time and resources, and must therefore be conducted in as efficient and effective a manner as possible. Recognizing the need to assist port facility operators in the Asia-Pacific Region, the Asia-Pacific Economic Cooperation (APEC) forum's Transportation Working Group developed a set of guidelines in the form of a manual. It provides a systematic and comprehensive approach to the planning, preparation for, conduct, debrief and reporting of maritime security drills and exercises. Workshops have been delivered to port security officials in several APEC member economies. To provide an appreciation of the scope of these practices, the manual's table of contents is shown in appendix 3.7 – APEC Manual of maritime security drills & exercises for port facilities: table of contents.

3.8.6 The accessible internet site containing the contents of the entire APEC Manual is referenced in appendix 3.6 – Examples of internet sources of guidance material on preparing, updating and implementing port facility security plans.

3.8.7 The conduct of drills and exercises may lead to amendments to the approved PFSP. Major amendments to an approved PFSP should be submitted to the Designated Authority for re-approval.

Reporting security incidents

3.8.8 PFSPs are required to document procedures for reporting security incidents and PFSOs are required to report them to relevant authorities.

3.8.9 Security incidents generally fall into two categories:

.1 those considered to be sufficiently serious that they should be reported to relevant authorities by the PFSO, including:

- unauthorized access to restricted areas within the port facility;
- unauthorized carriage or discovery of weapons or prohibited items in the port facility;
- incidents of which the media are aware;
- bomb warnings;
- unauthorized disclosure of a PFSP.

.2 those of a less serious nature but which require reporting to, and investigation by, the PFSO, including:

- breaches of screening points;
- inappropriate uses of passes;
- damage to security equipment through sabotage or vandalism;
- suspicious behaviour in or near the port facility;
- suspicious packages in or near the port facility;
- unsecured access points.

3.8.10 Experience to date indicates that some Designated Authorities have:

.1 specified the types of security incidents that must be immediately reported to them, as indicated below:

Type of security incident
Attack
Bomb warnings
Hijack
Armed robbery against a ship
Discovery of firearms
Discovery of other weapons
Discovery of explosives
Unauthorized access to a restricted area
Unauthorized access to the port facility
Media awareness

.2 with respect to bomb warnings, developed a checklist as a useful aid for anyone receiving a warning (which can be received in various ways, with a telephone call to a port authority, port facility operator or individual ship at the port facility being the most common).
www.cpni.gov.uk/security-planning/business-continuity-plan/bomb-threats

.3 designed standard forms for security incidents that must be reported to them and making them available on their internet sites. One such form – the Maritime security incident report online form developed by the Australian Government's Department of Infrastructure, Transport, Regional Development and local Government – may be downloaded from: www.infrastructure.gov.au. Although this form has been designed to fulfil incident reporting requirements prescribed in national legislation, it could be adapted by port facility operators to their particular reporting requirements. In such cases, the form's practical usefulness could be enhanced by:

- ensuring that its format is straightforward;
- allowing the PFSO to report the remedial action taken;
- ensuring that any associated reporting procedures are straightforward;
- specifying the situations when it is to be forwarded to the port facility's manager; and
- locating copies where they can be visible to, and easily accessed by, port facility personnel.

Information security

3.8.11 PFSPs are required to be protected from unauthorized access or disclosure and require the establishment of documented procedures for ensuring the security of information documented in them. Similar requirements apply to PFSAs and other security-sensitive information, including information on cargo movements and cargo.

3.8.12 Experience to date includes examples of Governments providing guidance to port facility operators on:

- **.1** ensuring that all sensitive information is password-protected;
- **.2** installing access–control and security systems in locations where sensitive information is stored (e.g., server rooms and control rooms);
- **.3** having effective data back-up procedures.

Shore access for seafarers and on-board visits to ships

3.8.13 The Maritime Security Measures require PFSPs to specify the procedures for facilitating:

- **.1** shore leave for ship's personnel or personnel changes;
- **.2** seafarer access to shore-based welfare and medical facilities; and
- **.3** on-board access by visitors, including representatives of seafarers' welfare and labour organizations.

3.8.14 In addition to access by visitors to ships, PFSPs should contain procedures at all security levels to cover access by shore-based ship support personnel, including those involved with the taking on board of ship's stores and bunkers.

3.8.15 From a practical perspective, it is important that port and port facility operators and security personnel seek a balance between the needs of security and the needs of the ship and its crew. Port facility operators and the PFSOs should ensure co-ordination of shore leave for ship personnel or crew change-out, as well as access through the port facility for visitors to the ship, including representatives of seafarers' welfare and labour organizations and those concerned with the maintenance of ships' equipment and safe operation, with the Company in advance of the ship's arrival.

3.8.16 A singular focus on the security of the port facility is contrary to the letter and spirit of Maritime Security Measures and has serious consequences for the international maritime transportation system that is a vital component of the global economy. The ILO/IMO Code of practice on security in ports also recommends that all port stakeholders work co-operatively to make such arrangements and advance plans.

3.8.17 Port States, while giving effect to security measures to prevent security incidents affecting ships or port facilities and to exercise control over access to their territories, have to recognize that shore leave for seafarers constitutes their right – not a privilege.

3.8.18 Access by authorized personnel to the ship is also a necessity. Wherever practicable, formalities, documentary requirements and procedures should be uniformly applied in order to provide for a consistent application of port facility security measures, provided that such uniformity does not bypass or eliminate the authority of Member States.

3.8.19 PFSOs and PSOs should ensure co-ordination of these requirements with SSOs, if possible, in advance of the ship's arrival at the port facility. The arrangements should strike a balance between the security needs of ports and port facilities with the needs of the ship and its crew. A single focus on port/port facility security is contrary to the letter and spirit of the Maritime Security Measures.

Conducting self-assessments

3.8.20 Checklists provide a useful way to assess and report progress in implementing PFSPs and, by extension, the Maritime Security Measures. Although they can be completed on an as-needed basis, it is a good management practice to conduct such an assessment at least once a year and to establish a link between any identified gaps and work-plan priorities. Appendix 3.8 – Implementation checklist for port facility operators, contains a checklist for port facility operators that can be used to assess progress in implementing the Maritime Security Measures. Except for minor modifications to its format and guidance material, it is identical to the Voluntary self-assessment tool for port facility security that was approved by IMO's Maritime Safety Committee in May 2006 and received widespread distribution.

3.8.21 PFSOs and PSOs are encouraged to modify the content and format of this checklist in order to ensure that they meet their specific assessment requirements (e.g., to identify when procedures were last reviewed or measures were tested).

Preventing unauthorized access

3.8.22 Experience to date indicates that some Governments require utilization of a credential checking system as a measure to prevent unauthorized access to the port facility:

.1 some systems are electronic in nature, while others are manual. As a general rule, for manual systems, these credentials are tamper-resistant, contain a photograph of the bearer, and have a specific expiration date;

.2 in both cases, the identity credential is required to be visibly displayed upon entry to the port facility and while on the port facility in order to facilitate verification by security personnel and duly authorized personnel conducting periodic inspections;

.3 every effort is made to limit the number of personnel in the port facility to only those personnel who have an immediate need to be there;

.4 in the case of visitor credentials, a document exchange is often utilized whereby the visitor (other than a duly authorized official) must surrender a Government-issued identity document in exchange for a visitor pass that must be displayed; and

.5 the facility should be monitored by patrols, electronic means such as CCTV, or a combination of the two, so as to ensure the integrity of the facility's security perimeter. In the case of patrols, they may be conducted on foot, by vehicle or via small vessels. Mechanisms should be in place to verify that the patrols occur as required.

Effective security of cargo and ship's stores

3.8.23 Experience to date indicates that some Governments require all cargo and ship's stores entering the port facility to have adequate and reliable documentation establishing that they are coming from a trustworthy and known source and are properly destined for the facility. Such documentation is normally standardized, matches the cargo with the conveyance transporting it to the port facility, is resistant to forgery and is consistently examined by security personnel prior to allowing admittance onto the port facility.

3.9 Port security

Introduction

3.9.1 The Maritime Security Measures apply to port facilities. Guidance on wider aspects of port security is contained in the ILO/IMO Code of practice on security in ports, which may be accessed at the following internet site: www.imo.org/OurWork/Security/Instruments/Pages/CoP.aspx. Several Governments have enacted legislation applying parts of the guidance.

3.9.2 There are broad similarities between the guidance offered on port facility security and port security. The significant differences relative to this section of the Guide are:

.1 establishing a port security committee;

.2 the appointment of PSOs;

.3 undertaking port security assessments (PSAs); and

.4 preparing port security plans (PSPs).

Port security committees

3.9.3 At the port level, the development and implementation of security procedures and measures can be enhanced through the establishment of a port security committee (PSC) comprising representatives from the port/harbour authority, the port facilities within the port, Government organizations operating in the port, local law-enforcement agencies, and those employed in the port as well as port users. Together, those represented on a PSC should have detailed knowledge of the security issues and patterns of criminality experienced at the particular port.

3.9.4 Many port operators have established PSCs to co-ordinate security procedures and measures across their port. Where established, committees have yielded significant security benefits through better co-ordination of security activities across the port and its port facilities.

3.9.5 Membership should be as broad as possible. In addition to the port security officer (PSO) – if appointed – and the PFSO of each facility in the port, it could comprise representatives from the:

- **.1** management of the port operator and each port facility operator;
- **.2** customs and immigration authorities operating at the port;
- **.3** law-enforcement and emergency services;
- **.4** port worker associations;
- **.5** associations for seafarers operating ships from the port;
- **.6** firms undertaking commercial activities at the port, e.g., storage, cargo handling;
- **.7** shipping companies operating at the port;
- **.8** shippers/cargo interests at the port;
- **.9** Designated Authority and Administration assigned to the port;
- **.10** municipal and regional Governments with jurisdictional interests; and
- **.11** community associations adjacent to the port.

3.9.6 Each PSC should have terms of reference, which could include:

- **.1** identifying security threats;
- **.2** reporting and assessing recent security incidents at the port;
- **.3** assessing the possible implications of security incidents at other ports;
- **.4** enhancing co-ordination in the application of security procedures and measures;
- **.5** planning, co-ordinating participation in and evaluating security drills and exercises;
- **.6** co-ordinating port facility security assessments with the port security assessment;
- **.7** co-ordinating, communicating and facilitating the implementation of applicable security measures specified in the port security plan;
- **.8** facilitating shore leave by seafarers;
- **.9** sharing best practices and experiences in the implementation of security plans;
- **.10** designing and evaluating security-awareness programmes.

3.9.7 Experience to date includes:

- **.1** the PSC being chaired by the senior manager in the port operator who has overall responsibility for port security (this is usually a position more senior than the PSO);
- **.2** the PSC appointing a deputy chair, usually the PSO, so as to ensure continuity of meetings;
- **.3** the terms of reference being approved by the port operator and available to all interested parties;
- **.4** the terms of reference specifying the meeting administration responsibilities of the chair and the meeting participation responsibilities of all members (i.e., to keep their organizations well informed of proceedings and raise their issues);

.5 meetings being held regularly (quarterly is often the minimum frequency – some PSCs meet weekly) so as to enable the timely handling of security matters, with decisions recorded and distributed to all members;

.6 in order for committees to conduct their business efficiently, consideration being given to limiting attendance to a single representative from each member organization. If necessary, smaller sub-committees could be established to address topics requiring multiple participation from organizations.

3.9.8 There is a need to balance the desired openness of an advisory/consultative committee with the need to protect the confidentiality of sensitive security information (e.g., intelligence on possible threats). In such instances, it may be necessary to establish a special subcommittee restricted to personnel with the necessary security clearances, e.g., security officers, police services and Government officials.

Port security officers

3.9.9 Many port authorities have appointed port security officers (PSOs). Their duties include co-ordination of security activities across the port, including liaison with PFSOs and membership of the port security committee.

3.9.10 At many ports the PSO can be the initial point of contact, on security matters, with the ships approaching the port and intending to use port facilities within the port.

3.9.11 PSOs can also have responsibility for the security of berths operated by the port authority or with responsibility for a PFSP which acts as a "master plan" for the port area. They can also have responsibility for the security of anchorages, waiting berths and approaches from seaward under the jurisdiction of the port authority.

3.9.12 PSOs can make a significant contribution to the co-ordination of security activities within port areas.

3.9.13 The competencies and training appropriate for PSOs are similar to those for PFSOs. Under the guidance in the ILO/IMO Code of practice, their duties could include:

.1 conducting a comprehensive security survey of the port, taking into account the approved PSA;

.2 ensuring the development and maintenance of the PSP;

.3 implementing and testing the PSP;

.4 undertaking regular security inspections of the port to ensure that appropriate measures are in place;

.5 recommending and incorporating, as appropriate, modifications to the PSP in order to correct deficiencies and take into account relevant changes to the port;

.6 enhancing security awareness and vigilance of port personnel;

.7 ensuring that adequate training has been provided to personnel responsible for the security of the port;

.8 reporting to the relevant authorities and maintaining records of security incidents that affect the security of the port;

.9 co-ordinating implementation of the PSP with appropriate persons or organizations;

.10 co-ordinating with security services, as appropriate;

.11 ensuring that standards for personnel responsible for security of the port are met;

.12 ensuring that security equipment is properly operated, tested, calibrated and maintained.

3.9.14 Experience to date provides examples of Governments requiring the appointment of a PSO for each port, including specifying their duties and responsibilities.

3.9.15 Designated authorities have generally endorsed the appointment of PSOs even when there is no obligation on a port authority to do so.

3.9.16 Other examples from experience to date include:

- **.1** the appointment of another port officer to undertake the duties of the PSO when necessary;
- **.2** PSOs and other port officers undertaking the duties possessing documentary evidence of their appointment and training;
- **.3** PSOs and other port officers undertaking their duties being port employees, not contracted resources from an external company (e.g., a security firm or consultant);
- **.4** PSOs having an approved documented list of security and non-security duties; non-security duties should not interfere with their ability to carry out their security duties;
- **.5** PSOs playing a key role on port security committees – on occasion acting as deputy chair or secretary; and
- **.6** ensuring that they report to the senior member of the port authority's management team who is the chair of the port security committee.

Port security assessments

3.9.17 Although the Maritime Security Measures do not require port security assessments (PSAs) to be conducted and submitted for approval, many Designated Authorities require their port authorities to do so.

3.9.18 The guidance provided on conducting PSAs is similar to the material provided for the conduct and approval of PFSAs.

3.9.19 However, using risk-assessment and management tools is much more of a challenge given the larger size of port areas (in some cases, with indistinct physical boundaries), the larger scale of potential impacts and vulnerabilities, and the greater number of countermeasures that need to be evaluated.

3.9.20 Experience to date provides examples of Designated Authorities recommending that port authorities establish a small team to conduct their PSAs . This can help ensure that the key personnel within a port area work together to conduct the assessment. However, given the confidential nature of the information being collated, membership would need to be restricted to those members of the port security committee with appropriate security clearances (e.g., the PSO, PFSOs and their counterparts in national authorities).

Port security plans

3.9.21 Although the Maritime Security Measures do not require port authorities to develop port security plans (PSPs), many do so. Several European States require PSAs and the preparation, submission and approval of PSPs. The guidance available for developing and maintaining PFSPs can be used for PSPs (refer to subsection 3.7). In instances where PSPs are required to be submitted for approval, PFSPs for facilities within the port area may be attached.

3.9.22 The guidance in subsection 3.8 on implementing PFSPs can also apply to PSPs.

3.10 Guidelines for non-SOLAS marinas, ports and harbours

3.10.1 Operators of marinas, ports and harbours which are not required to comply with the Maritime Security Measures may wish to consider taking the following steps:

- **.1** communicating information to users, such as:
  - the current security environment, including parts of the facility which are subject to security conditions and areas of restricted navigation;
  - areas where there might be interaction with SOLAS vessels; and
  - any local regulations produced for the guidance and direction of non-SOLAS vessels.

.2 if located in a complex of port facilities that are compliant with the Maritime Security Measures, regularly reviewing their security arrangements, in co-operation with the PFSOs;

.3 implementing physical security measures tailored to its size and complexity, such as:

- adequate illumination;
- passive monitoring controls and devices;
- segregation of visiting vessels in one particular area such that the visitors can be effectively monitored;
- holding transient vessels arriving at night in a specific area, with vessel details recorded; and
- installing radio frequency identification device (RFID) or similar systems to monitor the movements of vessels in and out of marinas, ports and harbours;

.4 implementing appropriate security procedures, such as training staff to be familiar with security operating procedures for their facility and for the safety of their customers and the public;

.5 implementing regular security patrols, which should include walking all pontoons/docks; checking that boats are moored normally; being alert for any suspicious activity; monitoring access gates, storage shed doors, overhead doors and fuel points; and inspecting restroom facilities;

.6 maintaining a security log of events, which should include:

- details of incidents and events that occurred while on patrol;
- the identity of anyone or any organization called in for emergencies and the time/results of the call;
- details of issues for referral to a supervisor; and
- any information which should be noted for the awareness of the personnel on the next shift.

Appendix 3.1

Declaration of Security form

Source: Part B of the ISPS Code

Name of ship: ....

Port of registry:..

IMO Number:..

Name of port facility: ....

This Declaration of Security is valid from .... until ...., for the following activities:

(list the activities with relevant details)

under the following security levels:

Security level(s) for the ship:..

Security level(s) for the port facility:..

The port facility and ship agree to the following security measures and responsibilities to ensure compliance with the requirements of the Maritime Security Measures.

	The affixing of the initials of the SSO or PFSO under these columns indicates that the activity will be done, in accordance with the relevant approved plan, by	
Activity	**The port facility:**	**The ship:**
Ensuring the performance of all security duties		
Monitoring restricted areas to ensure that only authorized personnel have access		
Controlling access to the port facility		
Controlling access to the ship		
Monitoring of the port facility, including berthing areas and areas surrounding the ship		
Monitoring of the ship, including berthing areas and areas surrounding the ship		
Handling of cargo		
Delivery of ship's stores		
Handling unaccompanied baggage		
Controlling the embarkation of persons and their effects		
Ensuring that security communication is readily available between the ship and the port facility		

The signatories to this agreement certify that security measures and arrangements for both the port facility and the ship during the specified activities meet the provisions of the Maritime Security Measures that will be implemented in accordance with the provisions already stipulated in their approved plans or the specific arrangements agreed to and set out in the attached annex.

Dated at .... on the ....

Signed for and on behalf of

the port facility:	the ship:
....	....
(Signature of port facility security officer)	*(Signature of master or ship security officer)*

Name and title of person who signed

Name:	Name: ....
Title:.	Title:. ....

Contact details

(to be completed as appropriate)

(indicate the telephone numbers or the radio channels or frequencies to be used)

for the port facility:	for the ship:
Port facility	Master
....	....
Port facility security officer	Ship security officer
....	....
	Company
	....
	Company security officer
	....

Note: This form is for use between a ship and a port facility. If the Declaration of Security is to cover two or more ships, or a port, this form should be appropriately modified.

Appendix 3.2
Competency matrix for port facility security officers

Source: MSC.1/Circ.1188, May 2006

Competence Knowledge requirement	**Method for demonstrating competence** Evaluation criterion
Develop, maintain and supervise the implementation of a PFSP • International maritime security policy and responsibilities of Governments, Companies and designated persons • The purpose for and the elements that make up a PFSP, related procedures and maintenance of records • Procedures to be employed in developing, maintaining and supervising the implementation, and the submission for approval, of a PFSP • Procedures for the initial and subsequent verification of the port facility's compliance • Security levels and consequential security measures and procedures aboard ship and in the port facility environment • Requirements and procedures for conducting internal audits, on-scene inspections, control and monitoring of security activities specified in a PFSP • Requirements and procedures for acting upon any deficiencies and non-conformities identified during internal audits, periodic reviews, and security inspections • Methods and procedures used to modify the PFSP • Security-related contingency plans and the procedures for responding to security threats or breaches of security, including provisions for maintaining critical operations of the ship/port interface • Procedures for facilitating shore leave for ship's personnel or personnel changes, as well as access of visitors to the ship, including representatives of seafarers' welfare and labour organizations • Procedures, instructions, and guidance for responding to ship security alerts • Maritime security terms and definitions (in the Maritime Security Measures)	**Assessment of evidence obtained from approved training or examination** • Procedures and actions are in accordance with the principles established by the Maritime Security Measures • Legislative requirements relating to security are correctly identified • Procedures achieve a state of readiness to respond to changes in security levels • Communications within the PFSO's area of responsibility are clear and understood
Assess security risk, threat, and vulnerability • Risk assessment and assessment tools • Security assessment documentation, including the Declaration of Security • Techniques used to circumvent security measures. • Recognition, on a non-discriminatory basis, of persons posing potential security risks • Recognition of weapons, dangerous substances, and devices and awareness of the damage they can cause • Crowd management and control techniques, where appropriate • Handling sensitive security-related information and security-related communications • Methods for implementing and co-ordinating searches • Methods for physical searches and non-intrusive inspections	**Assessment of evidence obtained from approved training or examination** • Procedures and actions are in accordance with the principles established by the Maritime Security Measures • Procedures achieve a state of readiness to respond to changes in security levels • Communications within the PFSO's area of responsibility are clear and understood

Competence Knowledge requirement	**Method for demonstrating competence** Evaluation criterion
Undertake regular inspections of the port facility to ensure that appropriate security measures are implemented and maintained • Requirements for designating and monitoring restricted areas • Controlling access to the port facility and to restricted areas in the port facility • Methods for effective monitoring of the port facility and areas surrounding the port facility • Methods for controlling the embarkation and disembarkation of persons and their effects aboard ships, including the confirmation of identity when requested by the SSO • Security aspects relating to the handling of cargo and ship's stores and co-ordinating these aspects with relevant SSOs and CSOs	**Assessment of evidence obtained from approved training or examination** • Procedures and actions are in accordance with the principles established by the Maritime Security Measures • Procedures achieve a state of readiness to respond to changes in security levels • Communications within the PFSO's area of responsibility are clear and understood
Ensure that security equipment and systems, if any, are properly operated, tested and calibrated • Various types of security equipment and systems and their limitations • Methods for testing, calibrating, and maintaining security systems and equipment	**Assessment of evidence obtained from approved training or examination** • Procedures and actions are in accordance with the principles established by the Maritime Security Measures
Encourage security awareness and vigilance • Training, drill and exercise requirements under relevant conventions and codes • Methods for enhancing security awareness and vigilance • Methods for assessing the effectiveness of drills and exercises • Instruction techniques for security training and education	**Assessment of evidence obtained from approved training or examination** • Procedures and actions are in accordance with the principles established by the Maritime Security Measures • Communications within the PFSO's area of responsibility are clear and understood

Appendix 3.3
Competency matrix for port facility personnel with security duties

Source: MSC.1/Circ.1341, May 2010

Competence Knowledge requirement	**Method for demonstrating competence** Evaluation criterion
Maintain the conditions set out in a PFSP • Maritime security terms and definitions • International maritime security policy and responsibilities of Governments/Designated Authorities, RSOs, PFSOs and designated persons • Maritime security levels and their impact on security measures and procedures in the port facility and aboard ships • Security reporting procedures • Procedures for drills and exercises • Procedures for conducting inspections and surveys and for the control and monitoring of security activities specified in a PFSP • Security-related contingency plans and the procedures for responding to security incidents, including provisions for maintaining critical operations of port facility and ship/port interface • Procedures for handling security-related information and communications • Security documentation, including the DoS	**Assessment of evidence obtained from approved instruction or during attendance at an approved course** • Procedures and actions are in accordance with the principles established by the Maritime Security Measures • Legislative requirements relating to security are correctly identified • Communications within the area of responsibility are clear and understood
Recognition of security threats • Techniques used to circumvent security measures • Recognition of weapons, dangerous substances, dangerous goods, and devices and awareness of damage they can cause • Security-related provisions for dangerous goods • Crowd management and control techniques, where appropriate • Methods for recognition, on a non-discriminatory basis, of behavioural characteristics of persons and patterns likely to threaten security	**Assessment of evidence obtained from approved instruction or during attendance at an approved course** • Procedures and actions are in accordance with the principles established by the Maritime Security Measures and the relevant provisions of the International Maritime Dangerous Goods Code
Inspection, control and monitoring activities • Controlling access to the port facility and its restricted areas • Techniques for monitoring restricted areas • Methods for effective monitoring of ship/port interface and areas surrounding the port facility • Inspection methods relating to the cargo and stores • Methods for physical searches and non-intrusive inspections	**Assessment of evidence obtained from approved instruction or during attendance at an approved course** • Procedures and actions are in accordance with the principles established by the Maritime Security Measures and the relevant provisions of the International Maritime Dangerous Goods Code
Proper usage of security equipment and systems • Various types of security equipment and systems, including their limitations • The need for testing, calibrating and maintaining security systems and equipment	**Assessment of evidence obtained from approved instruction or during attendance at an approved course** • Equipment and systems operations are carried out in accordance with established equipment operating instructions and taking into account the limitations of the equipment and systems • Procedures and actions are in accordance with the principles established by the Maritime Security Measures

Appendix 3.4
Competency matrix for port facility personnel without security duties

Source: MSC.1/Circ.1341, May 2010

Competence Basic knowledge requirement	**Method for demonstrating competence** Evaluation criterion
Contribute to the enhancement of maritime security through heightened awareness • Maritime security terms and definitions • International maritime security policy and responsibilities of Government/Designated Authority, PFSOs and designated persons • Maritime security levels and their impact on security measures and procedures in the port facility and aboard ships • Security reporting procedures • Security-related contingency plans • Security-related provisions for dangerous goods	**Assessment of evidence obtained from approved instruction or during attendance at an approved course** • Requirements relating to enhanced maritime security are correctly identified
Recognition of security threats • Recognition of potential security threats • Techniques used to circumvent security measures • Recognition of weapons, dangerous substances, dangerous goods, and devices and awareness of the damage they can cause • Procedures for security-related communications	**Assessment of evidence obtained from approved instruction or during attendance at an approved course** • Maritime security threats are correctly identified
Understanding the need for and methods of maintaining security awareness and vigilance • Training, drill and exercise requirements under relevant conventions and codes	**Assessment of evidence obtained from approved instruction or during attendance at an approved course** • Requirements relating to enhanced maritime security are correctly identified

Appendix 3.5
Example of a port facility security assessment and plan approval process

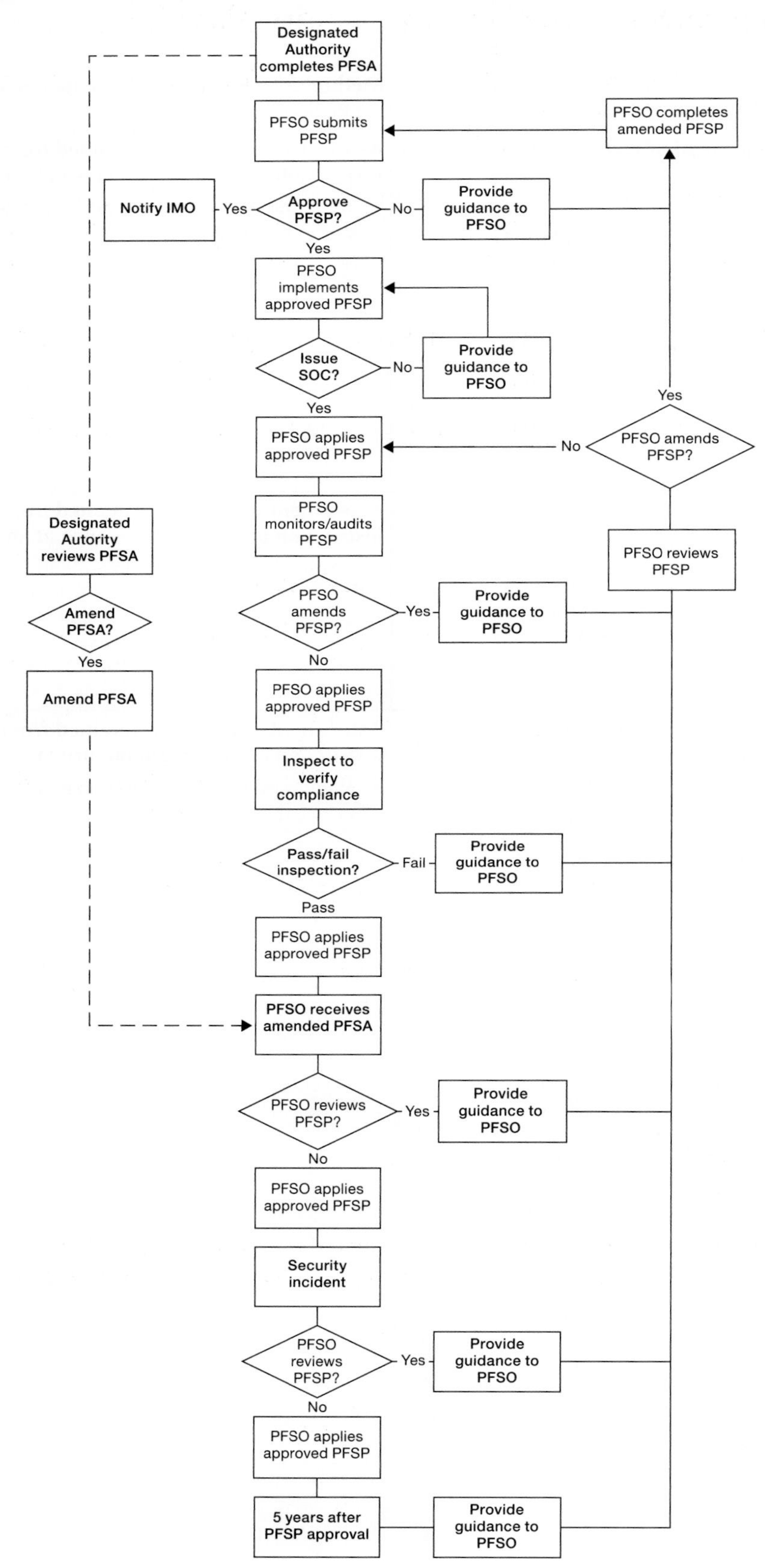

Appendix 3.6

Examples of internet sources of guidance material on preparing, updating and implementing port facility security plans

1. Australian Government, Department of Infrastructure and Transport: Guide to Preparing a maritime security plan for Port Facility Operators, April 2009.

Refer to: www.infrastructure.gov.au/transport/security/maritime

This 33-page guide has been developed to provide port facility operators covered by the Maritime Transport and Offshore Securities Act 2003 with a plan template so as to assist them with meeting all the requirements of an approved plan. It also contains a chart showing the plan approval. Similar guides exist for port operators and port service providers.

2. United Kingdom, Department for Transport: Port Facility Security plan, August 2008.

Refer to: www.dft.gov.uk/pgr/security/maritime

This 22-page document is a template showing port facility operators how to complete and submit their PFSP.

3. United States Coast Guard (USCG) Homeport Site.
Refer to: www.homeport.uscg.mil

Based on visits to ports in countries trading with the United States, this site documents port security best practices for compliance with the Maritime Security Measures. Its contents have the following characteristics:

- a single-page standardized format including description, discussion, potential downside, conclusion, cost and contact information for further details (including the website);
- emphasis is placed on low-cost or innovative practices that are judged to have a significant impact on port facility security;
- the ports listed are generally those where the practice was first observed and the country's national authority has expressed its willingness to share the information;
- the practices are grouped into nine categories:
  - Access control
  - Documents and forms
  - Perimeter control
  - Security infrastructure
  - Electronic surveillance
  - Guards and police
  - Communications
  - Lighting
  - Training and procedures

APEC's Manual of maritime security drills (volume 1) and exercises (volume 2) for port facilities is included on this site (refer to appendix 3.7 – APEC Manual of maritime security drills and exercises for port facilities: table of contents).

Appendix 3.7
APEC Manual of maritime security drills and exercises for port facilities: table of contents

Source: APEC, Transportation Working Group, August 2008

Module	Topics covered
Access control	Introduction Guidelines for the planning and conduct of maritime security drills Access control drills Person entering without permission Visitor seeking entry without means of identification Person seeking entry using false documents Entry by employees without their security pass Entry by contractor with expired long-term pass Entry by ship crew/shipping agency/seafarer organization representatives without prior notice Vehicle without authorized entry label Vehicle with suspicious person/item Vehicle parked in or in close proximity to a key area or restricted area Vehicle forcing entry
Contiguous zone security	Introduction Guidelines for the planning and conduct of maritime security drills Contiguous zone security Persons loitering outside the port facility Person taking photographs of the port facility Person on vessel engaged in suspicious activity Vehicle loitering near the port facility Vessel loitering offshore at the port facility
Materials handling	Introduction Guidelines for the planning and conduct of maritime security drills Materials handling Suspicious parcel/envelope Suspicious substances Suspicious items Vehicle delivering cargo without proper documents Cargo without proper seals Discovery of unauthorized cargo on board a ship alongside Vehicle delivering ship's stores without proper documents Delivery of ship's stores without prior notice Unauthorized item found in vehicle delivering ship's stores Unauthorized loading/unloading of cargo/ship's stores in a restricted area Unaccompanied baggage found in the port facility Unaccompanied baggage found within a restricted area Vehicle carrying unaccompanied baggage seeking entry to the port facility

Emergency response	Introduction Guidelines for the planning and conduct of maritime security drills Emergency response Security surveillance equipment malfunction Perimeter security compromised Activation of intrusion alarm Activation of ship security alert system Power failure Bomb threat Evacuation Changing the security level
Ship-shore interface	Introduction Guidelines for the planning and conduct of maritime security drills Shore interface Interface with non-ISPS-compliant vessel Exchange of Declaration of Security
Principal exercises	Introduction Guidelines for the planning and conduct of maritime security exercises Principal exercises State maritime security exercise Port facility security plan exercise
Port facility exercises	Introduction Guidelines for the planning and conduct of maritime security exercises Port facility exercises Response to security threats Handling unauthorized items Unauthorized access Cargo and ship's stores Interfacing with ship security activities Security incidents

Appendix 3.8
Implementation checklist for port facility operators

Source: MSC.1/Circ.1192, May 2006

This checklist may be used by port facility operators to examine the status of implementation of the Special Measures. The heading of each section is taken directly from paragraph A/14.2 of the ISPS Code.

Completion of the following section is recommended before using the checklist as it can be used to establish an overview of the port facility's operations.

1 Port facility overview:

Name of port facility	
Name of operator/authority	
Name of port, if applicable	
Name of PFSO	
Average number of SOLAS ships handled per annum	

2 Particular characteristics of the port facility, if any, including the vessel traffic, which may increase the likelihood of being the target of a security incident:

Passenger ships	☐	Other dangerous goods	☐
Ro–ro/container terminal	☐	Near military installation	☐
Explosives	☐	Military vessels	☐
Oil/gas refinery/terminal	☐	Embarkation of military personnel or cargo	☐
LPG, LNG or petrol storage	☐	Other (describe)	☐

3 Security agreements and arrangements:

Is the port facility covered by an alternative security agreement? If "Yes", provide relevant details.	
Has the port facility implemented any equivalent security arrangements allowed by the Contracting Government? If "Yes", provide relevant details.	
Is the port facility operating under any temporary security measures? If "Yes", have these been approved or authorized by the Contracting Government? If "Yes", provide relevant details.	

Guidance:

- For each question, one of the 'Yes/No/Other' boxes should be ticked. Whichever one is ticked, the 'Comments' box provides space for amplification.
- If the 'Yes' box is ticked, but the Measures/procedures are not documented in the PFSP, a short description of them should be included in the 'Comments' box. The 'Yes' box should be ticked only if all procedures or measures are in place. The 'comments' box may also be used to indicate when procedures were last reviewed and measures tested (e.g., drills and exercises).
- If the 'No' box is ticked, an explanation of why not should be included in the 'Comments' box along with details of any measures or procedures in place. Suggested actions should be recorded in the 'Recommendations' section at the end of the checklist.

- If the 'Other' box is selected, a short description should be provided in the 'Comments' box (e.g., it could include instances where alternative measures/procedures/agreements or equivalent arrangements have been implemented). If the reason is due to the question not being applicable, then it should be recorded in the 'Comments' box as "not applicable".
- If there is not enough space in the 'Comments' box, the explanation should be continued on a separate page (with the relevant question number and, in the case of questions with multiple options, the option added as a reference).
- The 'Recommendations' boxes at the end of the checklist should be used to record any identified deficiencies and how these could be mitigated. A schedule for their implementation should be included.
- The 'Outcomes' box at the end of the checklist should be used to provide a brief record of the assessment process. Along with the comments in the 'Recommendations' boxes, they form the basis for updating the PFSP.

1 Ensuring the performance of all port facility security duties (ISPS Code, sections A/14.2.1 and A/14.3)

Part A

.1 Does the port facility's means of ensuring the performance of all security duties meet the requirements set out in the PFSP for security levels 1 and 2? (ISPS Code, section A/14.2.1)	Yes ☐	No ☐	Other ☐
Comments:			

.2 Has the port facility established measures to prevent weapons or any other dangerous substances and devices intended for use against persons, ships, or the port from entering the facility? (ISPS Code, section A/16.3.1)	Yes ☐	No ☐	Other ☐
Comments:			

.3 Has the port facility established evacuation procedures in case of security threats or breaches of security? (ISPS Code, section A/16.3.5)	Yes ☐	No ☐	Other ☐
Comments:			

.4 Has the port facility established procedures for response to an activation of a ship security alert system? (ISPS Code, section A/16.3.14)	Yes ☐	No ☐	Other ☐
Comments:			

Part B – Organization of port facility security duties (ISPS Code, paragraph B/16.8)

.5 Has the port facility established the role and structure of the security organization? (ISPS Code, paragraph B/16.8.1)	Yes ☐	No ☐	Other ☐
Comments:			

.5 Has the port facility established the role and structure of the security organization? (ISPS Code, paragraph B/16.8.1)	Yes ☐	No ☐	Other ☐
Comments:			

.6 Has the port facility established the duties and responsibilities for personnel with security roles? (ISPS Code, paragraph B/16.8.2)	Yes ☐	No ☐	Other ☐
Comments:			

.7 Has the port facility established the training requirements for personnel with security roles? (ISPS Code, sections A/18.1, A/18.2, A/18.3 and paragraph B/16.8.2)	Yes ☐	No ☐	Other ☐
Comments:			

.8 Has the port facility established the performance measures needed to assess the individual effectiveness of personnel with security roles? (ISPS Code, paragraph B/16.8.2)	Yes ☐	No ☐	Other ☐
Comments:			

.9 Has the port facility established their security organization's link with other national or local authorities with security responsibilities? (ISPS Code, paragraph B/16.8.3)	Yes ☐	No ☐	Other ☐
Comments:			

.10 Has the port facility established procedures and practices to protect security-sensitive information held in paper or electronic format? (ISPS Code, paragraph B/16.8.6)	Yes ☐	No ☐	Other ☐
Comments:			

.11 Has the port facility established procedures to assess the continuing effectiveness of security measures and procedures? (ISPS Code, paragraph B/16.8.7)	Yes ☐	No ☐	Other ☐
Comments:			

.12 Has the port facility established procedures to assess security equipment, to include identification of, and response to, equipment failure or malfunction? (ISPS Code, paragraph B/16.8.7)	Yes ☐	No ☐	Other ☐
Comments:			

.13 Has the port facility established procedures governing submission and assessment of reports relating to possible breaches of security or security concerns? (ISPS Code, paragraph B/16.8.8)	Yes ☐	No ☐	Other ☐
Comments:			

.14 Has the port facility established procedures to maintain and update records of dangerous goods and hazardous substances, including their location within the port facility? (ISPS Code, paragraph B/16.8.11)	Yes ☐	No ☐	Other ☐
Comments:			

.15 Has the port facility established a means of alerting and obtaining the services of waterside patrols and search teams, to include bomb and underwater specialists? (ISPS Code, paragraph B/16.8.12)	Yes ☐	No ☐	Other ☐
Comments:			

.16 Has the port facility established procedures for assisting, when requested, ship security officers in confirming the identity of those seeking to board the ship? (ISPS Code, paragraph B/16.8.13)	Yes ☐	No ☐	Other ☐
Comments:			

.17 Has the port facility established the procedures for facilitating shore leave for ship's crew members or personnel changes? (ISPS Code, paragraph B/16.8.14)	Yes ☐	No ☐	Other ☐
Comments:			

.18 Has the port facility established the procedures for facilitating visitor access to the ship, to include representatives of seafarers' welfare and labour organizations? (ISPS Code, paragraph B/16.8.14)	Yes ☐	No ☐	Other ☐
Comments:			

2 Controlling access to the port facility (ISPS Code, sections A/14.2.1, A/14.2.2 and A/14.3)

Part A

.1 Does the port facility's means of controlling access to the port facility meet the requirements set out in the PFSP for security levels 1 and 2?	Yes ☐	No ☐	Other ☐
Comments:			

Part B – Establish facility security measures (ISPS Code, paragraphs B/16.10 to B/16.12, B/16.14, B/16.17 and B/16.19.1)

.2 Has the port facility identified the appropriate location(s) where security measures can be applied to restrict or prohibit access. These should include all access points identified in the PFSP at security levels 1 and 2? (ISPS Code, paragraphs B/16.11, B/16.19.1)	Yes ☐	No ☐	Other ☐
Comments:			

.3 Does the port facility specify the type of restrictions or prohibitions, and the means of enforcement to be applied at all access points identified in the PFSP at security levels 1 and 2? (ISPS Code, paragraphs B/16.11, B/16.19.2, B/16.19.3)	Yes ☐	No ☐	Other ☐
Comments:			

.4 Has the port facility established measures to increase the frequency of searches of people, personal effects, and vehicles at security level 2? (ISPS Code, paragraph B/16.19.4)	Yes ☐	No ☐	Other ☐
Comments:			

.5 Has the port facility established measures to deny access to visitors who are unable to provide verifiable justification for seeking access to the port facility at security level 2 (ISPS Code, paragraph B/16.19.5)	Yes ☐	No ☐	Other ☐
Comments:			

Part B – Establish security measures for individuals (ISPS Code, paragraphs B/16.12)

.6 Has the port facility established the means of identification required to access and remain unchallenged within the port facility? (ISPS Code, paragraph B/16.12)	Yes ☐	No ☐	Other ☐
Comments:			

.7 Does the port facility have the means to differentiate the identification of permanent, temporary, and visiting individuals? (ISPS Code, paragraph B/16.12)	Yes ☐	No ☐	Other ☐
Comments:			

.8 Does the port facility have the means to verify the identity and legitimacy of passenger boarding passes, tickets, etc? (ISPS Code, paragraph B/16.12)	Yes ☐	No ☐	Other ☐
Comments:			

.9 Has the port facility established provisions to ensure that the identification systems are regularly updated? (ISPS Code, paragraph B/16.12)	Yes ☐	No ☐	Other ☐
Comments:			

.10 Has the port facility established provisions to facilitate disciplinary action against those who abuse the identification system procedures? (ISPS Code, paragraph B/16.12)	Yes ☐	No ☐	Other ☐
Comments:			

.11 Has the port facility created procedures to deny access and report all individuals who are unwilling or unable to establish their identity or purpose for visit to the PFSO and to the national or local authorities? (ISPS Code, paragraph B/16.13)	Yes ☐	No ☐	Other ☐
Comments:			

Part B – Search locations (ISPS Code, paragraph B/16.14)

.12 Has the port facility identified a location(s) for searches of persons, personal effects, and vehicles that facilitates continuous operation, regardless of prevailing weather conditions? (ISPS Code, paragraph B/16.14)	Yes ☐	No ☐	Other ☐
Comments:			

.13 Does the port facility have procedures established to directly transfer persons, personal effects, or vehicles subjected to search to the restricted holding, embarkation, or vehicle loading area? (ISPS Code, paragraph B/16.14)	Yes ☐	No ☐	Other ☐
Comments:			

.14 Has the port facility established separate locations for embarking and disembarking passengers, ship's personnel, and their effects to ensure that unchecked persons do not come in contact with checked persons? (ISPS Code, paragraph B/16.15)	Yes ☐	No ☐	Other ☐
Comments:			

.15 Does the PFSP establish the frequency of application of all access controls? (ISPS Code, paragraph B/16.16)	Yes ☐	No ☐	Other ☐
Comments:			

Part B – Establish control points (ISPS Code, paragraph B/16.17)

.16 Does the PFSP establish control points for restricted areas bounded by fencing or other barriers to a standard which is approved by the national Government? (ISPS Code, paragraph B/16.17.1)	Yes ☐	No ☐	Other ☐
Comments:			

.17 Does the PFSP establish the identification of and procedures to control access points not in regular use, which should be permanently closed and locked? (ISPS Code, paragraph B/16.17.7)	Yes ☐	No ☐	Other ☐
Comments:			

3 Monitoring of the port facility, including anchoring and berthing area(s) (ISPS Code sections A/14.2.3 and A/14.3)

Part A

.1 Does the facility's means of monitoring the port facility, including berthing and anchorage area(s), meet the requirements set out in the PFSP for security levels 1 and 2?	Yes ☐	No ☐	Other ☐
Comments:			

Part B – Scope of security monitoring (ISPS Code, paragraph B/16.49)

.2 Does the port facility have the capability to continuously monitor, on land and water, the port facility and its nearby approaches? (ISPS Code, paragraph B/16.49)	Yes ☐	No ☐	Other ☐
Comments:			

.3 Which of the following means are employed to monitor the port facility and nearby approaches? (ISPS Code, paragraph B/16.49)	Yes	No	Other
A Patrols by security guards	A ☐	☐	☐
B Patrols by security vehicles	B ☐	☐	☐
C Patrols by watercraft	C ☐	☐	☐
D Automatic intrusion-detection devices	D ☐	☐	☐
E Surveillance equipment	E ☐	☐	☐
Comments:			

.4 If automatic intrusion-detection devices are employed, do they activate an audible and/or visual alarm(s) at a location(s) that is continuously monitored? (ISPS Code, paragraph B/16.50)	Yes ☐	No ☐	Other ☐
Comments:			

.5 Does the PFSP establish procedures and equipment needed at each security level? (ISPS Code, paragraph B/16.51)	Yes ☐	No ☐	Other ☐
Comments:			

.6 Has the port facility established measures to increase the security measures at security levels 1 and 2? (ISPS Code, paragraphs B/16.51, B/16.53.1, B/16.53.2 and B/16.53.3)	Yes	No	Other
A Increase intensity and coverage of lighting and surveillance equipment	A ☐	☐	☐
B Increase frequency of foot, vehicle and waterborne patrols	B ☐	☐	☐
C Assign additional personnel	C ☐	☐	☐
D Surveillance	D ☐	☐	☐
Comments:			

.7 Does the PFSP establish procedures and equipment necessary to ensure that monitoring equipment will be able to perform continually, including consideration of the possible effects of weather or power disruptions? (ISPS Code, paragraph B/16.51)	Yes ☐	No ☐	Other ☐
Comments:			

Part B – Illumination at port facility (ISPS Code, paragraph B/16.49.1)

.8 Does the port facility have adequate illumination, to allow for detection of unauthorized persons at or approaching access points, the perimeter, restricted areas and ships, at all times, including the night hours and periods of limited visibility? (ISPS Code, paragraph B/16.49.1)	Yes ☐	No ☐	Other ☐
Comments:			

4 Monitoring restricted areas to ensure that only authorized persons have access (ISPS Code, sections A/14.2.4 and A/14.3)

Part A

.1 Does the port facility's means of limiting and monitoring access to restricted areas meet the requirements of the PFSP for security levels 1 and 2? (ISPS Code, sections A/14.2.4 and A/14.3)	Yes ☐	No ☐	Other ☐
Comments:			

Part B – Establishment of restricted areas (ISPS Code, paragraph B/16.21)

.2 Are restricted areas identified within the port facility? (ISPS Code, paragraph B/16.21)	Yes	No	Other
Comments:	☐	☐	☐

.3 Which of the following elements are identified for restricted areas in the PFSP? (ISPS Code, paragraph B/16.21)	Yes	No	Other
A Extent of area	A ☐	☐	☐
B Times of application	B ☐	☐	☐
C Security measures to control access to areas	C ☐	☐	☐
D Security measures to control activities within areas	D ☐	☐	☐
E Measures to ensure restricted areas are swept before and after establishment	E ☐	☐	☐
Comments:			

Part B – Security measures (ISPS Code, paragraph B/16.22)

.4 Are restricted areas clearly marked, indicating that access to the area is restricted and that unauthorized presence constitutes a breach of security? (ISPS Code, paragraph B/16.23)	Yes ☐	No ☐	Other ☐
Comments:			

.5 Are measures established to control access by individuals to restricted areas? (ISPS Code, paragraph B/16.22.1)	Yes ☐	No ☐	Other ☐
Comments:			

.6 Does the port facility have the means to ensure that passengers do not have unsupervised access to restricted areas? (ISPS Code, paragraph B/16.12)	Yes ☐	No ☐	Other ☐
Comments:			

.7 Are measures established to control the entry, parking, loading, and unloading of vehicles? (ISPS Code, paragraph B/16.22.2)	Yes ☐	No ☐	Other ☐
Comments:			

.8 Are measures established to control movement and storage of cargo and ship's stores? (ISPS Code, paragraph B/16.22.3)	Yes ☐	No ☐	Other ☐
Comments:			

.9 Are measures established to control unaccompanied baggage or personal effects? (ISPS Code, paragraph B/16.22.4)	Yes ☐	No ☐	Other ☐
Comments:			

.10 If automatic intrusion-detection devices are installed, do they alert a control centre capable of responding to the alarm? (ISPS Code, paragraph B/16.24)	Yes ☐	No ☐	Other ☐
Comments:			

.11 Which of the following security measures are utilized to control access to restricted areas? (ISPS Code, paragraph B/16.27)	Yes	No	Other
A Permanent or temporary barriers to surround restricted area	A ☐	☐	☐
B Access points controlled by security guards when in use	B ☐	☐	☐
C Access points that can be locked or barred when not in use	C ☐	☐	☐
D Use of passes to indicate a person's authorization for access	D ☐	☐	☐
E Marking of vehicles that are allowed access	E ☐	☐	☐
F Use of guards and patrols	F ☐	☐	☐
G Use of automatic intrusion-detection devices or surveillance equipment and systems	G ☐	☐	☐
H Control of vessel movement in vicinity of ships using port facility	H ☐	☐	☐
Comments:			

.12 Has the port facility established measures to enhance the security of restricted areas for security level 2? (ISPS Code, paragraph B/16.28)	Yes	No	Other
A Enhance the effectiveness of barriers	A ☐	☐	☐
B Reduce access points	B ☐	☐	☐
C Enhance control of access points	C ☐	☐	☐
D Restrict parking	D ☐	☐	☐
E Control movement within restricted area	E ☐	☐	☐
F Continuously monitor restricted area	F ☐	☐	☐
G Enhance frequency of patrols	G ☐	☐	☐
H Limiting access to spaces adjacent to ship	H ☐	☐	☐
Comments:			

.13 Has the port facility established measures to enhance the effectiveness of barriers, reduce access points, and enhance access control for restricted areas at security level 2? (ISPS Code, paragraph B/16.28)	Yes ☐	No ☐	Other ☐
Comments:			

5 Supervising the handling of cargo (ISPS Code, sections A/14.2.5 and A/14.3)

Part A

.1 Does the port facility's means of supervising the handling of cargo meet the requirements identified in the PFSP for security levels 1 and 2?	Yes ☐	No ☐	Other ☐
Comments:			

Part B – Prevent tampering, the acceptance of unauthorized cargo, inventory control (ISPS Code, paragraph B/16.30.1, B/16.30.2, B/16.31)

.2 Are measures employed to routinely monitor the integrity of cargo, including the checking of seals, upon entry to the port facility and whilst stored in the port facility at security levels 1 and 2? (ISPS Code, paragraph B/16.32.1)	Yes ☐	No ☐	Other ☐
Comments:			

.3 Are measures employed to routinely monitor cargo transport units prior to and during cargo handling operations? (ISPS Code, paragraph B/16.32.1)	Yes ☐	No ☐	Other ☐
Comments:			

Question	Yes	No	Other
.4 Which of the following means are employed to conduct cargo checking? (ISPS Code, paragraph B/16.33)			
A Visual examinations	A ☐	☐	☐
B Physical examinations	B ☐	☐	☐
C Scanning or detection equipment	C ☐	☐	☐
D Other mechanical means	D ☐	☐	☐
E Dogs	E ☐	☐	☐
Comments:			

Question	Yes	No	Other
.5 Are restricted areas designated to perform inspections of cargo transport units if a container seal appears to have been compromised? (ISPS Code, paragraph B/16.32.4)	☐	☐	☐
Comments:			

Question	Yes	No	Other
.5 Are restricted areas designated to perform inspections of cargo transport units if a container seal appears to have been compromised? (ISPS Code, paragraph B/16.32.4)	☐	☐	☐
Comments:			

Question	Yes	No	Other
.6 Has the port facility established measures to intensify checks to ensure that only documented cargo enters the facility, and, if necessary, is only stored on a temporary basis at security level 2? (ISPS Code, paragraph B/16.35.2)	☐	☐	☐
Comments:			

Question	Yes	No	Other
.7 Has the port facility established measures to intensify vehicle searches, the frequency and detail of examining cargo seals, and other tampering-prevention methods at security level 2? (ISPS Code, paragraph B/16.35.3)	☐	☐	☐
Comments:			

Question	Yes	No	Other
.8 Are cargo delivery orders or equivalent cargo documentation verified before acceptance? (ISPS Code, paragraph B/16.32.2)	☐	☐	☐
Comments:			

Question	Yes	No	Other
.9 Are procedures utilized to randomly or selectively search vehicles at facility access points? (ISPS Code, paragraph B/16.32.3)	☐	☐	☐
Comments:			

Question	Yes	No	Other
.10 Are inventory control procedures employed at facility access points? (ISPS Code, paragraph B/16.31)	☐	☐	☐
Comments:			

.11 Are means of identification used to determine whether cargo inside the port facility awaiting loading has been either checked and accepted or temporarily stored in a restricted area? (ISPS Code, paragraph B/16.31)	Yes ☐	No ☐	Other ☐
Comments:			

6 Supervising the handling of ship's stores (ISPS Code, sections A/14.2.6 and A/14.3)

Part A

.1 Does the port facility's means of supervising the handling of ship's stores meet the requirements identified in the PFSP at security levels 1 and 2? (ISPS Code, section A/14.2.6)	Yes ☐	No ☐	Other ☐
Comments:			

Part B – Ship's stores security measures (ISPS Code, paragraph B/16.38)

.2 Are ship's stores examined to ensure package integrity at security levels 1 and 2? (ISPS Code, paragraphs B/16.38.1 and B/16.42.1)	Yes ☐	No ☐	Other ☐
Comments:			

.3 Are procedures established to ensure that no ship's stores are accepted into the port facility without checking at security levels 1 and 2? (ISPS Code, paragraphs B/16.38.2 and B/16.42.2)	Yes ☐	No ☐	Other ☐
Comments:			

.4 Which of the following means are employed to inspect ship's stores? (ISPS Code, paragraph B/16.41)	Yes	No	Other
A Visual examinations	A ☐	☐	☐
B Physical examinations	B ☐	☐	☐
C Scanning or detection equipment	C ☐	☐	☐
D Other mechanical means	D ☐	☐	☐
E Dogs	E ☐	☐	☐
Comments:			

.5 Are procedures established to prevent the tampering of ship's stores? (ISPS Code, paragraph B/16.38.3)	Yes ☐	No ☐	Other ☐
Comments:			

.6 Are ship's stores deliveries preceded with an advanced notification of load composition, driver information, and vehicle registration? (ISPS Code, paragraph B/16.40.2)	Yes ☐	No ☐	Other ☐
Comments:			

.7 Are unscheduled deliveries of ship's stores declined access to the port facility? (ISPS Code, paragraph B/16.38.4)	Yes ☐	No ☐	Other ☐
Comments:			

.8 Are there procedures in place to prevent ship's stores being accepted unless ordered? Are manifests and order documentation validated prior to allowing them into the port facility at security levels 1 and 2? (ISPS Code, paragraph B/16.38.4)	Yes ☐	No ☐	Other ☐
Comments:			

.9 Are searches of vehicles delivering ship's stores performed prior to entry into the port facility? (ISPS Code, paragraph B/16.38.5)	Yes ☐	No ☐	Other ☐
Comments:			

.10 Are escorts provided for ship's stores delivery vehicles within the port facility at security levels 1 and 2? (ISPS Code, paragraphs B/16.38.6 and B/16.42.4)	Yes ☐	No ☐	Other ☐
Comments:			

.11 Does the port facility increase the use of scanning/detection equipment, mechanical devices, or dogs at security level 2? (ISPS Code, paragraph B/16.43.2)	Yes ☐	No ☐	Other ☐
Comments:			

7 Ensuring security communication is readily available (ISPS Code, sections A/14.2.7 and A/14.3)

Part A

.1 Do the port facility's communication equipment and procedures meet the requirements identified in the PFSP at security level 1? (ISPS Code, section A/14.2.7)	Yes ☐	No ☐	Other ☐
Comments:			

Part B – Effectiveness and protection of communication equipment, procedures and facilities (ISPS Code, paragraphs B/16.8.4 and B/16.8.5)

.2 Is the port facility equipped with auxiliary communication systems for both internal and external communications that are readily available, regardless of security level, weather conditions or power disruptions, at security levels 1 and 2? (ISPS Code, paragraph B/16.8.4)	Yes ☐	No ☐	Other ☐
Comments:			

.3 Are security personnel trained on communication equipment to ensure efficiency? (ISPS Code, paragraph B/16.8.4)	Yes ☐	No ☐	Other ☐
Comments:			

.4 Are telephone numbers for key personnel accurate and routinely validated? (ISPS Code, paragraph B/16.8.4)	Yes ☐	No ☐	Other ☐
Comments:			

.5 Are procedures in place to ensure that port facility communication systems and equipment are serviced and maintained? (ISPS Code, paragraph B/16.8.4)	Yes ☐	No ☐	Other ☐
Comments:			

.6 Has the port facility established procedures and means for the PFSO to effectively disseminate changes in the security level at the port facility or with a vessel interfacing with the port? (ISPS Code, paragraph B/16.8.4)	Yes ☐	No ☐	Other ☐
Comments:			

.7 Are security procedures established to protect radio, telecommunication equipment and infrastructure, and computer systems? (ISPS Code, paragraph B/16.8.5)	Yes ☐	No ☐	Other ☐
Comments:			

.8 Are entry-control procedures established to restrict access to communication facilities and infrastructure? (ISPS Code, paragraph B/16.8.5)	Yes ☐	No ☐	Other ☐
Comments:			

8 Training, drills and exercises (ISPS Code, section A/18)

Part A

.1 Have the PFSO and appropriate port facility security personnel received sufficient training to perform their assigned duties as identified in the PFSP? (ISPS Code, sections A/18.1 and A/18.2)	Yes ☐	No ☐	Other ☐
Comments:			

.2 Has the port facility implemented drills and exercises? (ISPS Code, sections A/18.3 and A/18.4)	Yes ☐	No ☐	Other ☐
Comments:			

Part B – Training, drills, and exercises on port facility security (ISPS Code, paragraphs B/18.1, B/18.2, B/18.3, B/18.5 and B/18.6)

.3 Are the PFSO, personnel with security duties and all other port facility personnel familiar with the relevant provisions of the PFSP and have they received the appropriate levels of training? (ISPS Code, paragraphs B/18.1, B/18.2 and B/18.3)	Yes ☐	No ☐	Other ☐
Comments:			

.4 Are security drills conducted at least every three months and security exercises conducted at least once each calendar year, with no more than 18 months between the exercises? (ISPS Code, paragraphs B/18.5 and B/18.6)	Yes ☐	No ☐	Other ☐
Comments:			

9 Miscellaneous

.1 Has the port facility established procedures and adopted measures with respect to ships operating at a higher security level than the port facility? (ISPS Code, paragraph B/16.55)	Yes ☐	No ☐	Other ☐
Comments:			

.2 Has the port facility established procedures and adopted measures which can be applied when: (ISPS Code, paragraph 16.56):	Yes	No	Other
A it is interfacing with a ship which has been at a port of a State which is not a Contracting Government?	A ☐	☐	☐
B it is interfacing with a ship to which the ISPS Code does not apply?	B ☐	☐	☐
C service vessels covered by the PFSP are interfacing with fixed or floating platforms or mobile offshore drilling units on location?	C ☐	☐	☐
Comments:			

Recommendations

This section should be used to record any deficiencies identified by the checklist and how these could be mitigated. In essence this will provide an action plan for the CSO and/or SSO.

Recommendations/For action: Section 1: Ensuring the performance of all ship security duties.

Recommendations/For action: Section 2: Controlling access to the ship.

Recommendations/For action: Section 3: Controlling the embarkation of persons and their effects.

Recommendations/For action: Section 4: Monitoring of restricted areas.

Recommendations/For action: Section 5: Monitoring of deck areas and areas surrounding the ship.

Recommendations/For action: Section 6: Supervising the handling of cargo and ship's stores.

Recommendations/For action: Section 7: Ensuring security communication is readily available.

Recommendations/For action: Section 8: Training, drills and exercises.

Recommendations/For action: Section 9: Miscellaneous.

Outcomes

This section should be used to record the findings of the voluntary self-assessment and any other issues arising. These findings could be raised with ship or company personnel or be used as the basis to seek guidance from the Administration, as appropriate.

Signature of assessor:	Date of completion: . . . .
Name (*please print*):	
Title:	

Section 4 – Security responsibilities of ship operators

4.1 Introduction

4.1.1 This section provides guidance on the responsibilities of ship operators under the Maritime Security Measures. Following a general description of the security framework, guidance is offered on:

- **.1** security levels;
- **.2** ship security personnel;
- **.3** ship security communications;
- **.4** ship security assessments;
- **.5** ship Security plans;
- **.6** security-related documentation and information; and
- **.7** guidelines for non-SOLAS vessels.

4.1.2 Primarily addressed to those undertaking ship security responsibilities, the guidance is also relevant for those responsible for the security of the port facilities with which ships interface and for Government officials with regulatory responsibilities for shipping activities.

4.1.3 The Maritime Security Measures specify the responsibilities of Governments and, to a lesser extent, those of ship operators. To facilitate comparisons of the responsibilities of ship operators with those of Governments and their Administrations, the chart below references the equivalent sections and paragraphs in section 2.

Ship operator responsibilities	Maritime Security Measure	Cross-reference to responsibilities for Administrations
4.2.5	Participation on port security committees	2.8.17–2.8.18
4.2.6–4.2.8	Recognized security organizations	2.5
4.2.9–4.2.11	Alternative security agreements	2.13
4.2.12	Equivalent security arrangements	2.14
4.3	Changing security levels	2.6
4.4	Declarations of Security	2.7
4.5	Ship security personnel	2.9.1–2.9.11
4.6.1–4.6.10	Ship security alert systems	2.12.4–2.12.15
4.6.11–4.6.12	Automatic identification systems	2.12.16–2.12.19
4.6.13–4.6.15	Pre-arrival information	2.12.20–2.12.24
4.6.16–4.6.18	Long-range identification and tracking systems	2.12.25–2.12.37
4.7	Ship security assessments	2.9.12–2.9.14
4.8.1–4.8.11	Ship security plans	2.9.15–2.9.30
4.8.30–4.8.33	Shore leave and access to shore-based facilities by seafarers	2.17.5–2.17.10

Ship operator responsibilities	Maritime Security Measure	Cross-reference to responsibilities for Administrations
4.8.34–4.8.37	Reporting security incidents	2.9.37
4.8.38–4.8.39	Maintaining on-board records	2.9.38
4.9	International Ship Security Certificates	2.10
4.10.1–4.10.7	Control and compliance measures	2.11
4.11	Guidelines for non-SOLAS vessels	2.18.3–2.18.15

4.2 Security framework

Extent of application of the Maritime Security Measures

4.2.1 Ships falling under the Maritime Security Measures may be grouped into the following categories:

.1 *Passenger ships*, including high-speed passenger craft, carrying 12 or more passengers;

.2 *Cargo ships* of 500 gross tonnage and upwards, including high-speed craft, bulk carriers, chemical tankers, gas carriers and oil tankers;

.3 *Mobile offshore drilling units*, which are vessels capable of drilling for resources beneath the sea-bed. The Maritime Security Measures are only applicable when they are under way. When they are on site on the Continental Shelf, they are subject to any security requirements that the coastal State applies to its offshore activities;

.4 *Special-purpose ships* over 500 gross tonnage that are not Government-owned and that, by reason of their functions, carry on board more than 12 personnel other than normal crew who are engaged in special duties. These include research and survey ships, training ships, fish processing and factory ships, salvage ships, cable- and pipe-laying ships, diving ships and floating cranes.

4.2.2 The Maritime Security Measures do not apply to ships engaged in domestic voyages or the following types of ships engaged in international voyages:

.1 warships, naval auxiliaries or other ships operated by a Government and used only on Government non-commercial business;

.2 cargo ships of less than 500 gross tonnage;

.3 ships not propelled by mechanical means;

.4 wooden ships of primitive build;

.5 pleasure vessels not engaged in trade; or

.6 fishing vessels.

4.2.3 Experience to date indicates that some Administrations:

.1 have not fully applied the Maritime Security Measures to traditional sailing vessels although they fall into the category of special-purpose ships;

.2 have exempted ships that are not normally engaged as special-purpose ships but undertake an exceptional single special-purpose voyage (provided that they comply with the safety requirements judged to be adequate for the voyage by the Administration);

.3 using risk-based assessments, have extended the application of the Maritime Security Measures to certain categories of non-SOLAS vessels (such as ferries operating domestic services); and

.4 are actively encouraging owners and operators of non-SOLAS vessels to voluntarily apply some of the basic security practices and principles contained in the Maritime Security Measures as it helps to strengthen the overall maritime security framework (refer to subsection 4.10).

Overview of shipping company responsibilities

4.2.4 Shipping companies are required to ensure that:

.1 each ship security plan clearly states the master's overriding authority to:

- make decisions with respect to the safety and security of the ship;
- request assistance from the Company or Governments as may be necessary.

.2 CSOs, ships' masters and their SSOs are given the necessary support to fulfill their duties and responsibilities.

.3 for each ship, a security assessment is conducted and its documentation is retained.

.4 masters have information on board that allows authorized Government officials to establish:

- who is responsible for appointing crew members or other persons on board their ship to duties on the ship;
- who is responsible for deciding the employment of the ship;
- who are the parties to any charter that the ship is employed under.

Participation on port security committees

4.2.5 There is no requirement for shipping companies to participate on port security committees. However, there can be advantages if companies are represented, particularly on the Committees at their home port or other ports used frequently by their ships. Active participation helps to ensure that key aspects of the ship/shore interface (such as shore leave for crew members and access to ships) can be effectively dealt with. Guidance on port security committees is in paragraphs 2.8.17 to 2.8.18.

Recognized security organizations

4.2.6 Administrations may authorize recognized security organizations (RSOs) to act on their behalf (refer to subsection 2.5) to:

.1 approve SSPs;

.2 verify and certify compliance of ships with the provisions of the Maritime Security Measures.

4.2.7 Shipping companies may use RSOs or experts from other organizations to provide advice and assistance on SSAs and SSPs. However, RSOs should not approve SSPs if they have been involved in their preparation or in the conduct of the related SSAs.

4.2.8 Experience to date includes examples of shipping companies contracting the services of an RSO having a formal written agreement signed by both parties that, as a minimum:

.1 specifies the scope and duration of the work;

.2 identifies the main points of contact in both the Company and the RSO;

.3 details the data to be provided to the Company;

.4 identifies the legislation, policies, procedures and other work instruments to be provided to the RSO;

.5 specifies the records to be maintained by the RSO and made available as necessary;

.6 specifies any reports to be provided regularly, including changes in capability (e.g., loss of key personnel); and

.7 specifies a process for resolving performance-related issues.

Alternative security agreements

4.2.9 Alternative security agreements are agreements between national Governments on how to implement the Maritime Security Measures for short international voyages (refer to the definition in paragraph 1.8.1 using fixed routes between port facilities within their jurisdiction (refer to subsection 2.13). The majority of such agreements cover international ferry services and may address such topics as:

.1 acceptance of minor differences in regulatory requirements;

.2 alternative security arrangements to those in the Maritime Security Measures;

.3 a single security assessment for all ships covered by the agreement;

.4 how Declarations of Security are to be handled;

.5 how pre-arrival information is to be handled.

4.2.10 Ships covered by an alternative security agreement cannot conduct any ship-to-ship activities with ships not covered by that agreement.

4.2.11 Experience to date indicates that CSOs have actively participated in the security assessments and negotiations leading to the adoption of alternative security agreements.

Equivalent security arrangements

4.2.12 An Administration may allow a ship or group of ships entitled to fly its flag to implement security measures equivalent to those prescribed in the Maritime Security Measures (refer to subsection 2.14). Equivalent security arrangements could be included in the SSP.

4.3 Changing security levels

4.3.1 Governments are responsible for setting security levels and communicating changes rapidly to those who need to be informed, including shipping companies (refer to subsection 2.6). This requires Governments, through their Administrations, to compile and maintain an accurate set of contact details. In turn, this requires shipping companies to promptly communicate changes in contact details.

4.3.2 Ships intending to enter a port or port facility usually establish the security level applying at the port or port facility through direct contact with the port authority, or the PSO or PFSO, prior to entry. If a ship is operating at a higher security level than that applying at the port or port facility, the information should be passed to the port authority or the PSO or PFSO prior to entry.

4.3.3 A ship can never operate at a lower security level than the one being applied at the port or port facility that it is visiting.

4.3.4 A ship can, however, operate at a higher security level, when set by its Government, than that applying at the port or port facility it is in, or it intends to enter. The authorities at the port/port facility should not seek to have the ship reduce the security level set by the ship's Government.

4.3.5 If Governments have set higher security levels to a ship using a foreign port or port facility, entry procedures can be simplified if the decision is also communicated Government-to-Government.

4.3.6 In addition to security plans specifying the security measures in place at each security level, ship operators should ensure that their plans identify the Measures and procedures to be implemented when their ships are operating at a higher security level, set by its Administration, than that applying at the port or port facility which they are seeking to enter.

4.3.7 Experience to date includes examples of:

.1 CSOs being appointed as the point of contact for shipping companies;

.2 the appointment of a senior manager within the shipping company as an alternative contact point;

.3 the line of change of notification being a two-step process:
- Administration to CSOs;
- CSOs to key company personnel and SSOs;

.4 CSOs regularly testing lines of communication; and

.5 multiple means of communicating with contacts, i.e., by telephone, e-mail and fax.

4.4 Declarations of Security

4.4.1 The requirement for a ship to initiate, complete and retain a Declaration of Security (DoS) is determined by the ship's Administration (refer to subsection 2.7).

4.4.2 Details of how a port facility initiates or responds to a request for a DoS with a ship is documented in subsection 3.4. The Maritime Security Measures contain a model form for a DoS between a ship and a port facility (refer to appendix 3.1 – Declaration of Security form).

4.4.3 This model DoS form can be modified for a DoS between ships, as provided in appendix 4.1 – Sample of a Declaration of Security Form for a ship-to-ship interface. As well as including information on the name, port of registry and IMO Number of both ships, the DoS should specify the types of activity it covers, its duration and the security level applying to both ships. The activity should take place at the higher security level if the ships are operating at different security levels.

4.4.4 Normally, the DoS is completed by the ship's master or the SSO acting on his behalf. When completed, it must be signed and dated both by the ship's master or SSO and, in the case of a ship/port interface, by the PFSO (or an alternate designated by the Designated Authority). In the case of ship-to-ship activity, it must be signed and dated by both masters or their SSOs. Unless there are exceptional circumstances, the DoS only takes effect after it has been signed by both parties in a language common to both parties.

4.4.5 When a ship initiates a DoS, the port facility is required to acknowledge the request; however, it does not have to comply with the request.

4.4.6 When a port facility initiates a DoS, the request shall be acknowledged by the ship's master or SSO and the ship must comply with the request if the ship intends to continue its interface with the port facility.

4.4.7 The conditions under which a DoS may be requested are documented in paragraph 2.7.3. Those relevant to ships should be documented in the SSP.

4.4.8 The SSP should detail the procedures to be followed and the security measures and procedures to be implemented when responding to a request for a DoS or initiating a DoS. For a ship/port interface, these could include the respective responsibility accepted by the port facility and ship in accordance with their security plans to:

.1 ensure the performance of all security duties;

.2 monitor restricted areas to ensure that only authorized personnel have access;

.3 control access to the port facility and ship;

.4 monitor the port facility, including berthing areas and areas surrounding the ship;

.5 monitor the ship, including berthing areas and areas surrounding the ship;

.6 handle cargo and unaccompanied baggage;

.7 monitor the delivery of ship's stores;

.8 control the embarkation of persons and their effects;

.9 ensure that security communication is readily available between the ship and the port facility.

4.4.9 For a ship-to-ship activity, the respective responsibility accepted by each ship in accordance with its SSP is the same as above except that 'port facility' is replaced by 'ship'.

4.4.10 When an SSO on a SOLAS ship is unable to contact a person ashore with responsibility for shore-side security (including the completion of a DoS), the SSO can prepare a DoS indicating the security measures and procedures to be applied and maintained by the ship for the duration of the ship/port interface.

4.4.11 A SOLAS ship intending to undertake ship-to-ship activities with a non-SOLAS ship is normally required to complete a DoS with the non-SOLAS ship. Since the Maritime Security Measures were introduced, non-SOLAS ships have become accustomed to responding positively to such requests. If a DoS cannot be agreed between a SOLAS ship and a non-SOLAS ship, it is unlikely that ship-to-ship activity should take place.

4.4.12 Experience to date includes examples of:

.1 when the ship's security measures documented in the DoS are extracted from the SSP, care being taken to omit sensitive security information such as security standards;

.2 the SSO notifying the Designated Authority if a port facility:

- for any reason, refuses a request for a DoS
- requesting a DoS is at security level 3;

.3 the DoS being kept on file for three years (which may be longer than the minimum specified by the Administrations), so as to be aware of any trends in DoS requests; and

.4 SSPs including a requirement for ships to seek agreement of a DoS when using such a non-SOLAS port facility.

4.5 Ship security personnel

Introduction

4.5.1 This section provides guidance on the duties and security-related training required for company security officers (CSOs), ship security officers (SSOs) and all shipboard personnel (refer to paragraph 1.8.1 for the definition). Related guidance is in paragraphs 2.9.1 to 2.9.11.

4.5.2 The appointment of CSOs and SSOs is essentially a matter for the shipping company whose ships fall under the Maritime Security Measures.

4.5.3 As the CSOs and SSOs are likely to be entrusted with security-sensitive information, some Administrations require that they are subjected to security vetting before receiving such information. This requirement can extend to other company personnel who perform the responsibilities of a CSO and to the senior management of the company.

4.5.4 Certificates of proficiency issued by the ship's Administration to SSOs and shipboard personnel under the STCW Code can be one of the documents inspected by a duly authorized officer undertaking control and compliance measures under the Maritime Security Measures when the ship is in a foreign port (refer to paragraphs 4.10.1 to 4.10.7).

Company security officers

4.5.5 In shipping companies, the responsibility for the security of a ship rests with the CSO. Working together with their SSOs and with the PFSOs at the port facilities used by their ships, CSOs play a central, and essential, role in the implementation of the Maritime Security Measures. It is their responsibility to ensure that each of their ships meets the requirements of the Maritime Security Measures.

4.5.6 On security matters, CSOs are the main point of contact with both their ships and the Administration. CSOs can be the officers within companies who are directly notified of changes in security level for onward transmission to their ships.

4.5.7 Through their contact with their SSOs and PFSOs, they can ensure:

- **.1** that possible security threats are identified and appropriate action is taken to address them; and
- **.2** the continued effectiveness of the security measures and procedures on their ships.

4.5.8 Each shipping company is required to appoint one or more CSOs. The ship or ships that each CSO is responsible for should be clearly identified.

4.5.9 The duties of a CSO include:

- **.1** advising on the level of threats likely to be encountered by the ship, using appropriate security assessments and other relevant information;
- **.2** ensuring that SSAs are carried out;
- **.3** ensuring the development, submission for approval, implementation and maintenance of SSPs;
- **.4** ensuring that SSPs are modified, as appropriate, to correct deficiencies and satisfy the security requirements of individual ships;
- **.5** arranging for internal audits and reviews of security activities;
- **.6** arranging for the initial and subsequent verifications of ships by the Administration or RSOs authorized to act on their behalf;
- **.7** ensuring that deficiencies and non-conformities identified during internal audits, periodic reviews, security inspections and verifications of compliance are promptly dealt with;
- **.8** enhancing security awareness and vigilance;
- **.9** ensuring adequate training for personnel responsible for ship security;
- **.10** ensuring effective communication and co-operation between SSOs and relevant PFSOs;
- **.11** ensuring consistency between security requirements and safety requirements;
- **.12** ensuring that, if sister-ship or fleet security plans are used, the plan for each ship reflects the ship-specific information accurately;
- **.13** ensuring that any ASAs or ESAs approved for a particular ship or group of ships are implemented and maintained; and
- **.14** ensuring the effective co-ordination and implementation of SSPs by participating in exercises at appropriate intervals.

4.5.10 Each person performing the duties of a CSO should be able to satisfactorily demonstrate the competencies listed in appendix 4.2 – Competency matrix for company security officers. Persons who have satisfactorily completed a training course for CSOs which is recognized by the Administration should be considered to have met this requirement.

4.5.11 As CSOs are not serving on board ships, they do not fall under the provisions of the STCW Convention and its related Code.

4.5.12 Other shore-based personnel with security responsibilities are required to be able to demonstrate the same competencies.

4.5.13 Some of the practices observed in the appointment and certification of CSOs include:

- **.1** having to undergo an approval process based on training course certification and security clearances, particularly if they have access to sensitive security information provided by the Contracting Government (e.g., information on national threats);
- **.2** having to attend courses from training organizations approved by their Administration;
- **.3** designating an alternate to undertake the duties of the CSO when necessary;

.4 having documentary evidence of their appointment and training (this includes their alternate);

.5 being shipping company employees and not contracted from an external company such as a security firm or consultant (this includes their alternate);

.6 having an approved list of their security and non-security duties. Non-security duties should not interfere with their ability to carry out their security duties;

.7 wherever possible, being members of the port security committee at their home port; and

.8 reporting directly to a senior member of the shipping company's management team.

Ship security officers

4.5.14 On a ship, the ship security officer (SSO) is responsible for security. This responsibility gives SSOs a key role in ensuring the continued effectiveness of the Maritime Security Measures.

4.5.15 Responsible to the master of their ship and reporting to the CSOs ashore, SSOs:

.1 ensure that the ship and its shipboard personnel operate in accordance with the approved SSP;

.2 maintain security at all times;

.3 may have responsibility for shipboard personnel with designated security responsibilities;

.4 ensure that contact is established and maintained with the PFSOs at the port facilities that the ship uses; and

.5 liaise as necessary with PSOs/PFSOs or other officers and officials ashore with security responsibilities.

4.5.16 An SSO must be designated for every SOLAS ship. To allow for crew changes, a number of SSOs may be designated to serve on each ship. The duties of an SSO include:

.1 undertaking regular security inspections of the ship to ensure that appropriate security measures are maintained;

.2 maintaining and supervising the implementation of the SSP, including any amendments;

.3 co-ordinating the security aspects of the handling of cargo and ship's stores with other shipboard personnel and relevant PFSOs;

.4 proposing modifications to the SSP;

.5 reporting any deficiencies and non-conformities identified during internal audits, periodic reviews, security inspections and verifications of compliance to the CSO;

.6 implementing any corrective actions;

.7 enhancing security awareness and vigilance on board the ship;

.8 ensuring that adequate training has been provided to shipboard personnel, including security-related familiarization training;

.9 reporting all security incidents;

.10 co-ordinating implementation of the SSP with the CSO and relevant PFSOs;

.11 ensuring that security equipment is properly operated, tested, calibrated and maintained; and

.12 ensuring the effective implementation of the SSP by organizing drills at appropriate intervals.

4.5.17 Effective 1 January 2012, SSOs are required to hold a certificate of proficiency confirming that they:

.1 have approved seagoing service of not less than 12 months (or appropriate seagoing service and knowledge of ship operations); and

.2 meet the minimum standards of competency specified in the STCW Code, which are listed in appendix 4.3 – Competency matrix for ship security officers. They are similar to the guidance issued by IMO for CSOs in 2005 (refer to appendix 2.4 – Sample of a port facility security plan approval form).

4.5.18 The 2006 amendments to the STCW Convention, which came into force in 2008, included the insertion of a new regulation VI/5 on "Mandatory minimum requirements for the issue of certificates of proficiency for ship security officers". part A of the STCW Code stipulates the KUP requirements (Knowledge, Understanding and Proficiency) for certification of SSOs which can be found in appendix 4.3 – Competency matrix for ship security officers, so as to determine the need for updating their qualifications.

4.5.19 Some of the practices observed in the appointment and certification of SSOs include:

.1 undergoing an approval process based on training course certification and security clearances, particularly if they have access to sensitive security information provided by the Contracting Government (e.g., information on national threats);

.2 undertaking training courses by training providers approved by the Administration;

.3 having documentary evidence of their appointment and training;

.4 being shipping company employees, not contracted resources from an external company (e.g., a security firm or consultant);

.5 having an approved documented list of their security and non-security duties. Non-security duties should not interfere with their ability to carry out security duties;

.6 being given the opportunity, when newly appointed, to become familiar with the ship and its SSP before assuming the responsibilities; and

.7 being the master on ships with small crews. In such cases, the shipping company could consider deploying a trained crew member to assist the master in conducting security activities.

Shipboard personnel with designated security duties

4.5.20 Under the amended STCW Code, shipboard personnel with designated security duties (e.g., deck and gangway watch, including contract security guards) are required to hold a certificate of proficiency confirming that they meet the minimum standards of competency listed in appendix 4.4 – Competency matrix for shipboard personnel with designated security duties.

4.5.21 Given their responsibilities, ships' masters, if they are not also the SSO, should always be considered to have designated security duties.

4.5.22 As a transitional provision, the STCW Code provides that, until 1 January 2014, shipboard personnel with designated security duties who commence their seagoing service prior to 1 January 2012 should be able to demonstrate competence to undertake the tasks, duties and responsibilities listed in appendix 4.4 – Competency matrix for shipboard personnel with designated security duties, by:

.1 having seagoing service as shipboard personnel with designated security duties, for a period of at least six months in total during the preceding three years; or

.2 having performed security functions considered to be equivalent to the seagoing service referenced above; or

.3 passing an approved test; or

.4 successfully completing approved training.

4.5.23 Some of the practices observed in the appointment and certification of shipboard personnel with designated security duties include:

- **.1** receiving security-related familiarization training from the SSO (or equally qualified person) in their assigned duties, in accordance with the provisions specified in the SSP, before being assigned to their duties;
- **.2** having documentary evidence of their training; and
- **.3** being listed in the SSP (by category of personnel).

All shipboard personnel

4.5.24 Under the STCW Code, all shipboard personnel are required to receive approved security-related familiarization training before taking up their duties and to be able to:

- **.1** report a security incident, including a piracy or armed robbery threat or attack;
- **.2** know the procedures to follow when they recognize a security threat; and
- **.3** take part in security-related emergency and contingency procedures.

4.5.25 Before taking up their duties, all shipboard personnel are also required to:

- **.1** receive appropriate approved training or instruction in security awareness as set out in appendix 4.5 – Competency matrix on security awareness for all shipboard personnel;
- **.2** provide evidence of meeting the minimum standards of competency for security awareness listed in appendix 4.5 – Competency matrix on security awareness for all shipboard personnel. The required evidence may be achieved by demonstrating competence in terms of either the evaluation criteria specified in appendix 4.5 or through examination or continuous assessment as part of an approved training programme.

4.5.26 As a transitional provision, the STCW Code provides that, until 1 January 2014, seafarers who commence their seagoing service prior to 1 January 2012 are required to establish that they meet the requirements listed in appendix 4.5 – Competency matrix on security awareness for all shipboard personnel, by:

- **.1** having approved seagoing service as shipboard personnel, for a period of at least six months in total during the preceding three years; or
- **.2** having performed security functions considered to be equivalent to the seagoing service referenced above; or
- **.3** passing an approved test; or
- **.4** successfully completing approved training.

4.5.27 Experience to date includes examples of shipboard personnel:

- **.1** receiving security-awareness training at least once in their career from the SSO or equally qualified person; and
- **.2** having documentary evidence of their security-related training.

Security clearances

4.5.28 Shipping companies can be required to comply with any instructions issued by their flag State regarding the application of any security clearance procedures for their personnel.

4.5.29 Security clearances are the means of verifying that personnel whose duties require access to restricted areas or security-sensitive information do not pose a risk to maritime security. The vetting associated with these clearances is more stringent than the pre-employment background checks conducted by shipping companies.

4.5.30 Experience to date includes examples of flag States requiring security clearance for

- **.1** the senior managers of a shipping company; and
- **.2** the CSO and those appointed to undertake any of the duties of the CSO.

4.5.31 A number of Governments require security clearance for all those working in any capacity within port areas, including the employees of shipping companies.

4.6 Ship security communications

Ship security alert systems

4.6.1 All ships are required to be provided with a ship security alert system (SSAS) as described in paragraphs 2.12.4 to 2.12.15. Its intent is to send a covert signal or message from a ship that will not be obvious to anyone on the ship who is unaware of the alert mechanism.

4.6.2 When activated, the SSAS must:

- **.1** initiate and transmit a ship-to-shore security alert to a competent authority designated by the Administration, which in these circumstances may include the shipping company, identifying the ship and its location, and indicating that the security of the ship is under threat or has been compromised;
- **.2** not send the alert to any other ships;
- **.3** not raise any alarm on board the ship; and
- **.4** continue the alert until deactivated and/or reset.

4.6.3 The competent authority should be able to receive SSAS alerts on a 24/7 basis.

4.6.4 The SSAS must:

- **.1** be capable of being activated from the navigation bridge and in at least one other location;
- **.2** conform to performance standards not inferior to those adopted by IMO; and
- **.3** have its activation points designed so as to prevent the inadvertent initiation of an alert.

4.6.5 When an SSAS alert is received by the competent authority, either directly or via a service provider, it should include the following information:

- **.1** name of ship;
- **.2** IMO ship identification number;
- **.3** call sign;
- **.4** Maritime Mobile Service Identity (which is a series of 9 digits sent over a radio-frequency channel to provide a unique identifier used to call ships automatically);
- **.5** GNSS position of the ship; and
- **.6** date and time of the GNSS position.

4.6.6 The requirement for an SSAS may be met by using radio installations that have been approved by the Administration.

4.6.7 The competent authority is responsible for ascertaining whether a security alert is real or false.

4.6.8 The SSP must include the following, which Administrations may require to be kept in a document separate from the SSP to avoid compromising its confidentiality:

- **.1** the identification of the SSAS activation points; and
- **.2** procedures to be used, including testing, activation, deactivation and resetting to limit false alerts.

4.6.9 A master may use an overt alarm (i.e., one such as a VHF broadcast, which makes no attempt to deny knowledge of its activation) in addition to a covert alarm as a means of discouraging a security threat from becoming a security incident.

4.6.10 Experience to date of ship operators in establishing SSASs reveals examples of:

- **.1** procedures being included in SSPs using a standard template;
- **.2** the handling of false security alerts being included as a procedure;
- **.3** testing being performed at least annually;
- **.4** all concerned parties being notified by the shipping company when an SSAS is to be tested, so as to avoid any unintended emergency response actions;
- **.5** when an SSAS accidentally transmits in testing, the ship immediately notifying the shipping company or competent authority (if it is not the shipping company), so that all concerned parties can be made aware that the alert is false and that no emergency response action should be taken;
- **.6** a checklist being used when testing; and
- **.7** providing for an alternative power source.

Automatic identification systems

4.6.11 Further to the requirements documented in paragraphs 2.12.16 to 2.12.19, if the master believes that continual operation of AIS might compromise the safety of the ship, or where security incidents are imminent, the AIS may be switched off.

4.6.12 In doing so, masters should:

- **.1** bear in mind the possibility that attackers are monitoring ship-to-shore communications and using intercepted information to select their targets;
- **.2** be aware that switching off the AIS in high-risk areas reduces the ability of the supporting naval vessels to track and trace vessels which may require assistance;
- **.3** exercise caution when transmitting information on cargo or valuables on board by radio in areas where attacks occur;
- **.4** use professional judgement to decide whether the AIS should be switched off to avoid detection when entering areas where piracy is an imminent threat;
- **.5** balance the risk of attack against the need to maintain the safety of navigation;
- **.6** act in accordance with IMO guidance material;
- **.7** be aware that other ships operating in high-risk areas may have taken a decision to switch off their AIS system; and
- **.8** in the event of an attack, ensure, to the extent feasible, that the AIS is turned on again and transmitting to enable security forces to locate the ship.

Pre-arrival notification

4.6.13 If a ship intending to enter a port of another Contracting Government is requested to provide the information listed in paragraph 2.12.20, it should be provided by the master or on his behalf by the CSO, SSO or ship's agent at the port where entry is being sought. It may be submitted in the form of a standard data set such as the one shown in appendix 4.6 – Standard data set of security-related pre-arrival information. If submitted electronically, it may not be possible for a signature to be provided.

4.6.14 The master may decline to provide such information, but failure to do so may result in denial of entry into port.

4.6.15 The ship is required to keep records of the information provided for the last 10 calls at port facilities.

Long-range identification and tracking systems

4.6.16 Long-range identification and tracking (LRIT) system requirements are described in paragraphs 2.12.25 to 2.12.37.

4.6.17 The ship operator's obligation is to comply with these requirements by providing on board equipment for transmitting the identity of the ship, its position and the date and time of the position to the data centre (DC) nominated by the ship's Administration.

4.6.18 In exceptional circumstances and for the shortest duration possible, the LRIT system can be switched off if its operation is considered by the master to compromise the safety or security of the ship. In such instances, the master is required to inform the Administration without undue delay and to record the occurrence, with the reason for the decision and the duration of non-transmittal.

4.7 Ship security assessments

Introduction

4.7.1 A ship security assessment (SSA) must be undertaken for each ship as a prelude to the preparation of an SSP (refer to paragraphs 2.9.12 to 2.9.14).

4.7.2 An SSA may be considered to be a risk analysis of all aspects of a ship's operations in order to determine which parts of it are more vulnerable. It is an essential and integral part of developing or updating the SSP.

4.7.3 CSOs are responsible for ensuring that SSAs are carried out for each ship in their company's fleet by persons with appropriate skills to evaluate the security of a ship.

4.7.4 RSOs may carry out SSAs provided that they are not subsequently involved in the review and approval of the associated SSP (refer to paragraphs 4.2.6 to 4.2.8).

Conducting and documenting SSAs

4.7.5 The SSA is required to include the following elements:

- **.1** an on-scene security survey;
- **.2** identification of existing security measures, procedures and operations;
- **.3** identification and evaluation of important shipboard operations;
- **.4** identification of possible threats to important shipboard operations and the likelihood of their occurrence;
- **.5** identification of weaknesses, including human factors, in the infrastructure, policies and procedures.

4.7.6 As with PFSAs, Administrations could consider requiring SSAs to establish and prioritize countermeasures.

4.7.7 The SSA for each ship in the company's fleet is required to be documented, reviewed, accepted and retained by the shipping company.

Preparing SSA reports

4.7.8 A report must be prepared on completion of the SSA. It provides the means by which an SSA is accepted by the shipping company and is required to:

- **.1** summarize how the assessment was conducted;
- **.2** describe each vulnerability found during the assessment;
- **.3** describe the countermeasures that could address each vulnerability;
- **.4** be protected from unauthorized access or disclosure.

4.7.9 If used, a completed template could be attached to the SSA report as an annex.

4.7.10 If the SSA has not been carried out by the shipping company, the SSA report should be reviewed and accepted by the CSO.

4.7.11 Experience to date includes examples of CSOs:

.1 providing a numbered copy of an approved SSA report to a list of individuals within the Company;

.2 before commencing an SSA, seeking out available information on threat assessments at ports which will be visited by the ship; studying previous reports on similar security needs; and discussing how the SSA is to be conducted with appropriate persons on board ship and in the port facilities and ports to be visited;

.3 following the specific guidance offered by national authorities and seeking clarification when appropriate; and

.4 in conjunction with the SSO, when developing countermeasures, considering their effect in terms of:

- comfort and convenience;
- personal privacy; and
- the performance of duties by shipboard personnel who may have to remain on board for long periods.

Updating SSAs

4.7.12 An SSA should be reviewed and updated, as appropriate, when there has been:

.1 a significant security incident involving the ship;

.2 a change in the ship's trading pattern; or

.3 change of the owner or operator of the ship.

4.7.13 These changes could include changes to sea routes, particularly in instances where the change may result in new threat scenarios and increased probability of a security incident.

4.8 Ship security plans

Introduction

4.8.1 Each ship is required to carry on board a ship security plan (SSP) approved by the Administration. It must make provision for the three security levels (refer to subsection 4.3). The close inter-relationship between SSAs and SSPs is shown by the example of an SSA/SSP approval process illustrated in appendix 4.6 – Standard data set of security-related pre-arrival information.

4.8.2 RSOs may:

.1 prepare SSPs on behalf of CSOs (who are responsible for ensuring that the SSPs are prepared and submitted for approval);

.2 review and approve SSPs and their amendments on behalf of an Administration provided that they were not involved in the preparation of the SSP under review or of its related SSA.

4.8.3 When an SSP or its amendment is submitted for approval, it must be accompanied by the SSA on which the plan or amendment was based.

4.8.4 CSOs and their SSOs should retain records of any amendments made to an approved SSP.

Preparing and maintaining SSPs

4.8.5 SSPs should provide details of:

.1 measures designed to prevent weapons, dangerous substances and devices intended for use against persons, ships or ports from being taken on board;

.2 restricted areas and measures for the prevention of unauthorized access to them;

.3 measures and equipment for the prevention of unauthorized access to the ship, including boarding of a ship when in port or at sea;

.4 procedures for responding to security threats or breaches of security, including provisions for maintaining critical operations of the ship or ship/port interface;

.5 the minimum operational and physical security measures the ship shall take at all times, when operating at security level 1;

.6 the additional or intensified security measures the ship itself can take when moving to security level 2;

.7 procedures for promptly responding to any security instructions Governments may give at security level 3;

.8 procedures for evacuation in case of security threats or breaches of security;

.9 security-related duties of shipboard personnel assigned security responsibilities and other shipboard personnel;

.10 procedures for auditing the security activities;

.11 procedures for training, drills and exercises associated with the plan;

.12 procedures for interfacing with port facility security activities;

.13 procedures for the periodic review of the plan and updating;

.14 procedures for reporting security incidents;

.15 the SSO and CSO, including 24-hour contact details;

.16 procedures to ensure the inspection, testing, calibration and maintenance of any security equipment provided on board;

.17 frequency of testing or calibrating any security equipment provided on board; and

.18 procedures, instructions and guidance on SSAS usage, including the testing, activation, deactivation, resetting and limitation of false alerts.

4.8.6 In addition, SSPs should detail:

.1 the organizational structure of security for the ship;

.2 the ship's relationship with the Company, port facilities, other ships and relevant authorities with security responsibility;

.3 the communication systems to allow effective continuous communication within the ship and between the ship and others, including port facilities;

.4 the basic security measures for security level 1, both operational and physical, that will always be in place;

.5 the additional security measures that will allow the ship to progress without delay to security level 2 and, when necessary, to security level 3;

.6 provision for regular review, or audit, of the SSP and for its amendment in response to experience or to changing circumstances; and

.7 reporting procedures to the appropriate Contacting Government's contact points.

4.8.7 Due to considerations of conflict of interest, personnel conducting internal audits of the security measures specified in SSPs or evaluating their implementation are required to be independent of the Measures being audited unless this is impracticable, due to the size and nature of the shipping company or its fleet.

4.8.8 SSPs are required to be protected from unauthorized access or disclosure.

4.8.9 The relevant provisions of the SSP placed on board must be in the working language or languages of the ship. If the language or languages used is not English, French, or Spanish, a translation into one of these languages must be included.

4.8.10 If SSPs are kept in electronic format, procedures must be put in place to prevent their unauthorized deletion, destruction or amendment.

4.8.11 Experience to date indicates that several Administrations have developed model SSPs, pre-submission checklists and related guidance material. Those referenced on internet sites are listed in appendix 4.8 – Examples of internet sources of guidance material on preparing and validating ship security plans, along with a summary of their contents.

Planning and conducting ship security drills and exercises

4.8.12 The regular conduct of ship security drills and exercises is an important aspect of ensuring that ships comply with the requirements of the Maritime Security Measures.

4.8.13 Drills may be defined as supervised activities that are used to test a single measure or procedure in the SSP. Exercises are more complex activities which test several measures and procedures at the same time.

4.8.14 To ensure the effective implementation of the Measures and procedures specified in SSPs, drills should be conducted at least once every three months. These are usually organized by SSOs, who are responsible for ensuring that all shipboard personnel have received adequate training. In addition, in cases where more than 25% of the ship's personnel has been changed at any one time with personnel that have not previously participated in any drill on that ship within the last three months, a drill should be conducted within one week of the change.

4.8.15 As a minimum, SSOs should organize drills to cover such scenarios as:

- **.1** identification and search of unauthorized visitors on board the ship;
- **.2** recognition of materials that may pose a security threat;
- **.3** methods to deter attackers from approaching the ship;
- **.4** recognition of restricted areas;
- **.5** mustering for evacuation.

4.8.16 Further guidance on undertaking drills and exercises is available from https://homeport.uscg.mil/mycg/portal/ep/home.do

4.8.17 To ensure the effective implementation and co-ordination of SSPs, CSOs are required to participate in exercises at a recommended minimum interval of once each calendar year, with no more than 18 months between the exercises.

4.8.18 These exercises, which could test communications, co-ordination, resource availability, and response, may be:

- **.1** full-scale or live;
- **.2** tabletop simulation or seminar; or
- **.3** combined with other exercises organized by Government agencies to test search and rescue or emergency response capabilities.

4.8.19 Exercises may cover such on-board emergencies as searches for bombs, weapons and unauthorized personnel as well as responses to damage or destruction of ship infrastructure. They do not have to involve each ship within a fleet. If an exercise is carried out on board and/or involves one or more of a shipping company's ships then, as a minimum, the exercise details and lessons learnt can be circulated throughout the fleet by means of seminars on board each ship; also, any measures that are identified can be implemented on each ship.

4.8.20 Drills and exercises take up organizational time and resources, and must therefore be conducted in as efficient and effective a manner as possible. Recognizing the need to assist port facility operators in the Asia-Pacific Region, APEC's Transportation Working Group developed a set of guidelines in the form of a manual (refer to appendix 3.7 – APEC Manual of maritime security drills and exercises for port facilities: table of contents). Although its focus is on port operations, the Manual provides a systematic and comprehensive approach to the planning, preparation for, conduct, debrief and reporting of maritime security drills and exercises. Thus, it could provide a useful reference for planning and conducting drills and exercises on board ships.

4.8.21 The conduct of drills and exercises may require an approved SSP to be amended; for a major amendment, the SSP must be submitted to the Administration or authorized RSO for re-approval.

Access to ships by Government officials, emergency response services and pilots

4.8.22 Government officials entitled as part of their duties to board ships should carry appropriate identification documents issued by the Government. Identification documents should include a photograph of the holder of the document. They should also include the name of the holder or have a unique identification number. If the identity document is in a language other than English, French or Spanish, a translation into one of those languages should be provided.

4.8.23 Government officials should present their identification document when requested to do so when boarding a ship.

4.8.24 Ship security personnel should be able to verify the authenticity of identity documents issued to Government officials, and Governments should establish procedures, and provide contact details, to facilitate such validation.

4.8.25 Emergency response services and pilots should also carry appropriate identification documents and present them when boarding a ship. The authenticity of such identification documents should be capable of being verified.

4.8.26 Only the person in charge of an emergency response team need present an identification document when boarding a ship and should inform the relevant security personnel of the number of emergency response personnel entering or boarding.

4.8.27 Government officials, emergency response personnel and pilots should not be required to surrender their identity documents when boarding a ship. The issuance of visitor identification documents by a ship may not be appropriate when Government officials, emergency response personnel or pilots have produced an identity document which can be verified.

4.8.28 Government officials should not be subject to search by ship security personnel. Any search requirement in an SSP could be waived for emergency response personnel responding to an emergency or for a pilot boarding a ship once their identity has been verified.

4.8.29 SSOs should be able to secure the assistance of PFSOs to verify the identification of Government officials, emergency response personnel or pilots intending to board a ship.

Shore leave and access to shore-based facilities by seafarers

4.8.30 The obligations of Governments and their national authorities related to shore leave and access to shore-based facilities are addressed in paragraphs 2.17.5 to 2.17.10, while those of port and port facility operators are addressed in paragraphs 3.8.13 to 3.8.19.

4.8.31 Normally, the SSO contacts the PFSO before arrival at the port facility in order to co-ordinate shore access and arrangements for on-board visits. These arrangements must strike a balance between the need for port and port facility security and the needs of the ship and its crew.

4.8.32 Procedures to facilitate shore access by, or shore leave for, seafarers should be transparent, easy to follow and should not require involvement by the seafarers. The procedures should provide a system whereby seamen, pilots, welfare and labour organizations can board and depart vessels in a timely manner. These procedures should not impose undue costs upon the individual requiring passage to and from the vessel. Barriers such as excessive fees or restrictive hours of operation should not be imposed.

4.8.33 In instances where shore leave is denied to crew members, the SSO should immediately refer the matter to the CSO to raise with appropriate authorities.

Reporting security incidents

4.8.34 SSPs are required to document the procedures for reporting security incidents and threats to Administrations and other Government organizations (refer to paragraph 2.9.37).

4.8.35 Security incidents generally can fall into two categories:

.1 those considered to be sufficiently serious that they should be reported to relevant authorities by the CSO, including:

- unauthorized access to restricted areas within the ship for suspected threat-related reasons;
- unauthorized carriage or discovery of stowaways, weapons or explosives;
- incidents of which the media are aware;
- bomb warnings;
- attempted or successful boardings; and
- damage to the ship caused by explosive devices or arson.

.2 those of a less serious nature but which require reporting to, and investigation by, the SSO can include:

- unauthorized access to the ship caused by breaches of access control points or inappropriate use of passes;
- damage to equipment through sabotage or vandalism;
- unauthorized disclosure of an SSP;
- suspicious behaviour near the ship when at a port facility;
- suspicious packages near the ship when at a port facility; and
- unsecured access points to the ship.

4.8.36 If a security threat or incident develops which requires initiation of the security procedures and measures applying at a higher security level than the security level set for the port facility, the initiation of the appropriate response to the emerging threat by a port facility or ship should not, and cannot, await change of the security level by the relevant Government or its Administration. The response to the security threat or incident as it develops should be taken in accordance with the SSP. The ship should report the threat or incident – and the action taken – to the Government, Designated Authority or Administration at the earliest practicable opportunity.

4.8.37 Experience to date indicates that some Administrations have:

.1 specified the types of security incidents that must be immediately reported to them, as indicated below:

Type of security incident
Attack
Bomb warnings
Hijack
Armed robbery or piracy against a ship
Discovery of firearms
Discovery of other weapons
Discovery of explosives
Unauthorized access to a restricted area
Unauthorized access to the port facility
Media awareness

.2 with respect to bomb warnings, developed a checklist as a useful aid for anyone receiving a warning (which can be received in various ways, with a telephone call to a shipping agent, shipping company or individual ship being the most common). A sample checklist may be accessed at: http://www.cpni.gov.uk/documents/posters%20and%20checklists/bomb-threat-checklist.pdf

.3 designed standard forms for security incidents that must be reported to them and making them available on their internet sites. Examples of such forms may be downloaded from the following internet sites:

- www.infrastructure.gov.au/transport/security/maritime/MSIR_online_form.aspx
- http://www.mpa.gov.sg/sites/circulars_and_notices/pdfs/shipping_circulars/security_incident_form.pdf
- www.gibmaritime.com
- www.mcw.gov.cy/mcw/dms/dms.nsf/All/78C174E7BB90EA95C22575190042D23A?OpenDocument

.4 although these forms have been designed to fulfil incident reporting requirements prescribed in national legislation, they could be adapted by ship operators to their particular reporting requirements. In such cases, the form's practical usefulness could be enhanced by:

- ensuring that its format is straightforward;
- allowing the SSO to report the remedial action taken;
- ensuring that any associated reporting procedures are straightforward;
- establishing the situations when it is to be forwarded to the CSO; and
- locating copies where they can be visible to, and easily accessed by, shipboard personnel.

.5 specified the manner in which the reports should be made and the procedures for doing so, including the time period by which an incident must be reported and the recipients of such reports (e.g., local law-enforcement agencies when a ship is in a port facility or an adjacent coastal State).

Maintaining on-board records

4.8.38 Administrations should specify the security records that a ship is required to keep and be available for inspection, including the period for which they should be kept (subsection 2.9.38). The records must cover:

- **.1** DoS agreed with port facilities and other ships;
- **.2** security threats or incidents;
- **.3** breaches of security;
- **.4** changes in security level;
- **.5** communications relating to the direct security of the ship, such as specific threats to the ship or to port facilities where the ship is, or has, been;
- **.6** ship security training undertaken by the ship's personnel;
- **.7** security drills and exercises;
- **.8** maintenance of security equipment;
- **.9** internal audits and reviews;
- **.10** reviews of SSAs and SSPs; and
- **.11** any amendments to an approved SSP.

4.8.39 Records are required to be:

- **.1** kept in the working language(s) of the ship;
- **.2** protected by procedures aimed at preventing their unauthorized deletion, destruction or amendment if kept in an electronic format;
- **.3** protected from unauthorized access or disclosure;
- **.4** available to duly authorized officers of Contracting Governments to verify that the provisions of SSPs are being implemented, and
- **.5** kept on board for the period specified by the Administration.

Conducting self-assessments

4.8.40 Checklists can provide a useful way to assess and report progress in implementing SSPs and, by extension, the Maritime Security Measures.

4.8.41 Appendix 4.9 – Implementation checklist for ship security personnel, contains a checklist for ship security personnel that allows them to assess progress in implementing the Maritime Security Measures. Except for minor modifications to its format and guidance material, it is identical to the Voluntary Self-Assessment Tool for Ship Security that was approved by IMO's Maritime Safety Committee in May 2006 and received widespread distribution.

4.8.42 Appendix 4.10 – Implementation checklist for shipping companies and their CSOs contains a checklist for shipping companies and their CSOs to assess progress in implementing the Maritime Security Measures. It was issued by IMO in December 2006.

4.8.43 Experience to date reveals that:

- **.1** several Administrations have encouraged the annual use of these checklists as a good management practice; and
- **.2** CSOs and SSOs have modified its content and format to meet their specific assessment requirements (e.g., to identify when procedures were last reviewed or measures tested, or to establish a link between any identified gaps and work-plan priorities).

Reviewing and amending an approved SSP

4.8.44 Administrations should notify CSOs of amendments to an approved SSP that must be approved before they can be implemented. This notification can be provided on approval of the initial SSP or a subsequent amendment.

4.8.45 Similarly, Administrations should notify CSOs of the amendments to an approved SSP that do not require their prior approval.

4.8.46 Unless the Administration has allowed specified amendments to be made without their prior approval, proposed amendments to an approved SSP may not be implemented until authorized by the Administration.

4.8.47 The preparation of all amendments to an approved SSP is ultimately the responsibility of the CSO.

4.8.48 If the Administration allows a CSO or SSO to amend an SSP without its prior approval, the adopted amendments must be communicated to the Administration at the earliest opportunity.

4.8.49 Experience to date indicates that:

.1 SSPs are being reviewed annually and more frequently in response to incidents such as:

- changes in ship operations, ownership and structure;
- after an unsuccessful drill or exercise;
- after a security incident or threat involving the ship;
- completion of a review of the SSA;
- when an internal audit or inspection by the Administration has identified failings in the ship's security organization and operations, calling into question the continuing relevance of the approved SSP;

.2 amendment of an approved SSP also involves a review of the ship's SSA;

.3 Administrations have required the following types of proposed changes to be submitted for their approval prior to their implementation:

- procedures for acknowledging changes in security levels;
- measures or procedures at security levels 2 and 3;
- procedures for controlling access to the ships;
- procedures for reporting incidents;
- frequency of testing security equipment;
- procedures for maintaining security equipment;
- procedures for maintaining confidentiality of documents;
- frequency of conducting drills and exercises;
- SSAS procedures;
- the CSO's identity and contact details;

.4 some Administrations have shown flexibility in allowing minor amendments to an approved SSP without their prior approval. This can often relate to changes that can occur frequently, e.g., changing the SSO;

.5 it has proved convenient to format SSPs so as to facilitate the submission of amendments in the form of single pages rather than the whole document.

4.9 The International Ship Security Certificate

4.9.1 Ships falling under the Maritime Security Measures have to carry either the International Ship Security Certificate (ISSC) or, in limited circumstances, the Interim ISSC, both of which are issued by their Administration.

Details of their issuance, required verifications, duration of validity, loss of validity and remedial actions are provided in subsection 2.10.

4.9.2 Shipping companies are required to:

.1 ensure that verification of compliance with the Maritime Security Measures takes place:

– before their ships are put into service and the ISSC is issued (initial verification);

– at least once between the second and third anniversary of the issuance of the ISSC if the validity period is five years (intermediate verification); and

– before the ISSC is renewed (renewal verification); and

.2 notify the ship's Administration immediately when there is a failure of a ship's security equipment or system or the suspension of a security measure which compromises the ship's ability to operate at security levels 1 to 3. The notification should be accompanied by any proposed remedial actions; and

.3 notify the ship's Administration when the above circumstances do not compromise the ship's ability to operate at security levels 1 to 3. In such cases, the notification should be accompanied by an action plan, specifying the alternative security measure being applied until the failure or suspension is rectified, together with the timing of any repair or replacement.

4.9.3 Experience to date indicates that Administrations:

.1 provide guidance to their CSOs, reminding them of the cumulative effect that individual failures or suspensions of measures could have on the ability of their ships to operate at security levels 1 to 3;

.2 apply widely diverging interpretations of when a SOLAS ship is out of service or laid up; and of the circumstances and passage of time that could lead to consideration of suspension or withdrawal of the ship's ISSC. The Maritime Security Measures are silent on the specific issues.

4.10 Control and compliance measures

4.10.1 Governments can apply specific control and compliance measures to foreign-flagged SOLAS ships using, or intending to use, their ports when assessing their compliance with the Maritime Security Measures. Elements of these control and compliance measures are unique, including the authority to:

.1 require ships to provide security-related information prior to entering port;

.2 inspect ships intending to enter into port when there are clear grounds for doing this once the ship is within the territorial sea. The master has a right to refuse such an inspection and withdraw his intention to enter that port, and

.3 refuse to allow a ship to enter port; and

.4 expel a ship from port.

4.10.2 Details of the responsibilities, procedures and limitations of Administrations in exercising this authority, and the Measures that may be applied, are provided in subsection 2.11.

4.10.3 During interactions with duly authorized officers, ships' masters and their SSOs should be able to:

.1 communicate in English; and

.2 verify the identity of duly authorized officers intending to board their ship.

4.10.4 If requested to do so, a ship has to provide security-related information prior to entering into a port. IMO has developed a standard data set of the security-related information that a ship might be expected to provide (refer to appendix 4.6 – Standard data set of security-related pre-arrival information). The standard data set does not preclude a Government from requesting further security-related information on a regular basis or in specified circumstances. When Governments require additional information, the shipping industry should be appropriately advised.

4.10.5 If a ship has been advised of the intention to take control measures under the Maritime Security Measures, it can:

- **.1** decide to withdraw its intention to enter the port; or
- **.2** discuss ways of rectifying its non-compliance with the duly authorized officer.

4.10.6 If a ship is unduly delayed, the Maritime Security Measures provide for compensation to be claimed for loss or damage.

4.10.7 The Maritime Security Measures require documents to be carried on board ship, some of which can be inspected by duly authorized officers undertaking control and compliance measures when a ship is in, or is intending to enter, port. The documents which are required to be available for inspection include:

- **.1** the original of the valid ISSC or Interim ISSC;
- **.2** the current CSR and any amendment form;
- **.3** the certificates of proficiency for the SSO and shipboard personnel with designated security duties;
- **.4** parts of the SSP subject to authorization being received from the ship's Administration; and
- **.5** all DoS that the ship has agreed during the period covered by the ship's last 10 ports of call.

4.10.8 Information on the current CSR and any amendment form should include:

- **.1** the Administration, Government or RSO that issued the valid ISSC or Interim ISSC; or
- **.2** if different from above, the body that carried out the verification on which the certificate was issued.

4.10.9 Experience to date indicates that security-related deficiencies represent around 3–5% of the total number of deficiencies found on SOLAS ships, with the vast majority being safety-related.

4.11 Guidelines for non-SOLAS vessels

Introduction

4.11.1 As mentioned in paragraph 4.2.2, there is no requirement under the Maritime Security Measures for Contracting Governments to extend their application to non-SOLAS vessels. However, it has been generally recognized that voluntary application of security practices and principles contained in these Measures represents a desirable goal, one that helps to strengthen the overall maritime security framework.

4.11.2 The following sections provide general guidance that is relevant for all types of non-SOLAS vessels. Appendix 4.11 – General information on security practices for all non-SOLAS vessel operators, lists security practices for all non-SOLAS vessels as well as specific practices that are relevant to the following four types of non-SOLAS vessels:

- **.1** commercial non-passenger and special-purpose vessels
- **.2** passenger vessels
- **.3** fishing vessels
- **.4** pleasure craft.

General guidance

4.11.3 The implementation of appropriate security measures should be governed by the results of a risk assessment.

4.11.4 Operators of non-SOLAS vessels should consider maintaining an appropriate level of security awareness and incident response capability on-board their vessels by:

.1 providing all on-board personnel with information on how to reach appropriate officials and authorities in the event of security problems or if suspicious activity is observed. This information should include contact information for the officials responsible for emergency response, the national response centre(s) (if appropriate) and any authorities that may need to be notified;

.2 implementing security initiatives developed by national authorities with respect to education, information-sharing, co-ordination and outreach programmes;

.3 promoting links with Administrations' maritime security services;

.4 establishing a workplace culture that recognizes the need to balance security requirements with both the safe and efficient operation of the vessel and the rights and welfare of seafarers;

.5 developing security training policies and procedures to ensure that all personnel (including passengers, where appropriate) are familiar with basic security measures that are applicable;

.6 recommending basic security familiarization training for crew members, enabling them to have the capability to respond to security threats. In higher-risk environments, this training should cover the competencies required to implement any security measures that are in place.

4.11.5 Operators may also wish to adopt hiring practices such as background checks. However, when such practices are in place, it is important for there to be:

.1 provisions allowing seafarers and other workers to appeal adverse employment determinations based on disputed background information; and

.2 adequate protections for workers' rights to privacy.

4.11.6 Non-SOLAS vessels on international voyages may be required to declare arrival and departure information for purposes of obtaining a port clearance from the relevant authorities. This declaration may be required within a specified period, as determined by local authorities, following arrival and/or prior to departure. The information to be submitted may include the particulars of vessel, date/time of arrival, position in port, particulars of master/owner/shipping line/agent, purpose of call, amount of cargo on board, passenger and crew list, and emergency contact numbers.

4.11.7 Operators of non-SOLAS vessels on international voyages may be encouraged by their Administration to fit automated tracking equipment on their vessels. The benefits of such a system could include enhanced safety and security; more rapid emergency response to maritime accidents and casualties; better and more effective SAR capabilities; and better control of smuggling, human-trafficking attempts, and illegal, unregulated or unreported fishing.

4.11.8 Non-SOLAS vessel operators should be aware of the key aspects of the Maritime Security Measures relevant to their vessels, including:

.1 communication of changes in security levels and implications for their operations;

.2 requirements for interacting with ships and port facilities falling under the Maritime Security Measures.

4.11.9 If the operator of the non-SOLAS vessel is required to complete a Declaration of Security with a PFSO or SSO, the following procedures apply:

.1 the SSO or PFSO should contact the non-SOLAS vessel well in advance of the non-SOLAS vessel's interaction with the ship or port facility, giving the master of the non-SOLAS vessel reasonable time to prepare for those security measures that might be required;

.2 the SSO or PFSO should detail the security measures with which the non-SOLAS vessel is being asked to comply, using the appropriate DoS form;

.3 the DoS should be completed and signed by both parties.

4.11.10 It is important that all operators of non-SOLAS vessels are aware of the need to stay a reasonable distance from SOLAS ships when using shared waterways. The appropriate distance will vary due to navigational safety considerations. Non-SOLAS vessels should take care not to undertake any manoeuvres close to the vessel which may give the crew of the SOLAS ship cause for concern. Non-SOLAS vessels are encouraged to clearly indicate their intentions to the crew of SOLAS ships by radiotelephone or other means.

4.11.11 Some Administrations have issued guidance material for non-SOLAS vessels which are fitted with ship security alert systems. Vessel operators should check with their national authority to determine if guidelines have been issued.

Appendix 4.1
Sample of a Declaration of Security form for a ship-to-ship interface

Name of ship A:..

Port of registry:...

IMO Number:...

Name of ship B:..

Port of registry:...

IMO Number:...

This Declaration of Security is valid from .... until ...., for the following activities:

....

....

....

(list the activities with relevant details)

under the following security levels

Security level(s) for ship A: ....

Security level(s) for ship B: ....

Both ships agree to the following security measures and responsibilities to ensure compliance with the relevant requirements of their national maritime security legislation (or, if not enacted, of chapter 5 in part A of the ISPS Code).

	The initials of each SSO or master in these columns indicates that the activity will be done, in accordance with their approved ship security plan, by ship A and/ or ship B	
Activity	**Ship A:**	**Ship B:**
Ensuring the performance of all security duties		
Monitoring restricted areas to ensure that only authorized personnel have access		
Controlling access to ship A		
Controlling access to ship B		
Monitoring of ship A, including areas surrounding the ship		
Monitoring of ship B, including areas surrounding the ship		
Handling of cargo		
Delivery of ship's stores		
Handling unaccompanied baggage		
Controlling the embarkation of persons and their effects		
Ensuring that security communication is readily available between the ships		

The signatories to this agreement certify that security measures and arrangements for both ships during the specified activities meet the relevant provisions of their national maritime security legislation (or, if not enacted, of chapter 5 in part A of the ISPS Code) and will be implemented in accordance with the provisions already stipulated in their approved ship security plan(s) or with specific arrangements agreed to (as set out in the attached annex).

Dated at .... on the ....

Signed for and on behalf of	
Ship A:	Ship B:

(Signature of master or ship security officer) *(Signature of master or ship security officer)*

Name and title of person who signed	
Name:	Name:
Title:	Title

Contact details *(to be completed as appropriate)* *(indicate the telephone numbers, radio channels or frequencies to be used)*	
for ship A:	for ship B:
Master	Master ....
Ship security officer	Ship security officer ....
Company	Company ....
Company security officer	Company security officer ....

Appendix 4.2
Competency matrix for company security officers

Source: MSC/Circ.1154, May 2005

Competence Knowledge requirement	**Method for demonstrating competence** Evaluation criterion
Develop, maintain and supervise the implementation of an SSP • International maritime security policy and responsibilities of Governments, companies and designated persons. • Purpose for and the elements that make up an SSP. • Procedures to be employed in developing, maintaining, and supervising the implementation of, and the submission for approval of, an SSP. • Procedures for the initial and subsequent verification of the ship's compliance. • Maritime security levels and the consequential security measures and procedures aboard ship and in the port facility environment. • Requirements and procedures involved with arranging for internal audits and review of security activities specified in an SSP. • Requirements and procedures for acting upon reports by the SSO to the CSO concerning any deficiencies or non-conformities identified during internal audits, periodic reviews, and security inspections. • Methods and procedures used to modify the SSP. • Security-related contingency plans and the procedures for responding to security threats or breaches of security, including provisions for maintaining critical operations of the ship/port interface. • Maritime security terms and definitions used in the Maritime Security Measures.	**Assessment of evidence obtained from approved training or examination** • Procedures and actions are in accordance with the principles established by the Maritime Security Measures. • Legislative requirements relating to security are correctly identified. • Procedures achieve a state of readiness to respond to changes in security levels. • Communications within the CSO's area of responsibility are clear and understood.
Ensuring security equipment and systems, if any, are properly operated • Various types of security equipment and systems and their limitations.	**Assessment of evidence obtained from approved training or examination** • Procedures and actions are in accordance with the principles established by the Maritime Security Measures. • Procedures achieve a state of readiness to respond to changes in security levels. • Communications within the CSO's area of responsibility are clear and understood.

Competence Knowledge requirement	**Method for demonstrating competence** Evaluation criterion
Assess security risk, threat, and vulnerability • Risk assessment, assessment tools, and procedures for conducting security assessments. • Security assessment documentation, including the DoS. • Techniques used to circumvent security measures. • Recognition, on a non-discriminatory basis, of persons posing potential security risks. • Recognition of weapons, dangerous substances, and devices and awareness of the damage they can cause. • Crowd management and control techniques, where appropriate. • Handling sensitive security-related information and security-related communications. • Methods for implementing and co-ordinating searches. • Methods for physical searches and non-intrusive inspections.	**Assessment of evidence obtained from approved training or from examination and practical demonstration of ability to conduct physical searches and non-intrusive inspections** • Procedures and actions are in accordance with the principles established by the Maritime Security Measures. • Procedures achieve a state of readiness to respond to changes in the maritime security levels. • Communications within the CSO's area of responsibility are clear and understood.
Ensure appropriate security measures are implemented and maintained • Requirements and methods for designating and monitoring restricted areas. • Methods for controlling access to the ship and to restricted areas on board ship. • Methods for effective monitoring of deck areas and areas surrounding the ship. • Security aspects relating to the handling of cargo and ship's stores with other shipboard personnel and relevant PFSOs. • Methods for controlling the embarkation, disembarkation and access, while on board, of persons and their effects.	**Assessment of evidence obtained from approved training or examination** • Procedures and actions are in accordance with the principles established by the Maritime Security Measures • Procedures achieve a state of readiness to respond to changes in the maritime security levels. • Communications within the CSO's area of responsibility are clear and understood.
Encourage security awareness and vigilance • Training, drill and exercise requirements under relevant conventions and codes. • Methods for enhancing security awareness and vigilance on board. • Methods for assessing the effectiveness of drills and exercises. • Instructional techniques for security training and education.	**Assessment of evidence obtained from approved training or examination** • Procedures and actions are in accordance with the principles established by the SOLAS security measures. • Communications within the CSO's area of responsibility are clear and understood.

Appendix 4.3

Competency matrix for ship security officers

Source: Section A-VI/5 of the STCW Code, as amended, August 2010

Competence Knowledge requirement	**Method for demonstrating competence** Evaluation criterion
Maintain and supervise the implementation of a ship security plan • International maritime security policy and responsibilities of Governments, companies and designated persons, including elements that may relate to piracy and armed robbery. • Purpose for and the elements that make up an SSP, related procedures and maintenance of records, including those that may relate to piracy and armed robbery. • Procedures to be employed in implementing an SSP and reporting of security incidents. • Maritime security levels and the consequential security measures and procedures aboard ship and in the port facility environment. • Requirements and procedures for conducting internal audits, on-scene inspections, surveys and control and monitoring of security activities specified in an SSP. • Requirements and procedures for reporting to the CSO any deficiencies and non-conformities identified during internal audits, periodic reviews and security inspections. • Methods and procedures used to modify the SSP. • Security-related contingency plans and the procedures for responding to security threats or breaches of security, including provisions for maintaining critical operations of the ship/port interface and elements that may relate to piracy and armed robbery. • Maritime security terms and definitions, including elements that may relate to piracy and armed robbery.	**Assessment of evidence obtained from approved training or examination** • Procedures and actions are in accordance with the principles established by the Maritime Security Measures. • Legislative requirements relating to security are correctly identified. • Procedures achieve a state of readiness to respond to changes in security levels. • Communications within the SSO's area of responsibility are clear and understood.
Assess security risk, threat, and vulnerability • Risk assessment and assessment tools. • Security-assessment documentation, including the DoS. • Techniques used to circumvent security measures, including those used by pirates and armed robbers. • Recognition, on a non-discriminatory basis, of persons posing potential security risks. • Recognition of weapons, dangerous substances, and devices and awareness of the damage they can cause. • Crowd management and control techniques, where appropriate. • Handling sensitive security-related information and security-related communications. • Implementing and co-ordinating searches. • Methods for physical searches and non-intrusive inspections.	**Assessment of evidence obtained from approved training or approved experience and examination, including practical demonstration of competence to conduct physical searches and non-intrusive inspections** • Procedures and actions are in accordance with the principles established by the Maritime Security Measures. • Procedures achieve a state of readiness to respond to changes in security levels. • Communications within the SSO's area of responsibility are clear and understood.

Competence Knowledge requirement	**Method for demonstrating competence** Evaluation criterion
Undertake regular inspections of the ship to ensure that appropriate security measures are implemented and maintained • Requirements for designating and monitoring restricted areas. • Controlling access to the ship and to restricted areas on board ship. • Methods for effective monitoring of deck areas and areas surrounding the ship. • Security aspects relating to the handling of cargo and ship's stores with other shipboard personnel and relevant PFSOs. • Methods for controlling the embarkation, disembarkation and access while on board of persons and their effects.	**Assessment of evidence obtained from approved training or examination** • Procedures and actions are in accordance with the principles established by the Maritime Security Measures • Procedures achieve a state of readiness to respond to changes in the security levels. • Communications within the SSO's area of responsibility are clear and understood.
Ensure that security equipment and systems, if any, are properly operated, tested and calibrated • Various types of security equipment and systems and their limitations, including those that could be used in case of attacks by pirates and armed robbers. • Procedures, instructions and guidance on the use of SSASs. • Methods for testing, calibrating and maintaining security systems and equipment, particularly whilst at sea.	**Assessment of evidence obtained from approved training or examination** • Procedures and actions are in accordance with the principles established by the Maritime Security Measures • Procedures achieve a state of readiness to respond to changes in the security levels. • Communications within the SSO's area of responsibility are clear and understood.
Encourage security awareness and vigilance • Training, drill and exercise requirements under relevant conventions and codes and IMO circulars including those relevant to anti-piracy and anti-armed robbery. • Methods for enhancing security awareness and vigilance on board. • Methods for assessing the effectiveness of drills and exercises.	**Assessment of evidence obtained from approved training or examination.** • Procedures and actions are in accordance with the principles established by the Maritime Security Measures. • Communications within the SSO's area of responsibility are clear and understood.

Appendix 4.4

Competency matrix for shipboard personnel with designated security duties

Source: Section A-VI/6 of the STCW Code, as amended, August 2010

Competence Knowledge requirement	**Method for demonstrating competence** Evaluation criterion
Maintain the conditions set out in a ship security plan • Maritime security terms and definitions, including elements that may relate to piracy and armed robbery. • International maritime security policy and responsibilities of Governments, companies and persons, including elements that may relate to piracy and armed robbery. • Maritime security levels and their impact on security measures and procedures aboard ship and in the port facilities. • Security reporting procedures. • Procedures and requirements for drills and exercises under relevant conventions, codes and IMO circulars, including those that may relate to piracy and armed robbery. • Procedures for conducting inspections and surveys and for the control and monitoring of security activities specified in an SSP. • Security-related contingency plans and the procedures for responding to security threats or breaches of security, including provisions for maintaining critical operations of the ship/port interface and those that may relate to piracy and armed robbery.	**Assessment of evidence obtained from approved instruction or during attendance at an approved course** • Procedures and actions are in accordance with the principles established by the Maritime Security Measures. • Legislative requirements relating to security are correctly identified. • Communications within the area of responsibility are clear and understood.
Recognition of security risks and threats • Security documentation, including the DoS. • Techniques used to circumvent security measures, including those used by pirates and armed robbers. • Recognition of potential security threats. • Recognition of weapons, dangerous substances, and devices and awareness of the damage they can cause. • Crowd management and control techniques, where appropriate. • Handling security-related information and security-related communications. • Methods for physical searches and non-intrusive inspections.	**Assessment of evidence obtained from approved instruction or during attendance at an approved course** • Procedures and actions are in accordance with the principles established by the Maritime Security Measures.
Undertake regular security inspections of the ship • Techniques for monitoring restricted areas. • Controlling access to the ship and to restricted areas on board ship. • Methods for effective monitoring of deck areas and areas surrounding the ship. • Inspection methods relating to the cargo and ship's stores. • Methods for controlling the embarkation, disembarkation and access while on board of persons and their effects.	**Assessment of evidence obtained from approved instruction or during attendance at an approved course** • Procedures and actions are in accordance with the principles established by the Maritime Security Measures.

Competence Knowledge requirement	**Method for demonstrating competence** Evaluation criterion
Proper usage of security equipment and systems, if any • Various types of security equipment and systems and their limitations, including those that could be used in case of attacks by pirates and armed robbers. • The need for testing, calibrating and maintaining security systems and equipment, particularly whilst at sea.	**Assessment of evidence obtained from approved instruction or during attendance at an approved course** • Equipment and systems operations are carried out in accordance with established equipment operating instructions and taking into account the limitations of the equipment and systems. • Procedures and actions are in accordance with the principles established by the Maritime Security Measures.

Appendix 4.5

Competency matrix: Security awareness for all shipboard personnel

Source: Section A-VI/6 of the STCW Code, as amended, August 2010

Competence Knowledge requirement	**Method for demonstrating competence** Evaluation criterion
Contribute to the enhancement of maritime security through heightened awareness • Maritime security terms and definitions, including elements that may relate to piracy and armed robbery. • International maritime security policy and responsibilities of Governments, companies and persons. • Maritime security levels and their impact on security measures and procedures aboard ship and in port facilities. • Security reporting procedures. • Security-related contingency plans.	**Assessment of evidence obtained from approved instruction or during attendance at an approved course** • Requirements relating to enhanced maritime security are correctly identified.
Recognition of security threats • Techniques used to circumvent security measures. • Recognition of potential security threats, including elements that may relate to piracy and armed robbery. • Recognition of weapons, dangerous substances, and devices and awareness of the damage they can cause. • Handling security-related information and security-related communications.	**Assessment of evidence obtained from approved instruction or during attendance at an approved course** • Maritime security threats are correctly identified.
Understanding of the need for and methods of maintaining security awareness and vigilance • Training, drill and exercise requirements under relevant conventions, codes and IMO circulars including those relevant for anti-piracy and anti-armed robbery.	**Assessment of evidence obtained from approved instruction or during attendance at an approved course** • Requirements relating to enhanced maritime security are correctly identified.

Appendix 4.6

Standard data set of security-related pre-arrival information

Source: MSC.1/Circ.1305, June 2009

1 Particulars of the ship and contact details

1.1 IMO Number*

1.2 Name of ship*

1.3 Port of registry*

1.4 Flag State*

1.5 Type of ship

1.6 Call Sign

1.7 Inmarsat call numbers (if available)

1.8 Gross tonnage

1.9 Name of Company*

1.10 IMO Company identification number*

1.11 Name and 24-hour contact details of the company security officer (or designated duty officer)

2 Port and port facility information

2.1 Port of arrival and port facility where the ship is to berth, if known

2.2 Expected date and time of arrival of the ship in port

2.3 Primary purpose of call

3 Information required by SOLAS regulation XI-2/9.2.1

3.1 The ship is provided with a valid:

– International Ship Security Certificate ☐ Yes ☐ No

– Interim International Ship Security Certificate ☐ Yes ☐ No

3.2 The certificate indicated in 3.1 has been issued by (*enter name of the Contracting Government** *or the recognized security organization**) and which expires on (*enter date of expiry*).

3.3 If the ship is not provided with a valid International Ship Security Certificate or a valid Interim International Ship Security Certificate, explain why.

3.4 Does the ship have an approved ship security plan on board? ☐ Yes ☐ No

3.5 Current security level: ☐

3.6 Location of the ship at the time the report is made.

* No need to provide these details if a copy of the Continuous Synopsis Record has been submitted.

3.7 List the last ten calls, in chronological order with the most recent call first, at port facilities at which the ship conducted ship/port interface together with the security level at which the ship operated:

	Date			
No.*	**From**	**To**	**Port, country, port facility and UN LOCODE (if available)**	**Security level**
10				
9				
8				
7				
6				
5				
4				
3				
2				
1				

3.8 Did the ship, during the period specified in 3.7, take any special or additional security measures, beyond those specified in the approved ship security plan? ☐ Yes ☐ No

3.9 If the answer to 3.8 is YES, for each of such occasions please indicate the special or additional security measures which were taken by the ship:

	Date		
No.	**From**	**To**	**Special or additional security measures**

3.10 List the ship-to-ship activities, in chronological order with the most recent ship-to-ship activity first, which have been carried out during the period specified in 3.7:

☐ Not applicable

	Date			
No.	**From**	**To**	**Location or latitude and longitude**	**Ship-to-ship activity**

3.11 Have the ship security procedures, specified in the approved ship security plan, been maintained during each of the ship-to-ship activities specified in 3.10? ☐ Yes ☐ No

3.12 If the answer to 3.11 is NO, identify the ship-to-ship activities for which the ship security procedures were not maintained and indicate, for each, the security measures which were applied in lieu:

	Date			
No.	**From**	**To**	**Security measures applied**	**Ship-to-ship activity**

3.13 Provide a general description of cargo aboard the ship:

3.14 Is the ship carrying any dangerous substances (i.e., those covered by the International Maritime Dangerous Goods Code) as cargo? ☐ Yes ☐ No

* Port of call No. 10 is the last one before the port at which entry is being sought.

3.15 If the answer to 3.14 is YES, provide details or attach a copy of the Dangerous Goods Manifest (IMO FAL Form 7)

3.16 A copy of the ship's Crew List (IMO FAL Form 5) is attached ☐

3.17 A copy of the ship's Passenger List (IMO FAL Form 6) is attached ☐

4 Other security-related information

4.1 Is there any security-related matter you wish to report? ☐ Yes ☐ No

4.2 If the answer to 4.1 is YES, provide details (e.g., carriage of stowaways or persons rescued at sea).

5 Agent of the ship at the intended port of arrival

5.1 Name and contact details (telephone number) of the agent of the ship at the intended port of arrival:

6 Identification of the person providing the information

6.1 Name:

6.2 Title or position (master, SSO, CSO or ship's agent at intended port of arrival):

6.3 Signature:

This report is dated at (enter place) on (enter time and date).

Appendix 4.7
Example of a ship security assessment and plan approval process

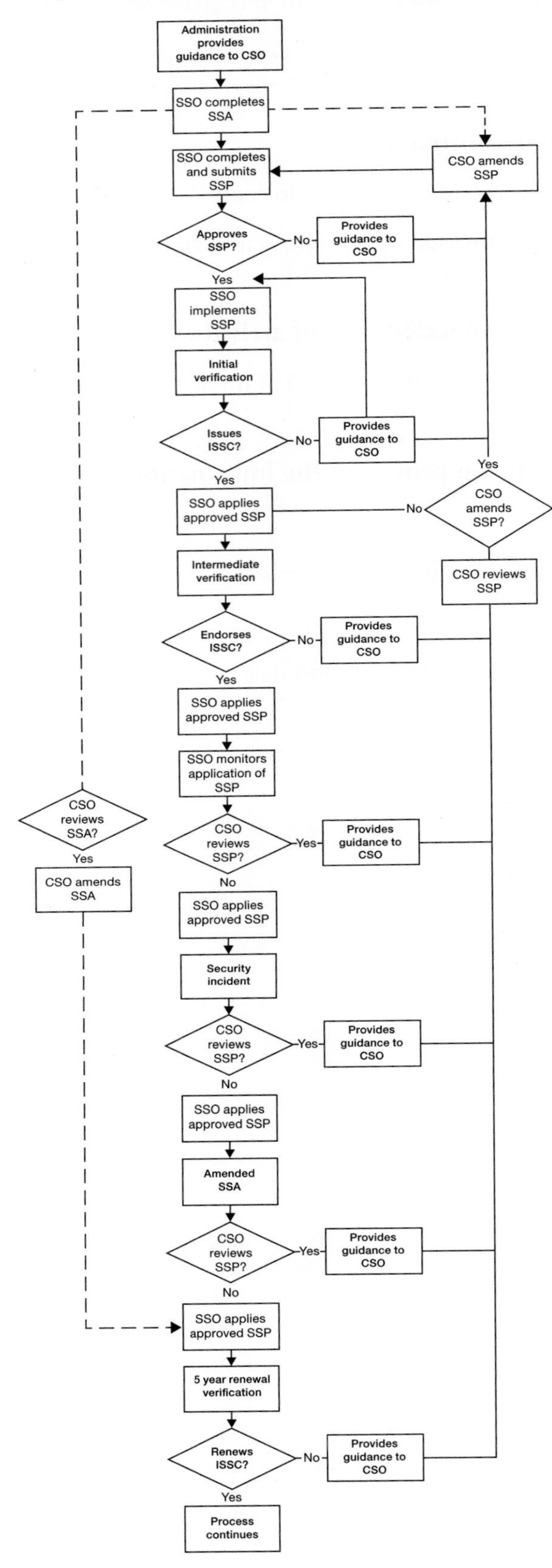

Appendix 4.8

Examples of internet sources of guidance material on preparing and validating ship security plans

1 Australian Government Department of Infrastructure and Transport, Guide to Preparing a Ship Security Plan, April 2009.

Refer to: www.infrastructure.gov.au/transport/security/maritime/

This 29-page guide has been developed to provide ship operators covered by the Maritime Transport and Offshore Securities Act 2003 with a plan template so as to assist them with meeting all the requirements of an approved plan. It also contains a chart showing the plan approval process. Also, the template may be downloaded in WORD format.

2 United Kingdom, Department for Transport, Model Ship Security Plan, September 2008.

Refer to: www.dft.gov.uk/pgr/security/maritime

This 31-page document is a template showing CSOs and SSOs how to compile and submit their SSPs, including a four-page template for the accompanying SSA Report. Also, the template may be downloaded in WORD format.

3 Commonwealth of Dominica Maritime Registry.

Refer to: www.dominica-registry.com

This site provides access to the following three documents, including a plan template and a checklist, both of which may be downloaded in WORD format:

- Model ship security plan guidance to accompany the security plan template, last modified in June 2004. The 38-page document is in the form of five guides:
  - Guide 1 – Developing threat assessments
  - Guide 2 – Ship initial security assessment (survey)
  - Guide 3 – How to identify and mitigate security vulnerabilities
  - Guide 4 – Guidance for establishing protective measures
  - Guide 5 – Developing final security assessment
- Ship security plan template, last modified July 2006. The 95-page template includes an SSO's security assessment form as an appendix; and
- Aid for reviewing compliance for ship security plans. The 15-page checklist was last modified in April 2008.

Appendix 4.9
Implementation checklist for ship security personnel

Source: MSC.1/Circ.1193, May 2006

This checklist may be used by ship security personnel to examine the status of implementation of the Special Measures. The heading of each section is taken directly from paragraph A/7.2 of the ISPS Code.

Completion of the following section is recommended before using the checklist. It can be used to establish an overview of the ship's operations.

1 Company and ship overview

Name of Administration	
Name of Company	
Name of ship	
IMO Ship identification Number	
Name of CSO	
Name of SSO	
Number of ships operated by the company	
Number of ships for which the CSO is responsible	

2 Total manning of the ship and crew with security duties on board at the time of this assessment

Total number of crew members	
Number of crew with security duties	

3 Ship security information in the last 12 months

Number of crew members assigned on first time to the ship	
Number of different SSOs	
Number of changes in the security level	
Number of security incidents	
Number of breaches of security	

4 Security agreements and arrangements

Is the ship operating between port facilities covered by an alternative security agreement? If "Yes", provide relevant details.	
Has the ship implemented any equivalent security arrangements allowed by the maritime Administration? If "Yes", provide relevant details.	
Is the ship operating under any temporary security measures? If "Yes", have these been approved or authorized by the maritime Administration? If "Yes", provide relevant details.	

Guidance:

For each question, one of the 'Yes/No/Other' boxes should be ticked. Whichever one is used, the 'Comments' box provides space for amplification.

If the 'Yes' box is ticked, but the Measures/procedures are not documented in the SSP, a short description of them should be included in the 'Comments' box. The 'Yes' box should be ticked only if all procedures and measures are in place. The 'Comments' box may also be used to indicate when procedures were last reviewed and measures tested (e.g., drills and exercises).

If the 'No' box is ticked, an explanation of why not should be included in the 'Comments' box along with details of any measures or procedures in place. Suggested actions should be recorded in the 'Recommendations' section at the end of the checklist.

If the 'Other' box is ticked, a short description should be provided in the 'Comments' box (e.g., it could include instances where alternative measures/procedures or equivalent arrangements have been implemented). If the reason is due to the question not being applicable, then it should be recorded in the 'Comments' box as "not applicable".

If there is not enough space in the 'Comments' box, the explanation should be continued on a separate page (with the relevant question number, and in the case of questions with multiple options, the option added as a reference).

The 'Recommendations' boxes at the end of the checklist should be used to record any identified deficiencies and how these could be mitigated. A schedule for their implementation should be included.

The 'Outcomes' box at the end of the checklist should be used to provide a brief record of the assessment process. Along with the comments in the 'Recommendations' boxes, they form the basis for updating the SSP.

1 Ensuring the performance of all ship security duties (ISPS Code, sections A/7.2.1, A/7.3 and A/9.4)

Part A

.1 Does the ship's means of ensuring the performance of all security duties meet the requirements set out in the SSP for security levels 1 and 2? (ISPS Code, section A/7.2.1)	Yes ☐	No ☐	Other ☐
Comments:			

.2 Has the ship established measures to prevent weapons, dangerous substances and devices intended for use against persons, ships or ports and the carriage of which is not authorized from being taken on board the ship? (ISPS Code, section A/9.4.1)	Yes ☐	No ☐	Other ☐
Comments:			

.3 Has the ship established procedures for responding to security threats or breaches of security, including provisions for maintaining critical operations of the ship or ship/port interface? (ISPS Code, section A/9.4.4)	Yes ☐	No ☐	Other ☐
Comments:			

.4 Has the ship established procedures for responding to any security instructions Contracting Governments may give at security level 3? (ISPS Code, section A/9.4.5)	Yes ☐	No ☐	Other ☐
Comments:			

.5 Has the ship established procedures for evacuation in case of security threats or breaches of security? (ISPS Code, section A/9.4.6)	Yes ☐	No ☐	Other ☐
Comments:			

.6 Have the duties of shipboard personnel assigned security responsibilities and other shipboard personnel on security aspects been specified? (ISPS Code, section A/9.4.7)	Yes ☐	No ☐	Other ☐
Comments:			

.7 Have procedures been established for auditing the security activities of the ship? (ISPS Code, section A/9.4.8)	Yes ☐	No ☐	Other ☐
Comments:			

.8 Has the ship established procedures for interfacing with port facility security activities? (ISPS Code, section A/9.4.10)	Yes ☐	No ☐	Other ☐
Comments:			

.9 Have procedures been established for the periodic review of the SSP and for its updating? (ISPS Code, section A/9.4.11)	Yes ☐	No ☐	Other ☐
Comments:			

.10 Has the ship established procedures for reporting security incidents? (ISPS Code, section A/9.4.12)	Yes ☐	No ☐	Other ☐
Comments:			

Part B – Organization and performance of ship security duties

.11 Has the ship implemented the organizational structure of security for the ships detailed in the SSP? (ISPS Code, paragraph B/9.2.1)	Yes ☐	No ☐	Other ☐
Comments:			

.12 Has the ship established the relationships with the Company, port facilities, other ships and relevant authorities with security responsibilities detailed in the SSP? (ISPS Code, paragraph B/9.2.2)	Yes ☐	No ☐	Other ☐
Comments:			

.13 Has the ship established the communication systems to allow effective continuous communication within the ship and between the ship and others, including port facilities, detailed in the SSP? (ISPS Code, paragraph B/9.2.3)	Yes ☐	No ☐	Other ☐
Comments:			

.14 Has the ship implemented the basic security measures for security level 1, both operational and physical, that will always be in place, detailed in the SSP? (ISPS Code, paragraph B/9.2.4)	Yes ☐	No ☐	Other ☐
Comments:			

.15 Has the ship implemented the additional security measures that will allow the ship to progress without delay to security level 2 and, when necessary, to security level 3 detailed in the SSP? (ISPS Code, paragraph B/9.2.5)	Yes ☐	No ☐	Other ☐
Comments:			

.16 Has the ship established procedures for regular review, or audit, of the SSP and for its amendment in response to experience or changing circumstances? (ISPS Code, paragraph B/9.2.6)	Yes ☐	No ☐	Other ☐
Comments:			

.17 Has the ship established reporting procedures to the appropriate Contracting Government's contact points? (ISPS Code, paragraph B/9.2.7)	Yes ☐	No ☐	Other ☐
Comments:			

.18 Has the ship established the duties and responsibilities of all shipboard personnel with a security role? (ISPS Code, paragraph B/9.7.1)	Yes ☐	No ☐	Other ☐
Comments:			

.19 Has the ship established the procedures or safeguards necessary to allow continuous communications to be maintained at all times? (ISPS Code, paragraph B/9.7.2)	Yes ☐	No ☐	Other ☐
Comments:			

.20 Has the ship established the procedures needed to assess the continuing effectiveness of security procedures and any security and surveillance equipment and systems, including procedures for identifying and responding to equipment or systems failure or malfunction? (ISPS Code, paragraph B/9.7.3)	Yes ☐	No ☐	Other ☐
Comments:			

.21 Has the ship established procedures and practices to protect security-sensitive information held in paper or electronic format? (ISPS Code, paragraph B/9.7.4)	Yes ☐	No ☐	Other ☐
Comments:			

.22 Has the ship established the type and maintenance requirements of security and surveillance equipment and systems, if any? (ISPS Code, paragraph B/9.7.5)	Yes ☐	No ☐	Other ☐
Comments:			

.23 Has the ship established the procedures to ensure timely submission and assessment of reports relating to possible breaches of security or security concerns? (ISPS Code, paragraph B/9.7.6)	Yes ☐	No ☐	Other ☐
Comments:			

.24 Has the ship put in place procedures to establish, maintain and update an inventory of any dangerous goods or hazardous substances carried on board, including their location? (ISPS Code, paragraph B/9.7.7)	Yes ☐	No ☐	Other ☐
Comments:			

2 Controlling access to the ship

Part A

.1 Does the ship's means of controlling access to the ship meet the requirements set out in the SSP for security level 1? (ISPS Code, section A/7.2.2)	Yes ☐	No ☐	Other ☐
Comments:			

.2 Has the ship established measures to prevent unauthorized access? (ISPS Code, section A/9.4.3)	Yes ☐	No ☐	Other ☐
Comments:			

Part B – Access to the ship

.3 Has the ship established security measures covering all means of access to the ship identified in the SSA? (ISPS Code, paragraph B/9.9)	Yes	No	Other
A Access ladders	A ☐	☐	☐
B Access gangways	B ☐	☐	☐
C Access ramps	C ☐	☐	☐
D Access doors, sidescuttles, windows and ports	D ☐	☐	☐
E Mooring lines and anchor chains	E ☐	☐	☐
F Cranes and hoisting gear	F ☐	☐	☐
Comments:			

.4 Has the ship identified appropriate locations where access restrictions or prohibitions should be applied for each of the security levels? (ISPS Code, paragraph B/9.10)	Yes ☐	No ☐	Other ☐
Comments:			

.5 Has the ship established for each security level the means of identification required to allow access to the ship and for individuals to remain on the ship without challenge? (ISPS Code, paragraph B/9.11)	Yes	No	Other
A Security level 1	A ☐	☐	☐
B Security level 2	B ☐	☐	☐
C Security level 3	C ☐	☐	☐
Comments:			

.6 Has the ship established the frequency of application of any access controls? (ISPS Code, paragraph B/9.13)	Yes ☐	No ☐	Other ☐
Comments:			

Security level 1

.7 Has the ship established security measures to check the identity of all persons seeking to board the ship and confirming their reasons for doing so? (ISPS Code, paragraph B/9.14.1)	Yes ☐	No ☐	Other ☐
Comments:			

.8 Has the ship established procedures to liaise with the port facility to ensure that designated secure areas are established in which inspections and searching of persons, baggage (including carry-on items), personal effects, vehicles and their contents can take place? (ISPS Code, paragraph B/9.14.2)	Yes ☐	No ☐	Other ☐
Comments:			

.9 Has the ship identified access points that should be secured or attended to prevent unauthorized access? (ISPS Code, paragraph B/9.14.6)	Yes ☐	No ☐	Other ☐
Comments:			

.10 Has the ship established security measures to secure, by locking or other means, access to unattended spaces adjoining areas to which passengers and visitors have access? (ISPS Code, paragraph B/9.14.7)	Yes ☐	No ☐	Other ☐
Comments:			

.11 Has the ship provided security briefings to all ship personnel on possible threats, the procedures for reporting suspicious persons, objects or activities and the need for vigilance? (ISPS Code, paragraph B/9.14.8)	Yes ☐	No ☐	Other ☐
Comments:			

.12 Has the ship established the frequency of searches, including random searches, of all those seeking to board the ship? (ISPS Code, paragraph B/9.15)	Yes ☐	No ☐	Other ☐
Comments:			

Security level 2

.13 Has the ship limited the number of access points to the ship, identifying those to be closed and the means of adequately securing them? (ISPS Code, paragraph B/9.16.2)	Yes ☐	No ☐	Other ☐
Comments:			

.14 Has the ship established a restricted area on the shore side of the ship, in close co-operation with the port facility? (ISPS Code, paragraph B/9.16.4)	Yes ☐	No ☐	Other ☐
Comments:			

.15 Has the ship arrangements to escort visitors on the ship? (ISPS Code, paragraph B/9.16.6)	Yes ☐	No ☐	Other ☐
Comments:			

.16 Has the ship provided additional specific security briefings to all ship personnel on any identified threats, re-emphasizing the procedures for reporting suspicious persons, objects, or activities and stressing the need for increased vigilance? (ISPS Code, paragraph B/9.16.7)	Yes ☐	No ☐	Other ☐
Comments:			

.17 Has the ship established procedures for carrying out a full or partial search of the ship? (ISPS Code, paragraph B/9.16.8)	Yes ☐	No ☐	Other ☐
Comments:			

3 Controlling the embarkation of persons and their effects

Part A

.1 Do the ship's measures for controlling the embarkation of persons and their effects meet the requirements set out in the SSP for security level 2? (ISPS Code, section A/7.3)	Yes ☐	No ☐	Other ☐
Comments:			

Part B – Access to the ship

Security level 1

.2 Has the ship established procedures to liaise with the port facility to ensure that vehicles destined to be loaded on board car carriers, ro–ro and other passenger ships are subjected to search prior to loading? (ISPS Code, paragraph B/9.14.3)	Yes ☐	No ☐	Other ☐
Comments:			

.3 Has the ship established security measures to segregate checked persons and their personal effects from unchecked persons and their personal effects? (ISPS Code, paragraph B/9.14.4)	Yes ☐	No ☐	Other ☐
Comments:			

.4 Has the ship established security measures to segregate embarking from disembarking passengers? (ISPS Code, paragraph B/9.14.5)	Yes ☐	No ☐	Other ☐
Comments:			

Security level 2

.5 Has the ship increased the frequency and detail of searches of persons, personal effects, and vehicles being embarked or loaded onto the ship? (ISPS Code, paragraph B/9.16.5)	Yes ☐	No ☐	Other ☐
Comments:			

Part B – Handling unaccompanied baggage

.6 Has the ship established security measures to be applied to ensure that unaccompanied baggage is identified and subject to appropriate screening, including searching, before it is accepted on board? (ISPS Code, paragraph B/9.38)	Yes ☐	No ☐	Other ☐
Comments:			

Security level 1

.7 Has the ship established security measures to be applied when handling unaccompanied baggage to ensure that unaccompanied baggage is screened or searched up to and including 100%, which may include use of x-ray screening? (ISPS Code, paragraph B/9.39)	Yes ☐	No ☐	Other ☐
Comments:			

Security level 2

.8 Has the ship established additional security measures to be applied when handling unaccompanied baggage, which should include 100% x-ray screening of all unaccompanied baggage? (ISPS Code, paragraph B/9.40)	Yes ☐	No ☐	Other ☐
Comments:			

4 Monitoring of restricted areas

Part A

.1 Do the ship's measures for monitoring access to restricted areas, to ensure that only authorized persons have access, meet the requirements set out in the SSP for security level 1? (ISPS Code, section A/7.2.4)	Yes ☐	No ☐	Other ☐
Comments:			

.2 Have restricted areas been identified and measures put in place to prevent unauthorized access to them? (ISPS Code, section A/9.4.2)	Yes ☐	No ☐	Other ☐
Comments:			

Part B – Restricted areas on the ship

.3 Has the ship clearly established policies and practices to control access to all restricted areas? (ISPS Code, paragraph B/9.19)	Yes ☐	No ☐	Other ☐
Comments:			

.4 Has the ship clearly marked all restricted areas, indicating that access to the area is restricted and that unauthorized presence in the area constitutes a breach of security? (ISPS Code, paragraph B/9.20)	Yes ☐	No ☐	Other ☐
Comments:			

.5 Which of the following have been identified as restricted areas? (ISPS Code, paragraph B/9.21)	Yes	No	Other
A Navigation bridge, machinery spaces of category A and other control stations	A ☐	☐	☐
B Spaces containing security and surveillance equipment and systems and their controls and lighting system controls	B ☐	☐	☐
C Ventilation and air-conditioning systems and other similar spaces	C ☐	☐	☐
D Spaces with access to potable water tanks, pumps or manifolds	D ☐	☐	☐
E Spaces containing dangerous goods or hazardous substances	E ☐	☐	☐
F Spaces containing cargo pumps and their controls	F ☐	☐	☐
G Cargo spaces and spaces containing ship's stores	G ☐	☐	☐
H Crew accommodation	H ☐	☐	☐
I Any other areas	I ☐	☐	☐
Comments:			

Security level 1

.6 Which of the following security measures have been applied to restricted areas on the ship? (ISPS Code, paragraph B/9.22)	Yes	No	Other
A Locking or securing access points	A ☐	☐	☐
B Using surveillance equipment to monitor the areas	B ☐	☐	☐
C Using guards or patrols	C ☐	☐	☐
D Using automatic intrusion-detection devices to alert the ship's personnel of unauthorized access	D ☐	☐	☐
Comments:			

Security level 2

.7 Which of the following additional security measures have been applied to restricted areas on the ship? (ISPS Code, paragraph B/9.23)	Yes	No	Other
A Establishing restricted areas adjacent to access points	A ☐	☐	☐
B Continuously monitoring surveillance equipment	B ☐	☐	☐
C Dedicating additional personnel to guard and patrol restricted areas	C ☐	☐	☐
Comments:			

5 Monitoring of deck areas and areas surrounding the ship

Part A

.1 Does the ship's means of monitoring deck areas and areas surrounding the ship meet the requirements identified in the SSP for security level 1? (ISPS Code, section A/7.2)	Yes ☐	No ☐	Other ☐
Comments:			

Part B – Access to the ship

Security level 2

.2 Has the ship assigned additional personnel to patrol deck areas during silent hours to deter unauthorized access? (ISPS Code, paragraph B/9.16.1)	Yes ☐	No ☐	Other ☐
Comments:			

.3 Has the ship established security measures to deter waterside access to the ship, including, for example, in liaison with the port facility, provision of boat patrols? (ISPS Code, paragraph B/9.16.3)	Yes ☐	No ☐	Other ☐
Comments:			

Part B – Monitoring the security of the ship (ISPS Code, paragraphs B/9.42 to B/9.48)

.4 Which of the following monitoring capabilities have been established by the ship to monitor the ship, the restricted areas on board and areas surrounding the ship? (ISPS Code, paragraph B/9.42)	Yes	No	Other
A Lighting	A ☐	☐	☐
B Watchkeepers, security guards and deck watches, including patrols	B ☐	☐	☐
C Automatic intrusion-detection devices and surveillance equipment	C ☐	☐	☐
Comments:			

.5 Do any automatic intrusion-detection devices on the ship activate an audible and/or visual alarm at a location that is continuously attended or monitored? (ISPS Code, paragraph B/9.43)	Yes ☐	No ☐	Other ☐
Comments:			

.6 Has the ship established the procedures and equipment needed at each security level and the means of ensuring that monitoring equipment will be able to perform continually, including consideration of the possible effects of weather conditions or power disruptions? (ISPS Code, paragraph B/9.44)	Yes ☐	No ☐	Other ☐
Comments:			

Security level 1

.7 Has the ship established the security measures to be applied, which may be a combination of lighting, watchkeepers, security guards or the use of security and surveillance equipment to allow ship's security personnel to observe the ship in general, and barriers and restricted areas in particular? (ISPS Code, paragraph B/9.45)	Yes ☐	No ☐	Other ☐
Comments:			

.8 Are the ship's deck and access points illuminated during hours of darkness and periods of low visibility while conducting ship/port interface activities or at a port facility or anchorage? (ISPS Code, paragraph B/9.46)	Yes ☐	No ☐	Other ☐
Comments:			

Security level 2

.9 Which of the following additional security measures have been established to enhance monitoring and surveillance activities? (ISPS Code, paragraph B/9.47)	Yes	No	Other
A Increasing the frequency and detail of security patrols	A ☐	☐	☐
B Increasing the coverage and intensity of lighting or the use of security and surveillance equipment	B ☐	☐	☐
C Assigning additional personnel as security look-outs	C ☐	☐	☐
D Ensuring co-ordination with water-side boat patrols, and foot or vehicle patrols on the shore side, when provided	D ☐	☐	☐
Comments:			

6 Supervising the handling of cargo and ship's stores

Part A

.1 Does the ship's means of supervising the handling of the following meet the requirements identified in the SSP at security levels 1 and 2? (ISPS Code, section A/7.2.6 and A/7.3): A cargo B ship's stores	Yes	No	Other
	A ☐	☐	☐
	B ☐	☐	☐
Comments:			

Part B – Handling of cargo

Security level 1

.2 Are measures employed to routinely check the integrity of cargo, including the checking of seals, during cargo handling? (ISPS Code, paragraphs B/9.27.1 and B/9.27.4)	Yes	No	Other
	☐	☐	☐
Comments:			

.3 Are measures employed to routinely check that cargo being loaded matches the cargo documentation? (ISPS Code, paragraph B/9.27.2)	Yes	No	Other
	☐	☐	☐
Comments:			

.4 Does the ship ensure, in liaison with the port facility, that vehicles to be loaded on car carriers, ro–ro and passenger ships are searched prior to loading, in accordance with the frequency required in the SSP? (ISPS Code, paragraph B/9.27.3)	Yes	No	Other
	☐	☐	☐
Comments:			

.5 Which of the following security measures are employed during cargo checking? (ISPS Code, paragraph B/9.28) A Visual examination B Physical examination C Scanning or detection equipment D Other mechanical devices E Dogs	Yes	No	Other
	A ☐	☐	☐
	B ☐	☐	☐
	C ☐	☐	☐
	D ☐	☐	☐
	E ☐	☐	☐
Comments:			

Security level 2

.6 Which of the following additional security measures are applied during cargo handling? (ISPS Code, paragraph B/9.30)	Yes	No	Other
A Detailed checking of cargo, cargo transport units and cargo spaces	A ☐	☐	☐
B Intensified checks to ensure that only the intended cargo is loaded	B ☐	☐	☐
C Intensified searching of vehicles	C ☐	☐	☐
D Increased frequency and detail in checking of seals or other methods used to prevent tampering	D ☐	☐	☐
Comments:			

Part B – Delivery of ship's stores (ISPS Code, paragraphs B/9.33 to B/9.36)

.7 Has the ship established security measures to ensure that stores being delivered match the order, prior to being loaded on board, and to ensure their immediate secure stowage at security level 1? (ISPS Code, paragraph B/9.35)	Yes ☐	No ☐	Other ☐
Comments:			

.8 Has the ship established additional security measures at security level 2 by exercising checks prior to receiving stores on board and intensifying inspections? (ISPS Code, paragraph B/9.36)	Yes ☐	No ☐	Other ☐
Comments:			

7 Ensuring security communication is readily available

Part A

.1 Do the ship's communication equipment and procedures meet the requirements identified in the SSP at security levels 1 and 2? (ISPS Code, section A/7.2)	Yes ☐	No ☐	Other ☐
Comments:			

.2 Has the ship security officer been identified? (ISPS Code, section A/9.4.13)	Yes ☐	No ☐	Other ☐
Comments:			

.3 Has the company security officer been identified and have 24 h contact details been provided? (ISPS Code, section A/9.4.14)	Yes ☐	No ☐	Other ☐
Comments:			

.4 Has the ship established procedures to ensure the inspection, testing, calibration and maintenance of any security equipment provided on board? (ISPS Code, section A/9.4.15)	Yes ☐	No ☐	Other ☐
Comments:			

.5 Has the frequency for testing or calibration of any security equipment provided on board been specified? (ISPS Code, section A/9.4.16)	Yes ☐	No ☐	Other ☐
Comments:			

.6 Have the locations on the ship been identified where the ship security alert system activation points are provided? (ISPS Code, section A/9.4.17)	Yes ☐	No ☐	Other ☐
Comments:			

.7 Have procedures, instructions and guidance been established and communicated on the use of the ship security alert system, including the testing, activation, deactivation and resetting, and to limit false alerts? (ISPS Code, section A/9.4.18)	Yes ☐	No ☐	Other ☐
Comments:			

8 Training, drills and exercises

Part A

.1 Have the:	Yes	No	Other
A CSO and appropriate shore-based security personnel received sufficient training to perform their assigned duties? (ISPS Code, section A/13.1)	A ☐	☐	☐
B SSO and appropriate shipboard personnel received sufficient training to perform their assigned duties? (ISPS Code, section A/13.2)	B ☐	☐	☐
Comments:			

.2 Do shipboard personnel having specific security duties and responsibilities understand their responsibilities for ship security and have sufficient knowledge and ability to perform their assigned duties? (ISPS Code, section A/13.3)	Yes ☐	No ☐	Other ☐
Comments:			

.3 Have the Company and ship implemented drills and participated in exercises? (ISPS Code, sections A/13.4 and A/13.5)	Yes ☐	No ☐	Other ☐
Comments:			

.4 Has the ship established procedures for training, drills and exercises associated with the ship security plan? (ISPS Code, section A/9.4.9)	Yes ☐	No ☐	Other ☐
Comments:			

Part B – Training, drills and exercises on ship security

.5 Have the CSO, appropriate shore-based Company personnel and the SSO received the appropriate levels of training? (ISPS Code, paragraphs B/13.1 and B/13.2)	Yes ☐	No ☐	Other ☐
Comments:			

.6 Do shipboard personnel with security responsibilities have sufficient knowledge and ability to perform their duties? (ISPS Code, paragraph B/13.3)	Yes ☐	No ☐	Other ☐
Comments:			

.7 Are security drills conducted:	Yes	No	Other
A at least every three months?	A ☐	☐	☐
B in cases where more than 25% of the ship's personnel has been changed, at any one time, with personnel that have not previously participated in any drill on that ship within the last three months? (ISPS Code, paragraph B/13.6)	B ☐	☐	☐
C to test individual elements of the ship security plan such as those security threats listed in ISPS Code, paragraph B/8.9? (ISPS Code, paragraph B/13.6)	C ☐	☐	☐
Comments:			

9 Miscellaneous

Part A

.1 Have different RSOs undertaken (a) the preparation of the SSA and SSP and (b) the review and approval of the SSP? (ISPS Code, sections A/9.2 and A/9.2.1)	Yes ☐	No ☐	Other ☐
Comments:			

.2 Has the master a contact point in the Administration to seek consent for the inspection of those provisions in the SSP that are considered confidential information, when access to them is requested by a duly authorized officer of another Contracting Government? (ISPS Code, section A/9.8.1)	Yes ☐	No ☐	Other ☐
Comments:			

.3 Has the ship established procedures to protect from unauthorized access or disclosure the records of activities addressed in the SSP which are required to be kept on board? (ISPS Code, section A/10.4)	Yes ☐	No ☐	Other ☐
Comments:			

.4 In which of the following circumstances does the ship request completion of a Declaration of Security (DoS)? (ISPS Code, section A/5.2)	Yes	No	Other
A When the ship is operating at a higher security level than the port facility or another ship it is interfacing with?	A ☐	☐	☐
B The ship is covered by an agreement on a DoS between Contracting Governments?	B ☐	☐	☐
C When there has been a security threat or a security incident involving the ship or port facility it is calling at?	C ☐	☐	☐
D When the ship is at a port which is not required to have and implement an approved PFSP?	D ☐	☐	☐
E When the ship is conducting ship-to-ship activities with another ship not required to have and implement an approved SSP?	E ☐	☐	☐
Comments:			

.5 Does the CSO or SSO periodically review the SSA for accuracy as part of the SSP review process? (ISPS Code, section A/10.1.7)	Yes	No	Other
	☐	☐	☐
Comments:			

.6 Does the ship adequately maintain the required security records and are they sufficiently detailed to allow the CSO and SSO to identify areas for improvement or change in the current security procedures and measures? (ISPS Code, section A/10.1)	Yes	No	Other
A Training, drills and exercises (ISPS Code, section A/10.1.1)	A ☐	☐	☐
B Security threats and security incidents (ISPS Code, section A/10.1.2)	B ☐	☐	☐
C Breaches of security (ISPS Code, section A/10.1.3)	C ☐	☐	☐
D Periodic review of the SSP (ISPS Code, section A/10.1.8)	D ☐	☐	☐
Comments:			

.7 Is the ship adequately manned and does its complement include the grades/capacities and number of persons required for the safe operation and the security of the ship and for the protection of the marine environment (IMO Assembly resolution A.890(21) as amended by Assembly resolution A.955(23), SOLAS regulation V/14.1 and ISPS Code, paragraph B/4.28):	Yes	No	Other
A When the ship is operating at security level 1?	A ☐	☐	☐
B When the ship is operating at security level 2?	B ☐	☐	☐
Comments:	C ☐	☐	☐
	D ☐	☐	☐

Part B – Miscellaneous

.8 Has the ship established procedures on handling requests for a Declaration of Security from a port facility? (ISPS Code, paragraph B/9.52)	Yes	No	Other
	☐	☐	☐
Comments:			

.9 Have procedures been established in the SSP as to how the CSO and SSO intend to audit the continued effectiveness of the SSP and to review, update or amend the SSP? (ISPS Code, paragraph B/9.53)	Yes	No	Other
	☐	☐	☐
Comments:			

.10 Has the ship established additional security procedures to be implemented when calling into a port facility which is not required to comply with the requirements of SOLAS chapter XI-2 and the ISPS Code? (ISPS Code, paragraph B/4.20)	Yes ☐	No ☐	Other ☐
Comments:			

Recommendations

This section should be used to record any deficiencies identified by the checklist and how these could be mitigated. In essence this will provide an action plan for the CSO and/or SSO.

Recommendations/For action: Section 1: Ensuring the performance of all ship security duties.

Recommendations/For action: Section 2: Controlling access to the ship.

Recommendations/For action: Section 3: Controlling the embarkation of persons and their effects.

Recommendations/For action: Section 4: Monitoring of restricted areas.

Recommendations/For action: Section 5: Monitoring of deck areas and areas surrounding the ship.

Recommendations/For action: Section 6: Supervising the handling of cargo and ship's stores.

Recommendations/For action: Section 7: Ensuring security communication is readily available.

Recommendations/For action: Section 8: Training, drills and exercises.

Recommendations/For action: Section 9: Miscellaneous.

Outcomes

This section should be used to record the findings of the voluntary self-assessment and any other issues arising. These findings could be raised with ship or company personnel or be used as the basis to seek guidance from the Administration, as appropriate.

Signature of assessor:	Date of completion: ....
Name (*please print*):	
Title:	

Appendix 4.10
Implementation checklist for shipping companies and their CSOs

Source: MSC.1/Circ.1217, December 2006

This checklist may be used by shipping companies and their CSOs to assess the status of implementation of the Maritime Security Measures within their company and on the ships they operate.

Completion of the following section is recommended before using the checklist. It can be used to establish an overview of company operations.

Company name	
Company address	
CSO name(s)	

Complete a separate table for each CSO as appropriate

Name of CSO	
Does the CSO hold an appropriate training certificate?	
Was this certificate submitted to the Administration for recognition?	

List of ship(s)

Name of ship	IMO Number	Type	Flag	SSP approved by, on	ISSC issued by, on
1)					
2)					
3)					
4)					
5)					
6)					
7)					
8)					
9)					
10)					

Guidance:

For each question, one of the 'Yes/No/Other' boxes should be ticked. Whichever one is used, the 'Comments' box provides space for amplification.

If the 'Yes' box is ticked, but the Measures/procedures are not documented in the SSP, a short description of them should be included in the 'Comments' box. The 'Yes' box should be ticked only if all procedures and measures are in place. The 'Comments' box may also be used to indicate when procedures were last reviewed and measures tested (e.g., drills and exercises).

If the 'No' box is ticked, an explanation of why not should be included in the 'Comments' box along with details of any measures or procedures in place. Suggested actions should be recorded in the 'Recommendations' section at the end of the checklist.

If the 'Other' box is ticked, a short description should be provided in the 'Comments' box (e.g., it could include instances where alternative measures/procedures or equivalent arrangements have been implemented). If the reason is due to the question not being applicable, then it should be recorded in the 'Comments' box as "not applicable".

If there is not enough space in the 'Comments' box, the explanation should be continued on a separate page (with the relevant question number, and in the case of questions with multiple options, the option added as a reference).

The 'Recommendations' boxes at the end of the checklist should be used to record any identified deficiencies and how these could be mitigated. A schedule for their implementation should be included.

The 'Outcomes' box at the end of the checklist should be used to provide a brief record of the assessment process. Along with the comments in the 'Recommendations' boxes, they form the basis for updating the SSP.

1 Continuous Synopsis Record (CSR) (SOLAS regulation XI-1/5)

.1 Has the Company ensured that all of its ships have been issued with an up-to-date CSR? (SOLAS regulation XI-1/5)	Yes ☐	No ☐	Other ☐
Comments:			

.2 Has the Company ensured that procedures are in place to notify the Administration when ships are transferred to the flag of another State? (SOLAS regulation XI-1/5.7)	Yes ☐	No ☐	Other ☐
Comments:			

2 Ship security alert system (SSAS)

.1 Has the Company ensured that an SSAS has been installed and that it operates as required? (SOLAS regulations XI-2/6.1 and XI-2/6.3)	Yes ☐	No ☐	Other ☐
Comments:			

.2 Has the Company been designated by each ship's Administration to receive ship-to-shore security alerts (a separate answer should be given for each flag under which the Company's ships are flying)? (SOLAS regulation XI-2/6.2.1)	Yes ☐	No ☐	Other ☐
Comments:			

.3 Does the CSO inform the Administration of SSAS implementation details and alterations? (SOLAS regulation XI-2/6.2.1)	Yes ☐	No ☐	Other ☐
Comments:			

.4 Does the Company have procedures in place to act upon receipt of a ship-to-shore security alert, including notification of the Administration? (SOLAS regulation XI-2/6.2.1)	Yes ☐	No ☐	Other ☐
Comments:			

3 Master's discretion for ship safety and security

.1 Has the Company adopted a clearly stated policy that nothing constrains the master from taking or executing any decision which, in his professional judgement, is necessary to maintain the safety and security of the ship? (SOLAS regulation XI-2/8.1)	Yes ☐	No ☐	Other ☐
Comments:			

4 Obligations of the Company

Part A

.1 Has the Company ensured that the master has available on board, at all times, information through which officers duly authorized by a Contracting Government can establish the following: (SOLAS regulation XI-2/5)	Yes	No	Other
A Who is responsible for appointing the members of the crew or other persons currently employed or engaged on board the ship in any capacity on the business of that ship?	A ☐	☐	☐
B Who is responsible for deciding the employment of the ship?	B ☐	☐	☐
C In cases where the ship is employed under the terms of charter party(ies), who are the parties to such charter party(ies)?	C ☐	☐	☐
Comments:			

.2 Has the Company established, in the ship security plan, that the master has the overriding authority and responsibility to make decisions with respect to the safety and the security of the ship and to request the assistance of the Company or of any Contracting Government as may be necessary? (ISPS Code, section A/6.1)	Yes ☐	No ☐	Other ☐
Comments:			

.3 Has the Company ensured that the CSO, the master and the ship security officer (SSO) are being given the necessary support to fulfil their duties and responsibilities in accordance with SOLAS chapter XI-2 and part A of the Code? (ISPS Code, section A/6.2)	Yes ☐	No ☐	Other ☐
Comments:			

Part B

.4 Has the Company provided the master of each ship with information to meet the requirements of the Company, under the provisions of SOLAS regulation XI-2/5, for each of the following: (ISPS Code, paragraph B/6.1)	Yes	No	Other
A Parties responsible for appointing shipboard personnel, such as ship management companies, manning agents, contractors, and concessionaries (for example, retail sales outlets, casinos, etc.)?	A ☐	☐	☐
B Parties responsible for deciding the employment of the ship, including time or bareboat charterer(s) or any other entity acting in such capacity?	B ☐	☐	☐
C In cases when the ship is employed under the terms of a charter party, the contact details of those parties, including time or voyage charterers?	C ☐	☐	☐
Comments:			

.5 Does the Company update and keep the information provided current as and when changes occur? (ISPS Code, paragraph B/6.2)	Yes	No	Other
	☐	☐	☐
Comments:			

.6 Is the information provided in the English, French or Spanish language? (ISPS Code, paragraph B/6.3)	Yes	No	Other
	☐	☐	☐
Comments:			

.7 If the ships were constructed before 1 July 2004, does this information reflect the actual condition on that date? (ISPS Code, paragraph B/6.4)	Yes	No	Other
	☐	☐	☐
Comments:			

.8 If the ships were constructed on or after 1 July 2004, or the ships were constructed before 1 July 2004 but were out of service on 1 July 2004, was the information provided as from the date of entry of the ship into service and does it reflect the actual condition on that date? (ISPS Code, paragraph B/6.5)	Yes	No	Other
	☐	☐	☐
Comments:			

.9 When a ship is withdrawn from service, is the information provided as from the date of re-entry of the ship into service and does it reflect the actual condition on that date? (ISPS Code, paragraph B/6.6)	Yes	No	Other
	☐	☐	☐
Comments:			

5 Control and compliance measures

.1 Does the Company provide, or has it ensured that its ships provide, confirmation to a Contracting Government, on request, of the information required in SOLAS regulation XI-2/9.2.1.1 to 9.2.1.6, using the standard data set detailed in MSC.1/Circ.1305? (SOLAS regulation XI-2/9.2.1)	Yes	No	Other
	☐	☐	☐
Comments:			

6 Verification and certification for ships

Part A

.1 Does the Company ensure that each ship to which SOLAS chapter XI-2 and the ISPS Code apply is covered by a valid International Ship Security Certificate (ISSC)? (ISPS Code, section A/19)	Yes ☐	No ☐	Other ☐
Comments:			

.2 Does the Company ensure that, when it assumes responsibility for a ship not previously operated by that Company, the existing ISSC is no longer used? (ISPS Code, section A/19.3.9.2)	Yes ☐	No ☐	Other ☐
Comments:			

.3 Does the Company, when it ceases to be responsible for the operation of a ship, transmit to the receiving Company, as soon as possible, copies of any information related to the ISSC or to facilitate the verifications required for an ISSC to be issued, as described in the ISPS Code, section A/19.4.2? (ISPS Code, section A/19.3.9.2)	Yes ☐	No ☐	Other ☐
Comments:			

7 Ship security assessment

Part A

.1 Does the CSO ensure that each ship security assessment is carried out by persons with appropriate skills to evaluate the security of a ship? (ISPS Code, sections A/2.1.7 and A/8.2 and paragraphs B/8.1 and B/8.4)	Yes ☐	No ☐	Other ☐
Comments:			

.2 Does the CSO ensure that the persons carrying out the ship security assessment take into account the guidance given in part B of the ISPS Code and, in particular, paragraphs B/8.2 to B/8.13 (see part B below)? (ISPS Code, section A/8.2 and paragraph B/8.1)	Yes ☐	No ☐	Other ☐
Comments:			

.3 Does the CSO ensure that ship security assessments include an on-scene security survey and at least the following elements: (ISPS Code, section A/8.4)	Yes	No	Other
A Identification of existing security measures, procedures and operations?	A ☐	☐	☐
B Identification and evaluation of key shipboard operations that it is important to protect?	B ☐	☐	☐
C Identification of possible threats to the key shipboard operations and the likelihood of their occurrence, in order to establish and prioritize security measures?	C ☐	☐	☐
D Identification of weaknesses, including human factors, in the infrastructure, policies and procedures?	D ☐	☐	☐
Comments:			

.4 Are ship security assessments documented, reviewed, accepted and retained by the Company? (ISPS Code, section A/8.5)	Yes ☐	No ☐	Other ☐
Comments:			

Part B – CSO requirements to conduct an assessment

.5 Has the CSO ensured that, prior to commencing the SSA, advantage was taken of information available on the assessment of threat for the ports at which the ship would call or at which passengers would embark or disembark and about the port facilities and their protective measures? (ISPS Code, paragraph B/8.2)	Yes ☐	No ☐	Other ☐
Comments:			

.6 Has the CSO studied previous reports on similar security needs? (ISPS Code, paragraph B/8.2)	Yes ☐	No ☐	Other ☐
Comments:			

.7 Has the CSO met with appropriate persons on the ship and in the port facilities to discuss the purpose and methodology of the assessment? (ISPS Code, paragraph B/8.2)	Yes ☐	No ☐	Other ☐
Comments:			

.8 Has the CSO followed any specific guidance offered by the Contracting Governments? (ISPS Code, paragraph B/8.2)	Yes ☐	No ☐	Other ☐
Comments:			

.9 Does the CSO obtain and record the information required to conduct an assessment, including the following: (ISPS Code, paragraph B/8.5)	Yes	No	Other
A The general layout of the ship?	A ☐	☐	☐
B The location of areas which should have restricted access, such as navigation bridge, machinery spaces of category A and other control stations as defined in chapter II-2 of SOLAS, etc?	B ☐	☐	☐
C The location and function of each actual or potential access point to the ship?	C ☐	☐	☐
D Changes in the tide which may have an impact on the vulnerability or security of the ship?	D ☐	☐	☐
E The cargo spaces and stowage arrangements?	E ☐	☐	☐
F The locations where ship's stores and essential maintenance equipment is stored?	F ☐	☐	☐
G The locations where unaccompanied baggage is stored?	G ☐	☐	☐
H The emergency and stand-by equipment available to maintain essential services?	H ☐	☐	☐
I The number of ship's personnel and existing security duties and any existing training requirement practices of the Company?	I ☐	☐	☐
J Existing security and safety equipment for the protection of passengers and ship's personnel?	J ☐	☐	☐
K Escape and evacuation routes and assembly stations which have to be maintained to ensure the orderly and safe emergency evacuation of the ship?	K ☐	☐	☐
L Existing agreements with private security companies providing ship-/water-side security services?	L ☐	☐	☐
M Existing security measures and procedures in effect, including inspection and control procedures, identification systems, surveillance and monitoring equipment, personnel identification documents and communication, alarms, lighting, access control and other appropriate systems?	M ☐	☐	☐
Comments:			

Part B – Content of the SSA

.10 Does the CSO ensure that the SSAs address the following elements on board or within the ship: (ISPS Code, paragraph B/8.3)	Yes	No	Other
A Physical security?	A ☐	☐	☐
B Structural integrity?	B ☐	☐	☐
C Personnel protection systems?	C ☐	☐	☐
D Procedural policies?	D ☐	☐	☐
E Radio and telecommunication systems, including computer systems and networks?	E ☐	☐	☐
F Other areas that may, if damaged or used for illicit observation, pose a risk to persons, property, or operations on board the ship or within a port facility?	F ☐	☐	☐
Comments:			

.11 Does the CSO ensure that those involved in conducting an SSA are able to draw upon expert assistance in relation to the following: (ISPS Code, paragraph B/8.4)	Yes	No	Other
A Knowledge of current security threats and patterns?	A ☐	☐	☐
B Recognition and detection of weapons, dangerous substances and devices?	B ☐	☐	☐
C Recognition, on a non-discriminatory basis, of characteristics and behavioural patterns of persons who are likely to threaten security?	C ☐	☐	☐
D Techniques used to circumvent security measures?	D ☐	☐	☐
E Methods used to cause a security incident?	E ☐	☐	☐
F Effects of explosives on ship's structures and equipment?	F ☐	☐	☐
G Ship security?	G ☐	☐	☐
H Ship/port interface business practices?	H ☐	☐	☐
I Contingency planning, emergency preparedness and response?	I ☐	☐	☐
J Physical security?	J ☐	☐	☐
K Radio and telecommunication systems, including computer systems and networks?	K ☐	☐	☐
L Marine engineering?	L ☐	☐	☐
M Ship and port operations?	M ☐	☐	☐
Comments:			

.12 Does the CSO ensure that SSAs examine each identified point of access, including open weather decks, and evaluate its potential for use by individuals who might seek to breach security? This includes points of access available to individuals having legitimate access as well as those who seek to obtain unauthorized entry. (ISPS Code, paragraph B/8.6)	Yes	No	Other
	☐	☐	☐
Comments:			

.13 Does the CSO ensure that SSAs consider the continuing relevance of the existing security measures and guidance, procedures and operations, under both routine and emergency conditions, and has the CSO determined security guidance, including the following: (ISPS Code, paragraph B/8.7)	Yes	No	Other
A The restricted areas?	A ☐	☐	☐
B The response procedures to fire or other emergency conditions?	B ☐	☐	☐
C The level of supervision of the ship's personnel, passengers, visitors, vendors, repair technicians, dock workers, etc?	C ☐	☐	☐
D The frequency and effectiveness of security patrols?	D ☐	☐	☐
E The access–control systems, including identification systems?	E ☐	☐	☐
F The security communications systems and procedures?	F ☐	☐	☐
G The security doors, barriers and lighting?	G ☐	☐	☐
H The security and surveillance equipment and systems, if any?	H ☐	☐	☐
Comments:			

.14 Does the CSO ensure that SSAs consider the persons, activities, services and operations that it is important to protect, which include the following: (ISPS Code, paragraph B/8.8)	Yes	No	Other
A The ship's personnel?	A ☐	☐	☐
B Passengers, visitors, vendors, repair technicians, port facility personnel, etc?	B ☐	☐	☐
C The capacity to maintain safe navigation and emergency response?	C ☐	☐	☐
D The cargo, particularly dangerous goods or hazardous substances?	D ☐	☐	☐
E The ship's stores?	E ☐	☐	☐
F The ship's security communication equipment and systems, if any?	F ☐	☐	☐
G The ship's security surveillance equipment and systems, if any?	G ☐	☐	☐
Comments:			

.15 Does the CSO ensure that SSAs consider all possible threats, which may include the following types of security incidents: (ISPS Code, paragraph B/8.9)	Yes	No	Other
A Damage to, or destruction of, the ship or of a port facility, e.g., by explosive devices, arson, sabotage or vandalism?	A ☐	☐	☐
B Hijacking or seizure of the ship or of persons on board?	B ☐	☐	☐
C Tampering with cargo, essential ship equipment or systems or ship's stores?	C ☐	☐	☐
D Unauthorized access or use, including presence of stowaways?	D ☐	☐	☐
E Smuggling weapons or equipment, including weapons of mass destruction?	E ☐	☐	☐
F Use of the ship to carry those intending to cause a security incident and/or their equipment?	F ☐	☐	☐
G Use of the ship itself as a weapon or as a means to cause damage or destruction?	G ☐	☐	☐
H Attacks from seaward whilst at berth or at anchor?	H ☐	☐	☐
I Attacks whilst at sea?	I ☐	☐	☐
Comments:			

.16 Does the CSO ensure that SSAs take into account all possible vulnerabilities, which may include the following: (ISPS Code, paragraph B/8.10)	Yes	No	Other
A Conflicts between safety and security measures?	A ☐	☐	☐
B Conflicts between shipboard duties and security assignments?	B ☐	☐	☐
C Watchkeeping duties, number of ship's personnel, particularly with implications on crew fatigue, alertness and performance?	C ☐	☐	☐
D Any identified security training deficiencies?	D ☐	☐	☐
E Any security equipment and systems, including communication systems?	E ☐	☐	☐
Comments:			

.17 Do the CSO and the SSO always have regard to the effect that security measures may have on ship's personnel who will remain on the ship for long periods? (ISPS Code, paragraph B/8.11)	Yes ☐	No ☐	Other ☐
Comments:			

.18 Does the CSO ensure that, upon completion of the SSA, a report is prepared consisting of a summary of how the assessment was conducted, a description of each vulnerability found during the assessment and a description of counter-measures that could be used to address each vulnerability? Is this report protected from unauthorized access or disclosure? (ISPS Code, paragraph B/8.12)	Yes ☐	No ☐	Other ☐
Comments:			

.19 Does the CSO review and accept the report of the SSA when the SSA has not been carried out by the Company? (ISPS Code, paragraph B/8.13)	Yes ☐	No ☐	Other ☐
Comments:			

8 Ship security plan

Part A

.1 Does the CSO ensure that a ship security plan (SSP) is carried on board every ship for which he/she is the CSO? (ISPS Code, section A/9.1)	Yes ☐	No ☐	Other ☐
Comments:			

.2 Does the SSP make provisions for the three security levels as defined in part A of the Code? (ISPS Code, section A/9.1)	Yes ☐	No ☐	Other ☐
Comments:			

.3 Does the CSO ensure that the SSP is written in the working language or languages of the ship? (ISPS Code, section A/9.4)	Yes ☐	No ☐	Other ☐
Comments:			

.4 Is an English, French or Spanish language version also available? (ISPS Code, section A/9.4)	Yes ☐	No ☐	Other ☐
Comments:			

.5 Does the SSP address, at least, the following: (ISPS Code, section A/9.4)	Yes	No	Other
A Measures designed to prevent weapons, dangerous substances and devices intended for use against persons, ships or ports, and the carriage of which is not authorized, from being taken on board the ship?	A ☐	☐	☐
B Identification of the restricted areas and measures for the prevention of unauthorized access to them?	B ☐	☐	☐
C Measures for the prevention of unauthorized access to the ship?	C ☐	☐	☐
D Procedures for responding to security threats or breaches of security, including provisions for maintaining critical operations of the ship or ship/port interface?	D ☐	☐	☐
E Procedures for responding to any security instructions Contracting Governments may give at security level 3?	E ☐	☐	☐
F Procedures for evacuation in case of security threats or breaches of security?	F ☐	☐	☐
G Duties of shipboard personnel assigned security responsibilities and of other shipboard personnel on security aspects?	G ☐	☐	☐
H Procedures for auditing the security activities?	H ☐	☐	☐
I Procedures for training, drills and exercises associated with the plan?	I ☐	☐	☐
J Procedures for interfacing with port facility security activities?	J ☐	☐	☐
K Procedures for the periodic review of the plan and for updating?	K ☐	☐	☐
L Procedures for reporting security incidents?	L ☐	☐	☐
M Identification of the ship security officer?	M ☐	☐	☐
N Identification of the CSO, including 24-hour contact details?	N ☐	☐	☐
O Procedures to ensure the inspection, testing, calibration, and maintenance of any security equipment provided on board?	O ☐	☐	☐
P Frequency for testing or calibration of any security equipment provided on board?	P ☐	☐	☐
Q Identification of the locations where the SSAS activation points are provided?	Q ☐	☐	☐
R Procedures, instructions and guidance on the use of the SSAS, including the testing, activation, deactivation and resetting and to limit false alerts?	R ☐	☐	☐
Comments:			

.6 Has the Company ensured that the personnel conducting internal audits of the security activities specified in the SSP, or evaluating its implementation, are independent of the activities being audited unless this is impracticable, due to the size and the nature of the Company or of the ship? (ISPS Code, section A/9.4.1)	Yes ☐	No ☐	Other ☐
Comments:			

.7 Where the SSP is kept in electronic format, has the Company established procedures aimed at preventing the unauthorized deletion, destruction or amendment of the SSP? (ISPS Code, section A/9.6)	Yes ☐	No ☐	Other ☐
Comments:			

.8 Has the Company established procedures to ensure the SSP is protected from unauthorized access or disclosure? (ISPS Code, section A/9.7)	Yes ☐	No ☐	Other ☐
Comments:			

Part B – Content of SSP

.9 Has the CSO taken into account whether the SSP is relevant for the ship it covers? (ISPS Code, paragraph B/9.1)	Yes ☐	No ☐	Other ☐
Comments:			

.10 Has the CSO complied with advice on the preparation and content of SSPs issued by the ship's Administration? (ISPS Code, paragraph B/9.1)	Yes ☐	No ☐	Other ☐
Comments:			

.11 Has the CSO taken into account that the SSP details those items listed in ISPS Code, paragraphs B/9.2.1 to 9.2.7?	Yes ☐	No ☐	Other ☐
Comments:			

.12 Does the CSO consider that all SSPs have been prepared having undergone a thorough assessment of all the issues relating to the security of the ship, including, in particular, a thorough appreciation of the physical and operational characteristics? (ISPS Code, paragraph B/9.3)	Yes ☐	No ☐	Other ☐
Comments:			

.13 Has the CSO developed the following procedures: (ISPS Code, paragraph B/9.5)	Yes	No	Other
A To assess the continuing effectiveness of the SSP?	A ☐	☐	☐
B To prepare amendments to the plan subsequent to its approval?	B ☐	☐	☐
Comments:			

9 Records

Part A

.1 Does the CSO ensure that records of the following activities addressed in the SSP are kept on board for at least the minimum period specified by the Administration, bearing in mind the provisions of SOLAS regulation XI-2/9.2.3: (ISPS Code, section A/10.1)	Yes	No	Other
A training, drills and exercises?	A ☐	☐	☐
B security threats and security incidents?	B ☐	☐	☐
C breaches of security?	C ☐	☐	☐
D changes in security level?	D ☐	☐	☐
E communications relating to the direct security of the ship, such as specific threats to the ship or to port facilities where the ship is or has been?	E ☐	☐	☐
F internal audits and reviews of security activities?	F ☐	☐	☐
G periodic review of the ship security assessment?	G ☐	☐	☐
H periodic review of the SSP?	H ☐	☐	☐
I implementation of any amendments to the plan?	I ☐	☐	☐
J maintenance, calibration and testing of any security equipment provided on board, including testing of the SSAS?	J ☐	☐	☐
Comments:			

.2 Does the CSO ensure that the records are kept in the working language or languages of the ship? (ISPS Code, section A/10.2)	Yes ☐	No ☐	Other ☐
Comments:			

.3 Is an English, French or Spanish language version of the records also available? (ISPS Code, section A/10.2)	Yes ☐	No ☐	Other ☐
Comments:			

.4 Where the records are kept in electronic format, has the Company established procedures aimed at preventing their unauthorized deletion, destruction or amendment? (ISPS Code, section A/10.3)	Yes ☐	No ☐	Other ☐
Comments:			

10 Company security officer

Part A

.1 Has the Company designated one or more CSOs? (ISPS Code, section A/11.1 and paragraph B/1.9)	Yes ☐	No ☐	Other ☐
Comments:			

.2 Where more than one CSO has been appointed, has it clearly been identified which ships each CSO is responsible for? (ISPS Code, section A/11.1)	Yes ☐	No ☐	Other ☐
Comments:			

.3 Do the CSO's duties and responsibilities include at least the following: (ISPS Code, section A/11.2)	Yes	No	Other
A Advising the level of threats likely to be encountered by the ship, using appropriate security assessments and other relevant information?	A ☐	☐	☐
B Ensuring that SSAs are carried out?	B ☐	☐	☐
C Ensuring the development, the submission for approval, and thereafter the implementation and maintenance of the SSP?	C ☐	☐	☐
D Ensuring that the SSP is modified, as appropriate, to correct deficiencies and satisfy the security requirements of the individual ship?	D ☐	☐	☐
E Arranging for internal audits and reviews of security activities?	E ☐	☐	☐
F Arranging for the initial and subsequent verifications of the ship by the Administration or the recognized security organization?	F ☐	☐	☐
G Ensuring that deficiencies and non-conformities identified during internal audits, periodic reviews, security inspections and verifications of compliance are promptly addressed and dealt with?	G ☐	☐	☐
H Enhancing security awareness and vigilance?	H ☐	☐	☐
I Ensuring adequate training for personnel responsible for the security of the ship?	I ☐	☐	☐
J Ensuring effective communication and co-operation between the SSO and the relevant PSOs?	J ☐	☐	☐
K Ensuring consistency between security requirements and safety requirements?	K ☐	☐	☐
L Ensuring that, if sister-ship or fleet security plans are used, the plan for each ship reflects the ship-specific information accurately?	L ☐	☐	☐
M Ensuring that any alternative or equivalent arrangements approved for a particular ship or group of ships, in accordance with SOLAS regulations XI-2/11 and XI-2/12, are implemented and maintained?	M ☐	☐	☐
Comments:			

.4 Has the CSO implemented a mechanism for receiving, from the SSO, reports of any deficiencies and non-conformities identified during internal audits, periodic reviews, security inspections and verifications of compliance, and any corrective actions taken? (ISPS Code, section A/12.2.5)	Yes ☐	No ☐	Other ☐
Comments:			

11 Training, drills and exercises on ship security

Part A

.1 Have the CSO and appropriate shore-based personnel received training, taking into account the guidance given in part B of ISPS Code? (ISPS Code, section A/13.1)	Yes ☐	No ☐	Other ☐
Comments:			

.2 Does the CSO ensure that drills are carried out at appropriate intervals, taking into account the ship type, ship personnel changes, port facilities to be visited and other relevant circumstances, and further taking into account the guidance in part B of ISPS Code? (ISPS Code, section A/13.4)	Yes ☐	No ☐	Other ☐
Comments:			

.3 Does the CSO ensure the effective co-ordination and implementation of SSPs by participating in exercises at appropriate intervals, taking into account the guidance given in part B of ISPS Code? (ISPS Code, section A/13.5)	Yes ☐	No ☐	Other ☐
Comments:			

Part B – Training, drills and exercises on ship security

.4 Have the CSO and appropriate shore-based Company personnel received training, in some or all of the following, as appropriate: (ISPS Code, paragraph B/13.1)	Yes	No	Other
A Security administrations?	A ☐	☐	☐
B Relevant international conventions, codes and recommendations?	B ☐	☐	☐
C Relevant Government legislation and regulations?	C ☐	☐	☐
D Responsibilities and functions of other security organizations?	D ☐	☐	☐
E Methodology of SSA?	E ☐	☐	☐
F Methods of ship security surveys and inspections?	F ☐	☐	☐
G Ship and port operations and conditions?	G ☐	☐	☐
H Ship and port facility security measures?	H ☐	☐	☐
I Emergency preparedness and response and contingency planning?	I ☐	☐	☐
J Instruction techniques for security training and education, including security measures and procedures?	J ☐	☐	☐
K Handling sensitive security-related information and security-related communications?	K ☐	☐	☐
L Knowledge of current security threats and patterns?	L ☐	☐	☐
M Recognition and detection of weapons, dangerous substances and devices?	M ☐	☐	☐
N Recognition, on a non-discriminatory basis, of characteristics and behavioural patterns of persons who are likely to threaten security?	N ☐	☐	☐
O Techniques used to circumvent security measures?	O ☐	☐	☐
P Security equipment and systems and their operational limitations?	P ☐	☐	☐
Q Methods of conducting audits, inspection, control and monitoring?	Q ☐	☐	☐
R Methods of physical searches and non-intrusive inspections?	R ☐	☐	☐
S Security drills and exercises, including drills and exercises with port facilities?	S ☐	☐	☐
T Assessment of security drills and exercises?	T ☐	☐	☐
Comments:			

.5 Does the CSO ensure that drills are conducted at least once every three months, with additional drills as recommended in ISPS Code, paragraph B/13.6?	Yes	No	Other
	☐	☐	☐
Comments:			

.6 Does the CSO ensure that exercises are conducted at least once each calendar year, with no more than 18 months between them? (ISPS Code, paragraph B/13.7)	Yes	No	Other
	☐	☐	☐
Comments:			

.7 Are these exercises: (ISPS Code, paragraph B/13.7)	Yes	No	Other
A full-scale or live?	A ☐	☐	☐
B tabletop simulation or seminar?	B ☐	☐	☐
C combined with other exercises held, such as search and rescue or emergency response exercises?	C ☐	☐	☐
D participated in by the CSO?	D ☐	☐	☐
Comments:			

.8 Has the Company participated in exercises with another Contracting Government? (ISPS Code, paragraph B/13.8)	Yes	No	Other
	☐	☐	☐
Comments:			

12 Information and co-operation (best practice)

.1 Is there a regular information exchange between the CSO and the Administration(s) responsible on best practices?	Yes ☐	No ☐	Other ☐
Comments:			

Recommendations

This section should be used to record any deficiencies identified by the checklist and how these could be mitigated. In essence this will provide an action plan for the CSO and/or SSO.

Recommendations/For action: Section 1: Continuous Synopsis Record

Recommendations/For action: Section 2: Ship security alert system

Recommendations/For action: Section 3: Master's discretion for ship safety and security

Recommendations/For action: Section 4: Obligations of the Company

Recommendations/For action: Section 5: Control and compliance measures

Recommendations/For action: Section 6: Verification and certification for ships

Recommendations/For action: Section 7: Ship security assessment

Recommendations/For action: Section 8: Ship security plan

Recommendations/For action: Section 9: Records

Recommendations/For action: Section 10: company security officer

Recommendations/For action: Section 11: Training, drills and exercises on ship security

Recommendations/For action: Section 12: Information and co-operation

Outcomes

This section should be used to record the findings of the assessment and any other issues arising. These findings could be raised with ship or company personnel or be used as the basis to seek guidance from the Administration, as appropriate.

Signature of assessor:	Date of completion: ....
Name (*please print*):	
Title:	

Appendix 4.11
General information on security practices for all non-SOLAS vessel operators

Source: MSC.1/Circ.1283, December 2008

A – Guidelines for non-SOLAS passenger vessels

1 Preventing unauthorized access

Members of the public and passengers should not be able to gain access to operational areas of the vessel or maintenance/storage facilities such as crew rest rooms, store rooms, cleaning cupboards, hatches and lockers. All doors leading into operational areas should be kept locked or controlled to prevent unauthorized access. The only exception to this should be where access is required to reach safety equipment or to use emergency escapes. Keys for doors should be kept in a secure location and controlled by a responsible person. If access is controlled by keypad, the code should only be given to people with a legitimate need to know. It is also recommended that codes are changed periodically. Where such access controls are in place, crew should be reminded of the importance of ensuring that nobody following can bypass the access controls. The following are suggested measures to deter unauthorized access to the vessel:

.1 over-the-side lighting which gives an even distribution of light on the whole hull and waterline;

.2 keeping a good watch from the deck;

.3 challenging all approaching boats. If unidentified, they should, where possible, be prevented from coming alongside.

2 Conducting a search

The vessel should be searched at the start of a voyage to ensure that nothing illegal or harmful has been placed on board and at the end of a voyage to ensure that nothing has been concealed or left behind. To the extent possible, checks should include any crew areas, stores, holds, underwater hull (if concern prevails) and areas that could conceal persons or articles that may be used for illegal purposes. There should be agreed procedures on how to isolate a suspect package if found and how to evacuate the vessel quickly and safely. The following are examples of good practice which should be implemented to assist crew undertaking patrolling duties when operating in a higher-risk environment:

.1 define the search area – crew members should be fully briefed and aware of what is required and have clearly defined start and finish points;

.2 plans – laminated plans of search areas should be produced in advance, highlighting the key features of the areas to be searched (such as storage bins and emergency exits);

.3 thoroughness – thorough searches help detect concealed items, and attention should be paid to vulnerable areas. Crew should not rely solely on visual checks, but should take note of unusual sounds, smells, etc;

.4 use of seals – unlockable equipment boxes, such as lifejacket boxes, can be fitted with tamper-evident seals, eliminating the need to search inside unless the seal is no longer intact.

Pre-planned action – crew members should be fully briefed on their expected actions in the event that a search identifies a security concern.

3 Verifying identity of persons on board a vessel

The following are examples of good practice which could be implemented to verify the identity of persons on board a vessel when operating in a higher-risk environment:

Visitors (other than passengers) should report to the master of the vessel, or other responsible person, to notify them of their arrival and departure. All visitors should have a form of identity, for example an ID card, passport or some other form of identification bearing the individual's photograph.

Passengers must present a valid ticket before boarding (except where tickets are bought on board the vessel) and where applicable have a form of identity such as an ID card, passport or some other form of identification bearing the individual's photograph. For chartered vessels where no tickets are required, the chartering party should give some thought as to how they will control access. This could be achieved through the provision of paper authorization, such as an invitation, to be shown or for names on a list to be checked off on presentation of identification.

It is recommended that passengers and visitors be advised on security procedures, such as the need to:

- **.1** be escorted at all times;
- **.2** wear a permit, if issued, at all times;
- **.3** be vigilant at all times when on the vessel. Should they find a suspicious item, they should not touch it but should contact a member of crew as soon as possible. Similarly, they should contact a member of crew if they see a person acting suspiciously; and
- **.4** secure all doors behind them when leaving, particularly those doors which lead to operational areas of the vessel. If they are leaving a work site, they must ensure that it is locked and that all equipment has been securely stored.

The vessel might maintain a security logbook at the point of entry/exit to the vessel, recording the identity of all persons boarding or disembarking.

4 Securing

With due regard to the need to facilitate escape in the event of an emergency, external doors and storage areas should be locked and portholes secured. If the vessel is to be left unattended for a lengthy period of time, such as overnight, it is recommended that the engine is disabled to prevent theft/unauthorized use and that the vessel is moored securely, in compliance with local port by laws. Masters should ensure that the gangway is raised when the vessel is left unattended.

5 Responding to bomb threats or discovery of suspicious items

Bomb threats are usually anonymous and communicated by telephone. While bomb threats are usually hoaxes intended to cause a nuisance, they must be taken seriously as a small number have been genuine and have preceded a terrorist or criminal act. It is recommended that advice is sought from local authorities on how to handle any genuine bomb threats that may be received.

Plans and procedures should be in place for dealing with health and safety alerts both on a vessel and at piers. These plans may be adapted to cover security alerts. Responsible individuals should consider appropriate responses for possible scenarios such as:

- **.1** suspect packages found on board a vessel or at a pier;
- **.2** individuals behaving suspiciously either on a vessel or at a pier;
- **.3** security alert at another pier or on another vessel, requiring suspension of operations;
- **.4** a direct attack against a vessel or pier by unknown persons, which could include ramming or the successful explosion of an improvised explosive device.

Responsible individuals should similarly consider how to isolate a suspect package, if found, without removing or touching it and how to evacuate the vessel and piers quickly and safely.

If a suspicious device or package is found while a vessel is at sea, the Master should take into account:

- **.1** the size and location of the device;
- **.2** the credibility of the threat;
- **.3** the vessel's location and the time it will take for security services and other assistance to arrive;
- **.4** the need to keep everyone well clear of the suspect device; and
- **.5** the need for all on board to keep clear of all doors, trunks and hatches leading from the space containing the device, to avoid possible blast injuries.

6 Maintaining a means for reporting security concerns

Vessel operators should implement procedures and processes for reporting and recording security incidents. In the event of a security incident occurring while the vessel is at sea, it should be reported to the master or SSO as appropriate. Depending on its seriousness, the master, in addition to activating an appropriate response, may alert the nearest coastal State or authorities and/or vessels in the vicinity and provide details of the incident.

Operators of non-SOLAS vessels should provide all personnel with contact information for authorities responsible for emergency response, the national response centre(s) (if appropriate) and any other authorities that may need to be notified. They should identify the actions that crew members should take in the event of a security incident, including how to notify authorities that a security incident is taking place (e.g., making radio calls, sounding alarms, etc.), and how to protect themselves, their vessel and the public.

All personnel should report suspicious activities to appropriate authorities. The report should include details of the activity and its location. The list below gives examples of activities which may by themselves constitute suspicious behaviour, any one of which may be considered suspicious by itself. However, those suspicions may warrant particular attention when one or more behaviour or a pattern of behaviour is observed or detected.

.1 unknown persons photographing vessels or facilities;

.2 unknown persons contacting, by any media, a ship or facility for the purpose of ascertaining security, personnel or standard operating procedures;

.3 unknown persons attempting to gain information about vessels or facilities by walking up to ship or facility personnel or associated individuals, or their families, and engaging them in conversation;

.4 theft or the unexplained absence of standard operating procedures documents;

.5 unknown or unauthorized workmen trying to gain access to facilities to repair, replace, service, install or remove equipment;

.6 inappropriate or unauthorized persons attempting to gain access to vessels or facilities;

.7 theft of facility vehicles, vehicle passes, personnel identification or personnel uniforms;

.8 inappropriate use of Global Maritime Distress Safety and Security procedures;

.9 suspicious individuals establishing ad hoc businesses or roadside stands either adjacent to or in proximity of port facilities;

.10 repeated or suspicious out-of-ordinary attempts at communication by voice media with duty personnel;

.11 vehicles or small vessels loitering in the vicinity of a facility without due cause, for extended periods of time;

.12 unknown persons loitering in the vicinity of a facility without due cause, for extended periods of time.

7 Prevention of trafficking in drugs and transportation of illicit cargoes

The following are general guidelines for precautionary measures which may be taken to safeguard a non-SOLAS vessel while in port, irrespective of whether at anchor or alongside a berth, to protect the vessel against trafficking in drugs and the transportation of illicit cargoes:

.1 the crew should be warned about the risks of knowingly transporting illicit cargoes and trafficking in drugs;

.2 crew going ashore should be advised that they should take care to ensure that persons they are meeting with are not connected with illegal activities;

.3 the vessel might maintain a security logbook at the point of entry/exit to the vessel, recording the identity of all persons boarding or disembarking. No unauthorized persons should be allowed to board;

.4 a permanent watch may be advisable in working areas. If appropriate, areas such as the forecastle, poop deck, main decks, etc., must be well lit during the hours of darkness;

.5 the vessel should maintain a good lookout for approaching small boats, or the presence of unauthorized divers, or other attempts by unauthorized persons to board the vessel;

.6 in the event that drugs or illicit cargoes are found on board, the crew should co-operate fully with the local authorities for the duration of the investigation.

8 Prevention of stowaways

For the purposes of the guidelines, a stowaway is defined as a person who is secreted on a vessel, or in cargo which is subsequently loaded onto a vessel, without the consent of the vessel owner or the master or other responsible person, and who is detected on board after the vessel has departed from a port and is reported as a stowaway by the master to the appropriate authorities.

The visible actions of the crew in implementing security measures will act as a deterrent to potential stowaways. Examples of general precautionary measures for the prevention of stowaways are set out below:

.1 prior to entering port, doors and hatchways should be securely fastened and locked, with due regard to the need to facilitate escape in the event of an emergency;

.2 fitting plates over anchor hawse pipes can prevent stowaways from boarding at anchorage or before a vessel is berthed;

.3 accommodation doors could also be secured and locked, leaving only one open entrance. In the interests of safety, keys to the locked doors should be placed in convenient positions so that doors can be opened in the event of an emergency;

.4 storerooms, equipment lockers on deck, the engine room and the accommodation should remain locked throughout a port call, only being opened for access and re-secured immediately thereafter;

.5 once alongside, a gangway watch is the first line of defence against stowaways, smugglers and theft. For this reason, it is important to ensure that an effective gangway watch is maintained at all times;

.6 at the commencement of loading, only the hold access doors of the compartments that are going to be used for the immediate loading of cargo should be opened. As soon as cargo operations cease, the compartment should be secured;

.7 the vessel's storerooms should also be kept locked at all times, only being opened when access is required;

.8 there may be some areas of the vessel that cannot be locked, for instance the funnel top. Any unlocked areas that can be accessed should be inspected on a regular basis;

.9 on completion of cargo loading operations and the disembarkation of all shore-based personnel, accessible areas of the vessel should be searched again;

.10 in high-risk ports, consideration should be given to anchoring in some convenient position outside the port and making a final stowaway search after tugs and pilots depart;

.11 if possible, the search should be conducted by two crew members. In the event that a stowaway is found, this will reduce the risk of the stowaway attacking or overpowering the searcher.

A detected stowaway should be reported immediately to the appropriate authorities. Any stowaways detected should be treated in accordance with humanitarian principles. However, as some stowaways may be violent, direct engagement is discouraged, as the safety and security of the vessel and its crew should not be compromised.

B – Specific guidelines for non-SOLAS passenger vessels

These guidelines are intended to complement the general guidelines contained above.

1 Searching

It is recommended that passengers are not permitted to board until the security check of the vessel has been completed. To the extent possible, checks should include all public areas, with special attention paid to underneath seating, toilets, and any storage areas, e.g., for luggage, on the vessel.

2 Control of passengers boarding and disembarking

Passengers must only be allowed to embark and disembark if crew or shore staff are present. Where ticket facilities exist for scheduled services, crew or shore staff should ensure that passengers present valid tickets before boarding. For chartered vessels where no tickets are required, the chartering party should seek to control access on to the boat, for example through the provision of an authorization card. If the vessel carries vehicles, special additional measures, including spot checks, may be required.

3 Passenger security awareness

Passengers should be reminded not to leave bags unattended and to report any unattended or suspect packages. Security messages should be displayed on posters and information screens and should be frequently delivered over public address systems, either as separate announcements or as part of the pre-sailing safety announcement.

C – Specific guidelines for pleasure craft

1 Introduction

Each national authority has its own definition of pleasure craft and may apply these guidelines as appropriate. Although they focus on pleasure craft engaged in international voyages or operating in waters where they might interact with or operate in close proximity to ISPS Code-compliant vessels or port facilities, they may have broader implementation as many pleasure craft are highly mobile, both via land and connecting waterways. The guidelines are intended to complement the general guidance contained in paragraphs 4.11.3 to 4.11.11.

Owners and operators of pleasure craft should remember that the overall safety and security of the vessel, crew and passengers is their responsibility. Prudent mariners are proactive in preventing incidents, planning in advance how best to respond to an incident, ensuring that all passengers and crew members know their roles, and being familiar with any particular directions that exist for an intended port or destination. Owners and operators should consider designating one crew member as responsible for all aspects of the security on the vessel. Some companies now offer courses specifically tailored for blue-water yachtsmen.

2 Security measures for unattended pleasure craft

Possible measures include:

- **.1** locking ignition switches and steering;
- **.2** fitting a small craft alarm system, possibly with an autodial facility to alert an operator to any unauthorized movement, or the activation of a variety of onboard security sensors, via mobile phone or e-mail. The alarm system could also be integrated with smoke and fire sensors for a complete vessel protection system;
- **.3** securing high-value items so that they are out of sight and in lockable compartments;
- **.4** not leaving anything valuable on display and preferably removing them, e.g., the ignition key;
- **.5** marking equipment, using approved property-marking equipment;

.6 etching the hull identification number onto windows and hatches;

.7 installing an engine immobilizer or a hidden device to shut off the fuel line;

.8 securing outboard motors with a strong case-hardened steel chain padlock, chain or some form of proprietary locking bar;

.9 covering the boat, as far as the design allows, and securing the cover;

.10 photographing the vessel and equipment (to assist authorities in returning stolen equipment);

.11 recording all available serial numbers and storing them in a safe place on and off the vessel;

.12 acquiring Radio Frequency Identification Tag (RFID) anti-theft systems (not only do such systems have the potential to reduce theft risk, but they also have been shown to increase recovery rates and in some instances to reduce insurance fees).

3 Higher-risk environments

Where safe and secure routes are not practicable, transits should be accomplished in the presence of other vessels, as expeditiously as possible, and prior notification made to the maritime authorities for the area, whose advice should be followed. A rigorous contact schedule should be maintained, preferably via satellite or mobile telephone or similar system which cannot be used to locate the vessel via radio direction-finding.

Consideration should be given to providing operator proficiency training for pleasure craft owners and operators that encompasses security-awareness familiarization.

4 Arrival and departure information

Pleasure craft departing a port could be required to submit voyage information when applying for port clearance, as has been implemented by Singapore. The voyage information may include the estimated time of departure, destination and the planned route of the trip. The additional information may be useful to the relevant authorities not only in monitoring and enforcement activities, but also when conducting search and rescue operations should the vessel run into trouble and require assistance. For more information refer to: *Declaration of Information by Pleasure Craft Departing Singapore*, Singapore, Maritime and Port Authority, Port Marine Circular No.17, 25 April 2003 , at the following website:

www.mpa.gov.sg/sites/circulars_and_notices/pdfs/port_marine_circulars/pc03-17.pdf

5 Registration

Some national authorities are encouraging operators of pleasure craft to register with their maritime Administration or delegated organization which could provide a database, available for authorized online access, to assist in both preventative and response activities related to both safety and security. One such example is the UK Ship Register, Part 3, Pleasure Craft/Small Ships, at the following website: http://www.dft.gov.uk/mca/

The registry is cheaper and simpler than full vessel registration and specifically aimed at pleasure craft. Owners benefit by having details of their craft's nationality and registered keeper recorded by an authoritative organization. It can be applied for online. However, it should be noted that registration in itself offers no protection against the misuse of a registered pleasure craft which may be stolen, hijacked or even legally acquired.

6 Information sharing

Some national authorities are seeking agreements to provide for information sharing, within the context of their individual laws and regulations, possibly as part of their individual coastal security initiatives. Pleasure craft engaged in international voyages present unique circumstances. Even when registered, information regarding vessel characteristics, ownership, etc., is often not shared between countries of departure and arrival. This can result in a lack of transparency for security and safety organizations, leading to, for example, complications in validating an arriving vessel's identity.

7 Ship security plan

Some national authorities have issued guidelines on developing effective security measures to address threats and other incidents at sea. One such authority is the International Merchant Marine Registry of Belize (IMMARBE), which has issued Guidelines for an effective ship security plan for yachts not required to hold ISPS Code certification. They may be accessed at the following website:

www.immarbe.com/yachts/guide_ship_security.html

Section 5 – Framework for conducting security assessments

5.1 Introduction

5.1.1 As noted in paragraphs 2.8.25 to 2.8.33 and 2.9.12 to 2.9.14, security assessments provide the foundation for the effective implementation of the Maritime Security Measures at port facilities and on board ships.

5.1.2 In December 2008, IMO issued guidance to assist national authorities in undertaking risk assessments. Although this guidance was aimed at non-SOLAS vessels, the methodology and the principles on which it is based are equally applicable to SOLAS port facilities and ships.

5.1.3 Although there are many different techniques, which range in the complexity of their application, the following six phases are common to all:

- **.1** pre-assessment;
- **.2** threat assessment;
- **.3** impact assessment;
- **.4** vulnerability assessment;
- **.5** risk scoring;
- **.6** risk management.

5.1.4 Each of the phases is discussed in turn below.

5.2 Pre-assessment phase

5.2.1 Effective project management is essential to the successful conduct of a security assessment. Before starting an assessment, the following steps should be considered.

Risk register

5.2.2 A useful first step is to establish a risk register that summarizes the assessment and scoring phases identified above. A sample format is shown below, along with accompanying explanations.

						Vulnerability			
Reference number – each threat scenario (TS) should be assigned a reference number so that it can be easily identified and its development tracked	Threat scenario – each possible threat scenario should be named, with a brief description of what it entails.	Lead organization – refer to subsection 5.3	Support organizations – refer to subsection 5.3	Threat (likelihood) – refer to subsection 5.3	Impact – refer to subsection 5.4	Key assets – refer to subsection 5.5	Mitigating controls – refer to subsection 5.5	Vulnerability score – refer to subsection 5.5	Risk score- refer to subsection 5.6
TS 1									
TS 2									

Establishing assessment teams

5.2.3 As the Maritime Security Measures identify both the conduct and approval of a PFSA to be a responsibility of the Designated Authority, the team leader should be a Government official appointed by the Designated Authority. However, as the conduct of a PFSA requires extensive input from the port facility, representatives of the facility operator (including the PFSO) should be team members.

5.2.4 Although the Maritime Security Measures do not require Government officials to be involved in the conduct or approval of an SSA, a small assessment team could still be established. Normally, its leader would either be the CSO or a suitably qualified member of the RSO to which the assessment has been delegated.

5.2.5 Where possible, assessment teams should follow project-management principles in planning and conducting an assessment.

5.2.6 Experience to date indicates that the assessment team should ensure that the operator is well-briefed on how the assessment is to be conducted. This is often achieved through the provision of an information package with follow-up, as required, by the team leader.

Process mapping

5.2.7 Where possible, the assessment process should be mapped as a basis for identifying critical-path items and responsibilities. The flowchart on the following page provides an example of the main steps in a typical assessment.

Inventory development

5.2.8 An inventory should be prepared of:

.1 assets and infrastructure;

.2 operating procedures;

.3 site or ship layout plans;

.4 previous security assessments;

.5 current security plan;

.6 previously reported security incidents;

.7 control measures in place; and

.8 risk-based classification based on type of facility or ship.

Methodology selection

5.2.9 The final pre-assessment task is to select the appropriate methodology. This is based in large part on the risk-based classification of the type of port, port facility or ship. The methodology used for a small, single-purpose port facility or small general cargo ship is likely to be less complex than the methodology required for a large, multi-purpose facility or cruise ship. The methodology described below should be suitable for most port facilities, small to medium-sized ports and most ships.

5.2.10 Internet sources of security assessment methodologies are shown in appendix 5.1 – Examples of internet sources of security assessment methodologies.

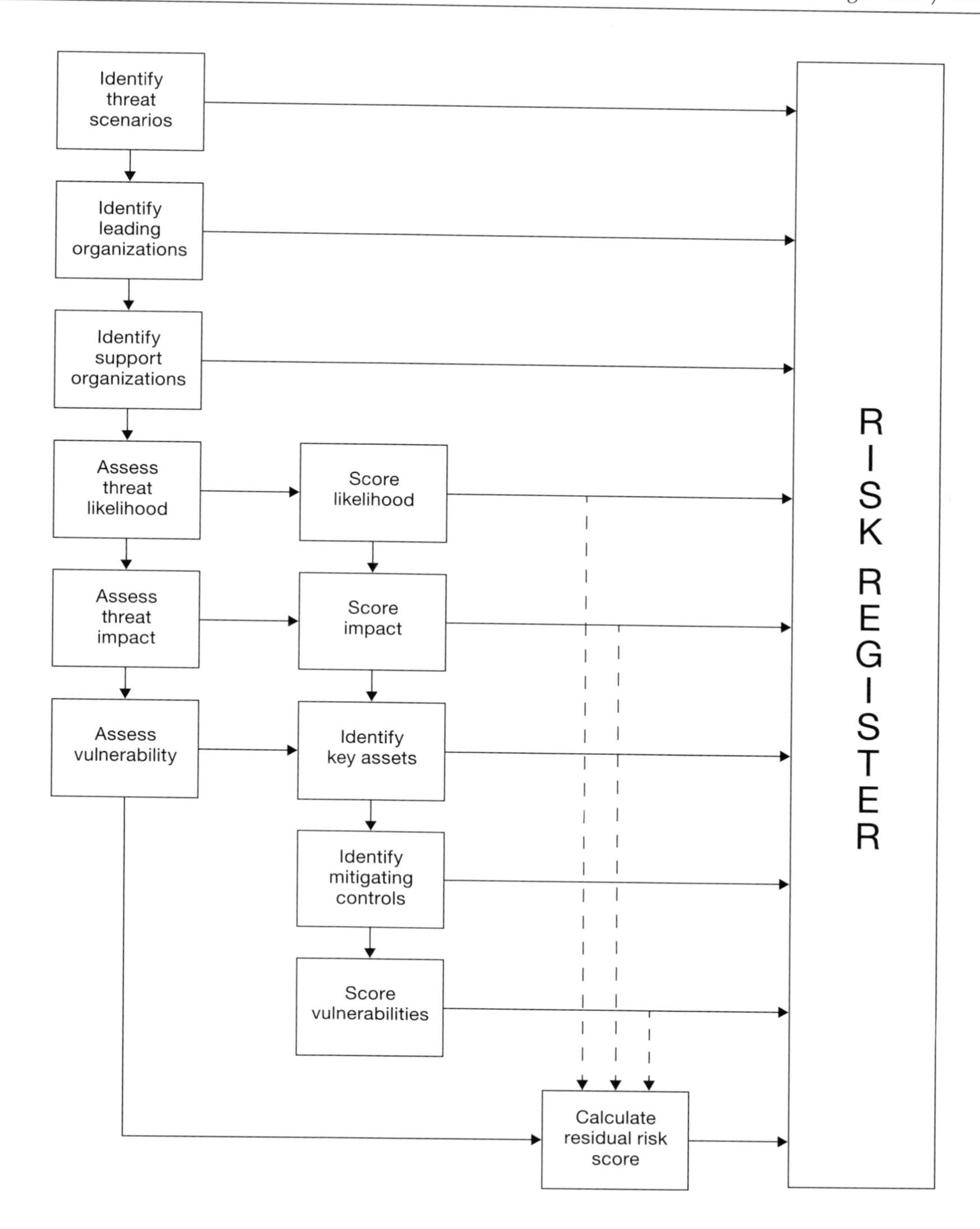

Risk-assessment flowchart

5.3 Threat-assessment phase

5.3.1 The first step is to list and agree on which threat scenarios could apply. Useful tips include:

.1 preparing an initial list of threat scenarios;

.2 having a "brainstorming" session where subject matter experts consider if there are any additional scenarios which should be listed and any refinements needed to develop the initial list;

.3 identifying potential perpetrators (e.g., terrorists, criminals, activists, disruptive passengers, employees);

.4 considering how they might operate (e.g., by reference to any precedents);

.5 considering their possible motivation and intent (e.g., financial gain, publicity, vengeance); and

.6 considering their capability to act (e.g., numbers, training, funding, weapons, track record, support).

5.3.2 The next step is to identify the lead organization or co-ordinating body, so that initial points of contact and responsibilities may be established for each scenario. Lead organizations should meet one of the following criteria:

.1 they own the assets;

.2 they set the policy for dealing with the threat;

.3 they have legal responsibility for, or have the major role in, mitigating or responding to a particular threat; or

.4 a combination of the above.

5.3.3 As there may be a different lead organization in instances where responsibilities vary depending on type of threat, location and method, distinctions should be made, where appropriate, between responsibilities for:

.1 preventive/protective security measures;

.2 contingency planning and reactive security measures to deal with and contain an incident;

.3 implementing the above measures.

5.3.4 Support organizations (e.g., first responders) should also be identified as they have a role in mitigating the threat but do not meet the criteria listed above for lead organizations. It may be decided that all stakeholders are support organizations through being vigilant, providing a deterring presence and sharing information with others.

5.3.5 For some threat scenarios, identifying lead and support organizations is not a simple task. If there are differing views, it is important that consensus is reached, particularly as lead organizations have a primary role in developing and delivering action plans. In situations where more than one lead organization is identified, it may be worth re-evaluating to minimize the potential for confusion and duplication.

5.3.6 The final step in this phase is to assign a score to each threat scenario. The score should reflect the likelihood of each threat scenario occurring if there were no security measures or mitigating controls in place to prevent them. To accurately score the threat, assessors should:

.1 consider local and international intelligence/knowledge about similar events which have or could have occurred;

.2 discuss how likely it would be for each threat scenario to occur if there were no security measures in place;

.3 read the definitions in the table below and decide which score best applies;

.4 use an alternative method of scoring if it produces a more logical and accurate assessment of the threats and risks;

.5 remember to apply any agreed rules around confidentiality.

Score	Likelihood	Criteria
4	PROBABLE	There have been previous reported incidents. There is intelligence to suggest that there are groups or individuals capable of causing the undesired event. There is specific intelligence to suggest that the port, port facility, ship or type of ship is a target.
3	LIKELY	There have been previous reported incidents. There is intelligence to suggest that there are groups or individuals currently capable of causing the undesired event. There is general intelligence to suggest that the port, port facility, ship or type of ship may be a likely target.
2	UNLIKELY	There is intelligence to suggest that there are groups or individuals capable of causing the undesired event. There is nothing to suggest that the port, port facility, ship or type of ship is a target.
1	IMPROBABLE	There have been no previously reported incidents anywhere worldwide. There is no intelligence to suggest that there are groups or individuals capable of causing the undesired event.

5.4 Impact assessment phase

5.4.1 The first step is to list examples of the type and magnitude of impact that might be expected if an undesired event happened. As the list of undesired events and their impacts in the table below is not exhaustive, assessors should consider modifying the table to meet their needs and to record discussions on the type and magnitude of impact associated with each listed undesired event.

Type of undesired event	Loss of life or personal injury	Loss or damage to ship and ship infrastructure	Loss of use of equipment	Disruption to services	Financial loss to vessel	Damage to reputation	Publicity to perpetrator
Improvised explosive device (IED)							
Sabotage							
Arson							
Unauthorized access							
Theft of vessels							

5.4.2 The second step is to assign a score to each impact. To score the impact accurately, assessors should:

.1 read the definitions in the table below and decide which one best applies to each undesired event in terms of its impact on the port facility or ship if the event occurs but without mitigating factors in place;

.2 consider how to record the scores allocated under each of the column headings for each undesired event. For simplicity, an average may be taken in most cases of the scores assigned to applicable impacts. If a particular impact is not applicable, it should be noted as such and excluded from the averaging process.

Score	Impact	Criteria – potential for:
4	SUBSTANTIAL	Multiple fatalities. Serious loss or damage to assets, infrastructure or ship. Economic cost of more than an agreed-on amount. Widespread media coverage, resulting in serious damage to reputation.
3	SIGNIFICANT	Loss of life. Significant but repairable loss or damage to assets, infrastructure or craft. Economic cost of less than an agreed-on amount. Adverse national media coverage.
2	MODERATE	Major injuries. Short-term minor loss or damage. Economic cost of less than an agreed-on amount. Major local damage to reputation.
1	MINOR	Minor injuries. Minimal operational disruption. Economic cost of less than an agreed-on amount. Minor damage to reputation.

5.5 Vulnerability assessment phase

5.5.1 The first step involves listing:

.1 the most important assets or targets, including infrastructure, which could be affected by the scenario, e.g., people (crew and passengers), objects, physical infrastructure and equipment; and

.2 their relevant characteristics and how they can be exploited.

5.5.2 Experience to date indicates that this is often achieved through an on-site or on-board survey by the assessment team.

5.5.3 The next step involves identifying the current mitigating controls (i.e., the security measures which are already in place to protect the key assets) and assessing their effectiveness and residual weaknesses. This is a vital step but, depending on the technique used, can be time-consuming, complex and intensive. As a minimum, assessors should undertake on-board or on-site inspections as a way of enhancing their understanding of the key assets or targets and the effectiveness of the mitigating controls in place. More sophisticated techniques (e.g., process mapping and event cause analysis) may provide for a more thorough assessment but should only be undertaken by individuals trained in their application.

5.5.4 Assessors may want to create a table similar to the one below to record their preliminary findings. This is a useful review tool to reconsider the effectiveness of control measures highlighted in the risk register and identify where there are weaknesses and gaps. Knowledge of those assets judged to be of high importance helps risk assessors to focus their review on what safeguards are in place and hence assess the vulnerability more accurately.

Assessment of security measures used to counter breaches of security

Security measures	Intended results
Security patrols Monitoring of security equipment Education and training of employees	Deterrence and detection Pre-empt breach or swift response Employee awareness
Possible weaknesses	**Follow-up actions**
Inadequate resources Gaps in security coverage Insufficient training	Discuss issues with relevant personnel Consider redeployment of resources Organize employee training programme

5.5.5 Assessors may also find it useful to ask the following questions and complete the table below, as they proceed through this phase:

- **.1** what are the key targets – people, critical infrastructure, communications and control, and support services?
- **.2** what are the systems designed to deter, detect, delay or deal with unlawful acts?
- **.3** what are the weaknesses in these systems, including consideration of predictability and opportunity?
- **.4** which assets are of high value?
- **.5** which stakeholders have a part to play in reducing the vulnerability of the target?
- **.6** how will this assist in defining "who" should work together on what?

Target – list of key assets (KA) grouped by category (e.g., infrastructure, communications and control assets, support services, people).	**KA1**	**KA2**
Strengths – systems designed to deter, detect, or deal with undesired events (e.g., vetting/pass systems, CCTV, restricted areas and police presence).		
Weaknesses – includes limited intelligence indicating the likelihood of a threat and the desirability of the target for the perpetrator (e.g., due to lack of search capability, poor surveillance, high traffic volumes, personnel shortages).		
Opportunities – opportunities for the perpetrator to exploit a loophole, conduct reconnaissance, etc.		
Predictability – the ways in which a target operates which make it predictable.		
Vulnerability – a High-Medium-Low rating based on a preliminary assessment of the net effect of the vulnerability factors identified above.		
Stakeholders involved in reducing vulnerability – includes members of port and ship security committees.		
Means of reducing vulnerability.		

5.5.6 The final step of this phase involves translating the vulnerability assessment into a vulnerability score. It requires consideration of, on the one hand, an evaluation of targets' characteristics and, on the other, the early warning indicators, embedded monitors and existing mitigating controls. The table below illustrates a possible scoring system to be used for assessing vulnerability, using the example of access to a sensitive area outside the boundary of a restricted area.

Score	**Extent of risk management**	**Countermeasures in place**
4	None	None
3	Limited	Some
2	Acceptable	Sufficient to manage the threat down to an acceptable level
1	Robust and effective	Complete set

5.6 Risk-scoring phase

5.6.1 All the information gathered on threat, impact and vulnerability should be used to identify and assess the residual risk. To score the risk accurately, assessors should use the formula:

RISK = THREAT × IMPACT × VULNERABILITY

5.6.2 For example, using an initial threat score of 2, an impact score of 4 and, where there are no mitigating measures in place (a vulnerability score of 4), the residual risk score would be 32 (2 × 4 × 4). Where measures are judged to reduce the vulnerability to some extent, but not to an acceptable level, the residual score would be 24. The threat and impact scores of 2 and 4 remain but the vulnerability score is now 3; hence $2 \times 4 \times 3 = 24$. And so on. As there is a presumption that no threat scenario can be managed totally out of existence, a score of 0 is not possible.

5.6.3 It should be noted that scenarios with differing individual threat, impact and vulnerability scores can have the same overall risk score. For instance, a particular scenario may have a threat score of 4, an impact score of 2 and a vulnerability score of 2, whereas another scenario may have a threat score of 1, an impact score of 4 and a vulnerability score of 4. Both scenarios produce a risk score of 16 despite having differing individual values for threat, impact and vulnerability.

5.6.4 Experience to date indicates that risk can be ranked into three broad categories – high, medium and low – as illustrated below:

.1 HIGH – a residual risk score of 27 or more.

.2 MEDIUM – a residual risk score of between 8 and 24.

.3 LOW – a residual risk score of 6 or less.

5.7 Risk-management phase

5.7.1 This phase considers how best to address the weaknesses identified during the vulnerability and risk-scoring stages and how to mitigate the risk effectively and practically on a sustainable long-term basis. This can be achieved by all stakeholders working together to agree joint tactical action plans, an example of which is shown below. The checklist below gives some pointers on how to work through the process:

.1 consider the overall risk profile from the risk register:

- High = Unacceptable risk – seek alternative and/or additional control measures,
- Medium = Manageable risk – requires management/monitoring,
- Low = Tolerable risk – no further control measures needed.

.2 reconsider the security measures review table in paragraph 5.5.4; the "possible weaknesses" and "follow-up actions" should assist in drawing up action plans.

.3 agree the priorities for action; these should be the "high" risks in the first instance.

.4 identify what actions can and need to be taken to bring the risk down to a "medium" (manageable) risk and from there to a "low" (tolerable) risk.

.5 agree on the lead agency in implementing changes.

.6 consider the resource implications.

.7 document recommendations, actions taken and link these back to the threats in the risk register.

.8 agreed actions should be recorded and progress monitored; such records are also evidence of decisions taken.

.9 determine the approval level for the action plan recommendations.

.10 consider the need to develop further systems for sharing information and intelligence.

.11 look for opportunities to share resources and assist others.

.12 establish a re-assessment schedule (e.g., as conditions change or on a regular schedule).

5.7.2 A sample action plan is illustrated on the following page.

Sample action plan

Ref number:		
THREAT:		
Current risk status:		
Area of port:		
Lead Agency:		
Current controls:		
Weaknesses:		
Review of residual risk:		
Date of review:		
AGREED ACTIONS:		CURRENT STATUS
		GREEN
Timescale for review:		
CURRENT STATUS KEY		
RED	Behind schedule, no remedial action in place	
AMBER	Falling behind schedule, remedial action in place	
GREEN	On track	

Appendix 5.1
Examples of internet sources of security assessment methodologies

1 Threat and risk analysis matrix (TRAM)

Source: ILO/IMO Code of practice on security in ports

Purpose: To provide smaller ports with few significant facilities and ports located in isolated areas with a practical risk-assessment and management tool.

Summary description: TRAM is a simplified version of the tool described above. It is a 10-step methodology which produces a risk score for each identified threat scenario as a basis for assigning priorities to security measures identified in an action plan. The tool is demonstrated by an example based on a specific threat scenario – destruction of a port authority's communication tower by explosives.

Internet site: www.imo.org/OurWork/Security/Instruments/Pages/CoP.aspx.

2 Port security risk assessment tool (PSRAT)

Source: United States Coast Guard (USCG) International Port Security Program

Purpose: To provide US Coast Guard captains of the port with a methodology for performing a risk-based analysis of assets and infrastructure within their area of responsibility.

Summary description: PSRAT is a more sophisticated version of the TRAM tool described above. It is a multi-step automated tool, created as a Microsoft Access 2000-based application. Input screens are used to capture the data needed for the analysis; all data is stored in the Access database. Unlike manual techniques, PSRAT facilitates the ability to update risk scenarios and their associated risks, identify the key drivers of risk scores and estimate the effectiveness of countermeasures.

Internet site: www.homeport.uscg.mil

PSRAT has been placed on the best practices website to enable its use as a template by other Government entities for their own risk-assessment purposes. The site contains two supporting documents, including the PSRAT User's Manual.

Resolutions of the Conference of Contracting Governments to the International Convention for the Safety of Life at Sea, 1974, adopted in December 2002

Conference resolution 1

(adopted on 12 December 2002)

Adoption of amendments to the Annex to the International Convention for the Safety of Life at Sea, 1974

THE CONFERENCE,

BEARING IN MIND the purposes and principles of the Charter of the United Nations concerning the maintenance of international peace and security and the promotion of friendly relations and co-operation among States,

DEEPLY CONCERNED about the world-wide escalation of acts of terrorism in all its forms, which endanger or take innocent human lives, jeopardize fundamental freedoms and seriously impair the dignity of human beings,

BEING AWARE of the importance and significance of shipping to the world trade and economy and, therefore, being determined to safeguard the world-wide supply chain against any breach resulting from terrorist attacks against ships, ports, offshore terminals or other facilities,

CONSIDERING that unlawful acts against shipping jeopardize the safety and security of persons and property, seriously affect the operation of maritime services and undermine the confidence of the peoples of the world in the safety of maritime navigation,

CONSIDERING that the occurrence of such acts is a matter of grave concern to the international community as a whole, while also recognizing the importance of the efficient and economic movement of world trade,

BEING CONVINCED of the urgent need to develop international co-operation between States in devising and adopting effective and practical measures, additional to those already adopted by the International Maritime Organization (hereinafter referred to as "the Organization"), to prevent and suppress unlawful acts directed against shipping in its broad sense,

RECALLING the United Nations Security Council resolution 1373(2001), adopted on 28 September 2001, requiring States to take measures to prevent and suppress terrorist acts, including calling on States to implement fully anti-terrorist conventions,

HAVING NOTED the Co-operative G8 Action on Transport Security (in particular, the Maritime Security section thereof), endorsed by the G8 Leaders during their Summit in Kananaskis, Alberta (Canada) in June 2002,

RECALLING article VIII(c) of the International Convention for the Safety of Life at Sea, 1974, as amended (hereinafter referred to as "the Convention"), concerning the procedure for amending the Convention by a Conference of Contracting Governments,

NOTING resolution A.924(22) entitled "Review of measures and procedures to prevent acts of terrorism which threaten the security of passengers and crew and the safety of ship", adopted by the Assembly of the Organization on 20 November 2001, which, *inter alia*:

(a) recognizes the need for the Organization to review, with the intent to revise, existing international legal and technical measures, and to consider appropriate new measures, to prevent and suppress terrorism against ships and to improve security aboard and ashore in order to reduce the risk to passengers, crew and port personnel on board ships and in port areas and to the vessels and their cargoes; and

(b) requests the Organization's Maritime Safety Committee, the Legal Committee and the Facilitation Committee under the direction of the Council to undertake, on a high-priority basis, a review to ascertain whether there is a need to update the instruments referred to in the preambular paragraphs of the aforesaid resolution and any other relevant IMO instrument under their scope and/or to adopt other security measures and, in the light of such a review, to take action as appropriate,

HAVING IDENTIFIED resolution A.584(14) entitled "Measures to prevent unlawful acts which threaten the safety of ships and the security of their passengers and crew", MSC/Circ.443 on "Measures to prevent unlawful acts against passengers and crew on board ships" and MSC/Circ.754 on "Passenger ferry security" among the IMO instruments relevant to the scope of resolution A.924(22),

RECALLING resolution 5 entitled "Future amendments to chapter XI of the 1974 SOLAS Convention on special measures to enhance maritime safety", adopted by the 1994 Conference of Contracting Governments to the International Convention for the Safety of Life at Sea, 1974,

HAVING CONSIDERED amendments to the Annex of the Convention proposed and circulated to all Members of the Organization and to all Contracting Governments to the Convention,

1. ADOPTS, in accordance with article VIII(c)(ii) of the Convention, amendments to the Annex of the Convention, the text of which is given in the Annex to the present resolution;

2. DETERMINES, in accordance with article VIII(b)(vi)(2)(bb) of the Convention, that the aforementioned amendments shall be deemed to have been accepted on 1 January 2004, unless, prior to that date, more than one third of the Contracting Governments to the Convention or Contracting Governments the combined merchant fleets of which constitute not less than 50% of the gross tonnage of the world's merchant fleet have notified their objections to the amendments;

3. INVITES Contracting Governments to the Convention to note that, in accordance with article VIII(b)(vii)(2) of the Convention, the said amendments shall enter into force on 1 July 2004 upon their acceptance in accordance with paragraph 2 above;

4. REQUESTS the Secretary-General of the Organization, in conformity with article VIII(b)(v) of the Convention, to transmit certified copies of the present resolution and the text of the amendments contained in the annex to all Contracting Governments to the Convention;

5. FURTHER REQUESTS the Secretary-General to transmit copies of this resolution and its Annex to all Members of the Organization which are not Contracting Governments to the Convention.

Annex

Amendments to the Annex to the International Convention for the Safety of Life at Sea, 1974 as amended

Chapter V
Safety of navigation

Regulation 19
Carriage requirements for shipborne navigational systems and equipment

1 *The existing subparagraphs .4, .5 and .6 of paragraph 2.4.2 are replaced by the following:*

"**.4** in the case of ships, other than passenger ships and tankers, of 300 gross tonnage and upwards but less than 50,000 gross tonnage, not later than the first safety equipment survey* after 1 July 2004 or by 31 December 2004, whichever occurs earlier; and"

2 *The following new sentence is added at the end of the existing subparagraph .7 of paragraph 2.4:*

"Ships fitted with AIS shall maintain AIS in operation at all times except where international agreements, rules or standards provide for the protection of navigational information."

Chapter XI
Special measures to enhance maritime safety

3 *The existing chapter XI is renumbered as chapter XI-1.*

Regulation 3
Ship identification number

4 *The following text is inserted after the title of the regulation:*

"(Paragraphs 4 and 5 apply to all ships to which this regulation applies. For ships constructed before 1 July 2004, the requirements of paragraphs 4 and 5 shall be complied with not later than the first scheduled dry-docking of the ship after 1 July 2004)"

5 *The existing paragraph 4 is deleted and the following new text is inserted:*

"**4** The ship's identification number shall be permanently marked:

.1 in a visible place either on the stern of the ship or on either side of the hull, amidships port and starboard, above the deepest assigned load line or either side of the superstructure, port and starboard or on the front of the superstructure or, in the case of passenger ships, on a horizontal surface visible from the air; and

.2 in an easily accessible place either on one of the end transverse bulkheads of the machinery spaces, as defined in regulation II-2/3.30, or on one of the hatchways or, in the case of tankers, in the pump-room or, in the case of ships with ro–ro spaces, as defined in regulation II-2/3.41, on one of the end transverse bulkheads of the ro–ro spaces.

* *The first safety equipment survey* means the first annual survey, the first periodical survey or the first renewal survey for safety equipment, whichever is due first after 1 July 2004, and, in addition, in the case of ships under construction, the initial survey.

5.1 The permanent marking shall be plainly visible, clear of any other markings on the hull and shall be painted in a contrasting colour.

5.2 The permanent marking referred to in paragraph 4.1 shall be not less than 200 mm in height. The permanent marking referred to in paragraph 4.2 shall not be less than 100 mm in height. The width of the marks shall be proportionate to the height.

5.3 The permanent marking may be made by raised lettering or by cutting it in or by centre-punching it or by any other equivalent method of marking the ship identification number which ensures that the marking is not easily expunged.

5.4 On ships constructed of material other than steel or metal, the Administration shall approve the method of marking the ship identification number."

6 *The following new regulation 5 is added after the existing regulation 4:*

"Regulation 5
Continuous Synopsis Record

1 Every ship to which chapter I applies shall be issued with a Continuous Synopsis Record.

2.1 The Continuous Synopsis Record is intended to provide an on-board record of the history of the ship with respect to the information recorded therein.

2.2 For ships constructed before 1 July 2004, the Continuous Synopsis Record shall, at least, provide the history of the ship as from 1 July 2004.

3 The Continuous Synopsis Record shall be issued by the Administration to each ship that is entitled to fly its flag and it shall contain, at least, the following information:

.1 the name of the State whose flag the ship is entitled to fly;

.2 the date on which the ship was registered with that State;

.3 the ship's identification number in accordance with regulation 3;

.4 the name of the ship;

.5 the port at which the ship is registered;

.6 the name of the registered owner(s) and their registered address(es);

.7 the name of the registered bareboat charterer(s) and their registered address(es), if applicable;

.8 the name of the Company, as defined in regulation IX/1, its registered address and the address(es) from where it carries out the safety-management activities;

.9 the name of all classification society(ies) with which the ship is classed;

.10 the name of the Administration or of the Contracting Government or of the recognized organization which has issued the Document of Compliance (or the Interim Document of Compliance), specified in the ISM Code as defined in regulation IX/1, to the Company operating the ship and the name of the body which has carried out the audit on the basis of which the Document was issued, if other than that issuing the Document;

.11 the name of the Administration or of the Contracting Government or of the recognized organization that has issued the Safety Management Certificate (or the Interim Safety Management Certificate), specified in the ISM Code as defined in regulation IX/1, to the ship and the name of the body which has carried out the audit on the basis of which the Certificate was issued, if other than that issuing the Certificate;

.12 the name of the Administration or of the Contracting Government or of the recognized security organization that has issued the International Ship Security Certificate (or the Interim International Ship Security Certificate), specified in part A of the ISPS Code as defined in regulation XI-2/1, to the ship and the name of the body which has carried out the verification on the basis of which the Certificate was issued, if other than that issuing the Certificate; and

.13 the date on which the ship ceased to be registered with that State.

4.1 Any changes relating to the entries referred to in paragraphs 3.4 to 3.12 shall be recorded in the Continuous Synopsis Record so as to provide updated and current information together with the history of the changes.

4.2 In case of any changes relating to the entries referred to in paragraph 4.1, the Administration shall issue, as soon as is practically possible but not later than three months from the date of the change, to the ships entitled to fly its flag either a revised and updated version of the Continuous Synopsis Record or appropriate amendments thereto.

4.3 In case of any changes relating to the entries referred to in paragraph 4.1, the Administration, pending the issue of a revised and updated version of the Continuous Synopsis Record, shall authorize and require either the Company as defined in regulation IX/1 or the master of the ship to amend the Continuous Synopsis Record to reflect the changes. In such cases, after the Continuous Synopsis Record has been amended, the Company shall, without delay, inform the Administration accordingly.

5.1 The Continuous Synopsis Record shall be in English, French or Spanish language. Additionally, a translation of the Continuous Synopsis Record into the official language or languages of the Administration may be provided.

5.2 The Continuous Synopsis Record shall be in the format developed by the Organization and shall be maintained in accordance with guidelines developed by the Organization. Any previous entries in the Continuous Synopsis Record shall not be modified, deleted or, in any way, erased or defaced.

6 Whenever a ship is transferred to the flag of another State or the ship is sold to another owner (or is taken over by another bareboat charterer) or another Company assumes the responsibility for the operation of the ship, the Continuous Synopsis Record shall be left on board.

7 When a ship is to be transferred to the flag of another State, the Company shall notify the Administration of the name of the State under whose flag the ship is to be transferred so as to enable the Administration to forward to that State a copy of the Continuous Synopsis Record covering the period during which the ship was under its jurisdiction.

8 When a ship is transferred to the flag of another State the Government of which is a Contracting Government, the Contracting Government of the State whose flag the ship was flying hitherto shall transmit to the Administration, as soon as possible after the transfer takes place, a copy of the relevant Continuous Synopsis Record covering the period during which the ship was under their jurisdiction together with any Continuous Synopsis Records previously issued to the ship by other States.

9 When a ship is transferred to the flag of another State, the Administration shall append the previous Continuous Synopsis Records to the Continuous Synopsis Record the Administration will issue to the ship so to provide the continuous history record intended by this regulation.

10 The Continuous Synopsis Record shall be kept on board the ship and shall be available for inspection at all times."

7 *The following new chapter XI-2 is inserted after the renumbered chapter XI-1:*

"Chapter XI-2
Special measures to enhance maritime security

Regulation 1
Definitions

1 For the purpose of this chapter, unless expressly provided otherwise:

.1 *Bulk carrier* means a bulk carrier as defined in regulation IX/1.6.

.2 *Chemical tanker* means a chemical tanker as defined in regulation VII/8.2.

.3 *Gas carrier* means a gas carrier as defined in regulation VII/11.2.

.4 *High-speed craft* means a craft as defined in regulation X/1.2.

.5 *Mobile offshore drilling unit* means a mechanically propelled mobile offshore drilling unit, as defined in regulation IX/1, not on location.

.6 *Oil tanker* means an oil tanker as defined in regulation II-1/2.12.

.7 *Company* means a Company as defined in regulation IX/1.

.8 *Ship/port interface* means the interactions that occur when a ship is directly and immediately affected by actions involving the movement of persons, goods or the provisions of port services to or from the ship.

.9 *Port facility* is a location, as determined by the Contracting Government or by the Designated Authority, where the ship/port interface takes place. This includes areas such as anchorages, waiting berths and approaches from seaward, as appropriate.

.10 *Ship-to-ship activity* means any activity not related to a port facility that involves the transfer of goods or persons from one ship to another.

.11 *Designated Authority* means the organization(s) or the administration(s) identified, within the Contracting Government, as responsible for ensuring the implementation of the provisions of this chapter pertaining to port facility security and ship/port interface, from the point of view of the port facility.

.12 *International Ship and Port Facility Security (ISPS) Code* means the International Code for the Security of Ships and of Port Facilities consisting of part A (the provisions of which shall be treated as mandatory) and part B (the provisions of which shall be treated as recommendatory), as adopted, on 12 December 2002, by resolution 2 of the Conference of Contracting Governments to the International Convention for the Safety of Life at Sea, 1974 as may be amended by the Organization, provided that:

.1 amendments to part A of the Code are adopted, brought into force and take effect in accordance with article VIII of the present Convention concerning the amendment procedures applicable to the Annex other than chapter I; and

.2 amendments to part B of the Code are adopted by the Maritime Safety Committee in accordance with its Rules of Procedure.

.13 *Security incident* means any suspicious act or circumstance threatening the security of a ship, including a mobile offshore drilling unit and a high-speed craft, or of a port facility or of any ship/port interface or any ship-to-ship activity.

.14 *Security level* means the qualification of the degree of risk that a security incident will be attempted or will occur.

.15 *Declaration of Security* means an agreement reached between a ship and either a port facility or another ship with which it interfaces, specifying the security measures each will implement.

.16 *Recognized security organization* means an organization with appropriate expertise in security matters and with appropriate knowledge of ship and port operations authorized to carry out an assessment, or a verification, or an approval or a certification activity, required by this chapter or by part A of the ISPS Code.

2 The term "ship", when used in regulations 3 to 13, includes mobile offshore drilling units and high-speed craft.

3 The term "all ships", when used in this chapter, means any ship to which this chapter applies.

4 The term "Contracting Government", when used in regulations 3, 4, 7 and 10 to 13, includes a reference to the Designated Authority.

Regulation 2
Application

1 This chapter applies to:

.1 the following types of ships engaged on international voyages:

.1.1 passenger ships, including high-speed passenger craft;

.1.2 cargo ships, including high-speed craft, of 500 gross tonnage and upwards; and

.1.3 mobile offshore drilling units; and

.2 port facilities serving such ships engaged on international voyages.

2 Notwithstanding the provisions of paragraph 1.2, Contracting Governments shall decide the extent of application of this chapter and of the relevant sections of part A of the ISPS Code to those port facilities within their territory which, although used primarily by ships not engaged on international voyages, are required, occasionally, to serve ships arriving or departing on an international voyage.

2.1 Contracting Governments shall base their decisions, under paragraph 2, on a port facility security assessment carried out in accordance with the provisions of part A of the ISPS Code.

2.2 Any decision which a Contracting Government makes, under paragraph 2, shall not compromise the level of security intended to be achieved by this chapter or by part A of the ISPS Code.

3 This chapter does not apply to warships, naval auxiliaries or other ships owned or operated by a Contracting Government and used only on Government non-commercial service.

4 Nothing in this chapter shall prejudice the rights or obligations of States under international law.

Regulation 3
Obligations of Contracting Governments with respect to security

1 Administrations shall set security levels and ensure the provision of security level information to ships entitled to fly their flag. When changes in security level occur, security-level information shall be updated as the circumstance dictates.

2 Contracting Governments shall set security levels and ensure the provision of security-level information to port facilities within their territory, and to ships prior to entering a port or whilst in a port within their territory. When changes in security level occur, security-level information shall be updated as the circumstance dictates.

Regulation 4
Requirements for Companies and ships

1 Companies shall comply with the relevant requirements of this chapter and of part A of the ISPS Code, taking into account the guidance given in part B of the ISPS Code.

2 Ships shall comply with the relevant requirements of this chapter and of part A of the ISPS Code, taking into account the guidance given in part B of the ISPS Code, and such compliance shall be verified and certified as provided for in part A of the ISPS Code.

3 Prior to entering a port or whilst in a port within the territory of a Contracting Government, a ship shall comply with the requirements for the security level set by that Contracting Government, if such security level is higher than the security level set by the Administration for that ship.

4 Ships shall respond without undue delay to any change to a higher security level.

5 Where a ship is not in compliance with the requirements of this chapter or of part A of the ISPS Code, or cannot comply with the requirements of the security level set by the Administration or by another Contracting Government and applicable to that ship, then the ship shall notify the appropriate competent authority prior to conducting any ship/port interface or prior to entry into port, whichever occurs earlier.

Regulation 5
Specific responsibility of Companies

The Company shall ensure that the master has available on board, at all times, information through which officers duly authorized by a Contracting Government can establish:

- **.1** who is responsible for appointing the members of the crew or other persons currently employed or engaged on board the ship in any capacity on the business of that ship;
- **.2** who is responsible for deciding the employment of the ship; and
- **.3** in cases where the ship is employed under the terms of charter party(ies), who are the parties to such charter party(ies).

Regulation 6
Ship security alert system[*]

1 All ships shall be provided with a ship security alert system, as follows:

- **.1** ships constructed on or after 1 July 2004;
- **.2** passenger ships, including high-speed passenger craft, constructed before 1 July 2004, not later than the first survey of the radio installation after 1 July 2004;
- **.3** oil tankers, chemical tankers, gas carriers, bulk carriers and cargo high-speed craft, of 500 gross tonnage and upwards constructed before 1 July 2004, not later than the first survey of the radio installation after 1 July 2004; and
- **.4** other cargo ships of 500 gross tonnage and upward and mobile offshore drilling units constructed before 1 July 2004, not later than the first survey of the radio installation after 1 July 2006.

[*] Refer to the performance standards for a ship security alert system adopted by resolution MSC.136(76).

2 The ship security alert system, when activated, shall:

.1 initiate and transmit a ship-to-shore security alert to a competent authority designated by the Administration, which in these circumstances may include the Company, identifying the ship, its location and indicating that the security of the ship is under threat or it has been compromised;

.2 not send the ship security alert to any other ships;

.3 not raise any alarm on board the ship; and

.4 continue the ship security alert until deactivated and/or reset.

3 The ship security alert system shall:

.1 be capable of being activated from the navigation bridge and in at least one other location; and

.2 conform to performance standards not inferior to those adopted by the Organization.

4 The ship security alert system activation points shall be designed so as to prevent the inadvertent initiation of the ship security alert.

5 The requirement for a ship security alert system may be complied with by using the radio installation fitted for compliance with the requirements of chapter IV, provided all requirements of this regulation are complied with.

6 When an Administration receives notification of a ship security alert, that Administration shall immediately notify the State(s) in the vicinity of which the ship is presently operating.

7 When a Contracting Government receives notification of a ship security alert from a ship which is not entitled to fly its flag, that Contracting Government shall immediately notify the relevant Administration and, if appropriate, the State(s) in the vicinity of which the ship is presently operating.

Regulation 7
Threats to ships

1 Contracting Governments shall set security levels and ensure the provision of security-level information to ships operating in their territorial sea or having communicated an intention to enter their territorial sea.

2 Contracting Governments shall provide a point of contact through which such ships can request advice or assistance and to which such ships can report any security concerns about other ships, movements or communications.

3 Where a risk of attack has been identified, the Contracting Government concerned shall advise the ships concerned and their Administrations of:

.1 the current security level;

.2 any security measures that should be put in place by the ships concerned to protect themselves from attack, in accordance with the provisions of part A of the ISPS Code; and

.3 security measures that the coastal State has decided to put in place, as appropriate.

Regulation 8
Master's discretion for ship safety and security

1 The master shall not be constrained by the Company, the charterer or any other person from taking or executing any decision which, in the professional judgement of the master, is necessary to maintain the safety and security of the ship. This includes denial of access to persons (except those identified as duly authorized by a Contracting Government) or their effects and refusal to load cargo, including containers or other closed cargo transport units.

2 If, in the professional judgement of the master, a conflict between any safety and security requirements applicable to the ship arises during its operations, the master shall give effect to those requirements necessary to maintain the safety of the ship. In such cases, the master may implement temporary security measures and shall forthwith inform the Administration and, if appropriate, the Contracting Government in whose port the ship is operating or intends to enter. Any such temporary security measures under this regulation shall, to the highest possible degree, be commensurate with the prevailing security level. When such cases are identified, the Administration shall ensure that such conflicts are resolved and that the possibility of recurrence is minimized.

Regulation 9
Control and compliance measures

1 Control of ships in port

1.1 For the purpose of this chapter, every ship to which this chapter applies is subject to control when in a port of another Contracting Government by officers duly authorized by that Government, who may be the same as those carrying out the functions of regulation I/19. Such control shall be limited to verifying that there is on board a valid International Ship Security Certificate or a valid Interim International Ship Security Certificate issued under the provisions of part A of the ISPS Code ("Certificate"), which if valid shall be accepted, unless there are clear grounds for believing that the ship is not in compliance with the requirements of this chapter or part A of the ISPS Code.

1.2 When there are such clear grounds, or when no valid Certificate is produced when required, the officers duly authorized by the Contracting Government shall impose any one or more control measures in relation to that ship as provided in paragraph 1.3. Any such measures imposed must be proportionate, taking into account the guidance given in part B of the ISPS Code.

1.3 Such control measures are as follows: inspection of the ship, delaying the ship, detention of the ship, restriction of operations, including movement within the port, or expulsion of the ship from port. Such control measures may additionally or alternatively include other lesser administrative or corrective measures.

2 Ships intending to enter a port of another Contracting Government

2.1 For the purpose of this chapter, a Contracting Government may require that ships intending to enter its ports provide the following information to officers duly authorized by that Government to ensure compliance with this chapter prior to entry into port with the aim of avoiding the need to impose control measures or steps:

- **.1** that the ship possesses a valid Certificate and the name of its issuing authority;
- **.2** the security level at which the ship is currently operating;
- **.3** the security level at which the ship operated in any previous port where it has conducted a ship/port interface within the timeframe specified in paragraph 2.3;
- **.4** any special or additional security measures that were taken by the ship in any previous port where it has conducted a ship/port interface within the timeframe specified in paragraph 2.3;
- **.5** that the appropriate ship security procedures were maintained during any ship-to-ship activity within the timeframe specified in paragraph 2.3; or
- **.6** other practical security-related information (but not details of the ship security plan), taking into account the guidance given in part B of the ISPS Code.

If requested by the Contracting Government, the ship or the Company shall provide confirmation, acceptable to that Contracting Government, of the information required above.

2.2 Every ship to which this chapter applies intending to enter the port of another Contracting Government shall provide the information described in paragraph 2.1 on the request of the officers duly authorized by that Government. The master may decline to provide such information on the understanding that failure to do so may result in denial of entry into port.

2.3 The ship shall keep records of the information referred to in paragraph 2.1 for the last 10 calls at port facilities.

2.4 If, after receipt of the information described in paragraph 2.1, officers duly authorized by the Contracting Government of the port in which the ship intends to enter have clear grounds for believing that the ship is in non-compliance with the requirements of this chapter or part A of the ISPS Code, such officers shall attempt to establish communication with and between the ship and the Administration in order to rectify the non-compliance. If such communication does not result in rectification, or if such officers have clear grounds otherwise for believing that the ship is in non-compliance with the requirements of this chapter or part A of the ISPS Code, such officers may take steps in relation to that ship as provided in paragraph 2.5. Any such steps taken must be proportionate, taking into account the guidance given in part B of the ISPS Code.

2.5 Such steps are as follows:

.1 a requirement for the rectification of the non-compliance;

.2 a requirement that the ship proceed to a location specified in the territorial sea or internal waters of that Contracting Government;

.3 inspection of the ship, if the ship is in the territorial sea of the Contracting Government the port of which the ship intends to enter; or

.4 denial of entry into port.

Prior to initiating any such steps, the ship shall be informed by the Contracting Government of its intentions. Upon this information the master may withdraw the intention to enter that port. In such cases, this regulation shall not apply.

3 Additional provisions

3.1 In the event:

.1 of the imposition of a control measure, other than a lesser administrative or corrective measure, referred to in paragraph 1.3; or

.2 any of the steps referred to in paragraph 2.5 are taken,

an officer duly authorized by the Contracting Government shall forthwith inform in writing the Administration specifying which control measures have been imposed or steps taken and the reasons thereof. The Contracting Government imposing the control measures or steps shall also notify the recognized security organization which issued the Certificate relating to the ship concerned and the Organization when any such control measures have been imposed or steps taken.

3.2 When entry into port is denied or the ship is expelled from port, the authorities of the port State should communicate the appropriate facts to the authorities of the State of the next appropriate ports of call, when known, and any other appropriate coastal States, taking into account guidelines to be developed by the Organization. Confidentiality and security of such notification shall be ensured.

3.3 Denial of entry into port, pursuant to paragraphs 2.4 and 2.5, or expulsion from port, pursuant to paragraphs 1.1 to 1.3, shall only be imposed where the officers duly authorized by the Contracting Government have clear grounds to believe that the ship poses an immediate threat to the security or safety of persons, or of ships or other property and there are no other appropriate means for removing that threat.

3.4 The control measures referred to in paragraph 1.3 and the steps referred to in paragraph 2.5 shall only be imposed, pursuant to this regulation, until the non-compliance giving rise to the control measures or steps has been corrected to the satisfaction of the Contracting Government, taking into account actions proposed by the ship or the Administration, if any.

3.5 When Contracting Governments exercise control under paragraph 1 or take steps under paragraph 2:

.1 all possible efforts shall be made to avoid a ship being unduly detained or delayed. If a ship is thereby unduly detained, or delayed, it shall be entitled to compensation for any loss or damage suffered; and

.2 necessary access to the ship shall not be prevented for emergency or humanitarian reasons and for security purposes.

Regulation 10
Requirements for port facilities

1 Port facilities shall comply with the relevant requirements of this chapter and part A of the ISPS Code, taking into account the guidance given in part B of the ISPS Code.

2 Contracting Governments with a port facility or port facilities within their territory, to which this regulation applies, shall ensure that:

.1 port facility security assessments are carried out, reviewed and approved in accordance with the provisions of part A of the ISPS Code; and

.2 port facility security plans are developed, reviewed, approved and implemented in accordance with the provisions of part A of the ISPS Code.

3 Contracting Governments shall designate and communicate the measures required to be addressed in a port facility security plan for the various security levels, including when the submission of a Declaration of Security will be required.

Regulation 11
Alternative security agreements

1 Contracting Governments may, when implementing this chapter and part A of the ISPS Code, conclude in writing bilateral or multilateral agreements with other Contracting Governments on alternative security arrangements covering short international voyages on fixed routes between port facilities located within their territories.

2 Any such agreement shall not compromise the level of security of other ships or of port facilities not covered by the agreement.

3 No ship covered by such an agreement shall conduct any ship-to-ship activities with any ship not covered by the agreement.

4 Such agreements shall be reviewed periodically, taking into account the experience gained as well as any changes in the particular circumstances or the assessed threats to the security of the ships, the port facilities or the routes covered by the agreement.

Regulation 12
Equivalent security arrangements

1 An Administration may allow a particular ship or a group of ships entitled to fly its flag to implement other security measures equivalent to those prescribed in this chapter or in part A of the ISPS Code, provided such security measures are at least as effective as those prescribed in this chapter or part A of the ISPS Code. The Administration which allows such security measures shall communicate to the Organization particulars thereof.

2 When implementing this chapter and part A of the ISPS Code, a Contracting Government may allow a particular port facility or a group of port facilities located within its territory, other than those covered by an agreement concluded under regulation 11, to implement security measures equivalent to those prescribed in this chapter or in part A of the ISPS Code, provided such security measures are at least as effective as those prescribed in this chapter or part A of the ISPS Code. The Contracting Government which allows such security measures shall communicate to the Organization particulars thereof.

Regulation 13
Communication of information

1 Contracting Governments shall, not later than 1 July 2004, communicate to the Organization and shall make available for the information of Companies and ships:

- **.1** the names and contact details of their national authority or authorities responsible for ship and port facility security;
- **.2** the locations within their territory covered by approved port facility security plans;
- **.3** the names and contact details of those who have been designated to be available at all times to receive and act upon the ship-to-shore security alerts referred to in regulation 6.2.1;
- **.4** the names and contact details of those who have been designated to be available at all times to receive and act upon any communications from Contracting Governments exercising control and compliance measures referred to in regulation 9.3.1; and
- **.5** the names and contact details of those who have been designated to be available at all times to provide advice or assistance to ships and to whom ships can report any security concerns referred to in regulation 7.2

and thereafter update such information as and when changes relating thereto occur. The Organization shall circulate such particulars to other Contracting Governments for the information of their officers.

2 Contracting Governments shall, not later than 1 July 2004, communicate to the Organization the names and contact details of any recognized security organizations authorized to act on their behalf together with details of the specific responsibility and conditions of authority delegated to such organizations. Such information shall be updated as and when changes relating thereto occur. The Organization shall circulate such particulars to other Contracting Governments for the information of their officers.

3 Contracting Governments shall, not later than 1 July 2004, communicate to the Organization a list showing the approved port facility security plans for the port facilities located within their territory together with the location or locations covered by each approved port facility security plan and the corresponding date of approval and thereafter shall further communicate when any of the following changes take place:

- **.1** changes in the location or locations covered by an approved port facility security plan are to be introduced or have been introduced. In such cases the information to be communicated shall indicate the changes in the location or locations covered by the plan and the date as of which such changes are to be introduced or were implemented;
- **.2** an approved port facility security plan, previously included in the list submitted to the Organization, is to be withdrawn or has been withdrawn. In such cases, the information to be communicated shall indicate the date on which the withdrawal will take effect or was implemented. In these cases, the communication shall be made to the Organization as soon as is practically possible; and

.3 additions are to be made to the list of approved port facility security plans. In such cases, the information to be communicated shall indicate the location or locations covered by the plan and the date of approval.

4 Contracting Governments shall, at five year intervals after 1 July 2004, communicate to the Organization a revised and updated list showing all the approved port facility security plans for the port facilities located within their territory together with the location or locations covered by each approved port facility security plan and the corresponding date of approval (and the date of approval of any amendments thereto) which will supersede and replace all information communicated to the Organization, pursuant to paragraph 3, during the preceding five years.

5 Contracting Governments shall communicate to the Organization information that an agreement under regulation 11 has been concluded. The information communicated shall include:

.1 the names of the Contracting Governments which have concluded the agreement;

.2 the port facilities and the fixed routes covered by the agreement;

.3 the periodicity of review of the agreement;

.4 the date of entry into force of the agreement; and

.5 information on any consultations which have taken place with other Contracting Governments

and thereafter shall communicate, as soon as practically possible, to the Organization information when the agreement has been amended or has ended.

6 Any Contracting Government which allows, under the provisions of regulation 12, any equivalent security arrangements with respect to a ship entitled to fly its flag or with respect to a port facility located within its territory shall communicate to the Organization particulars thereof.

7 The Organization shall make available the information communicated under paragraphs 3 to 6 to other Contracting Governments upon request."

Conference resolution 2

(adopted on 12 December 2002)

Adoption of the International Code for the Security of Ships and of Port Facilities

THE CONFERENCE,

HAVING ADOPTED amendments to the International Convention for the Safety of Life at Sea, 1974, as amended (hereinafter referred to as "the Convention"), concerning special measures to enhance maritime safety and security,

CONSIDERING that the new chapter XI-2 of the Convention makes a reference to an International Ship and Port Facility Security (ISPS) Code and requires ships, companies and port facilities to comply with the relevant requirements of part A of the International Ship and Port Facility Security (ISPS) Code, as specified in part A of the ISPS Code,

BEING OF THE OPINION that the implementation by Contracting Governments of the said chapter will greatly contribute to the enhancement of maritime safety and security and safeguarding those on board and ashore,

HAVING CONSIDERED a draft of the International Code for the Security of Ships and of Port Facilities prepared by the Maritime Safety Committee of the International Maritime Organization (hereinafter referred to as "the Organization"), at its seventy-fifth and seventy-sixth sessions, for consideration and adoption by the Conference,

1. ADOPTS the International Code for the Security of Ships and of Port Facilities (hereinafter referred to as "the Code"), the text of which is set out in the annex to the present resolution;

2. INVITES Contracting Governments to the Convention to note that the Code will take effect on 1 July 2004 upon entry into force of the new chapter XI-2 of the Convention;

3. REQUESTS the Maritime Safety Committee to keep the Code under review and amend it, as appropriate;

4. REQUESTS the Secretary-General of the Organization to transmit certified copies of the present resolution and the text of the Code contained in the annex to all Contracting Governments to the Convention;

5. FURTHER REQUESTS the Secretary-General to transmit copies of this resolution and its annex to all Members of the Organization which are not Contracting Governments to the Convention.

Annex

International Code for the Security of Ships and of Port Facilities

Preamble

1 The Diplomatic Conference on Maritime Security held in London in December 2002 adopted new provisions in the International Convention for the Safety of Life at Sea, 1974 and this Code* to enhance maritime security. These new requirements form the international framework through which ships and port facilities can co-operate to detect and deter acts which threaten security in the maritime transport sector.

2 Following the tragic events of 11th September 2001, the twenty-second session of the Assembly of the International Maritime Organization ("the Organization"), in November 2001, unanimously agreed to the development of new measures relating to the security of ships and of port facilities for adoption by a Conference of Contracting Governments to the International Convention for the Safety of Life at Sea, 1974 (known as the Diplomatic Conference on Maritime Security) in December 2002. Preparation for the Diplomatic Conference was entrusted to the Organization's Maritime Safety Committee (MSC) on the basis of submissions made by Member States, intergovernmental organizations and non-governmental organizations in consultative status with the Organization.

3 The MSC, at its first extraordinary session, held also in November 2001, in order to accelerate the development and the adoption of the appropriate security measures, established an MSC Intersessional Working Group on Maritime Security. The first meeting of the MSC Intersessional Working Group on Maritime Security was held in February 2002 and the outcome of its discussions was reported to, and considered by, the seventy-fifth session of the MSC in May 2002, when an *ad hoc* Working Group was established to further develop the proposals made. The seventy-fifth session of the MSC considered the report of that Working Group and recommended that work should be taken forward through a further MSC Intersessional Working Group, which was held in September 2002. The seventy-sixth session of the MSC considered the outcome of the September 2002 session of the MSC Intersessional Working Group and the further work undertaken by the MSC Working Group held in conjunction with the Committee's seventy-sixth session in December 2002, immediately prior to the Diplomatic Conference, and agreed the final version of the proposed texts to be considered by the Diplomatic Conference.

4 The Diplomatic Conference (9 to 13 December 2002) also adopted amendments to the existing provisions of the International Convention for the Safety of Life at Sea, 1974 (SOLAS 74) accelerating the implementation of the requirement to fit automatic identification systems and adopted new regulations in chapter XI-1 of SOLAS 74 covering marking of the Ship Identification Number and the carriage of a Continuous Synopsis Record. The Diplomatic Conference also adopted a number of Conference resolutions, including those covering implementation and revision of this Code, technical co-operation, and co-operative work with the International Labour Organization and World Customs Organization. It was recognized that review and amendment of certain of the new provisions regarding maritime security may be required on completion of the work of these two Organizations.

5 The provisions of chapter XI-2 of SOLAS 74 and this Code apply to ships and to port facilities. The extension of SOLAS 74 to cover port facilities was agreed on the basis that SOLAS 74 offered the speediest means of ensuring the necessary security measures entered into force and given effect quickly. However, it was further agreed that the provisions relating to port facilities should relate solely to the ship/port interface. The wider issue of the security of port areas will be the subject of further joint work between the International Maritime Organization and the International Labour Organization. It was also agreed that the provisions should not extend to the actual response to attacks or to any necessary clear-up activities after such an attack.

* The complete name of the Code is the International Code for the Security of Ships and of Port Facilities. The abbreviated name of this Code, as referred to in regulation XI-2/1 of SOLAS 74 as amended, is the International Ship and Port Facility Security (ISPS) Code, or in short, the ISPS Code.

6 In drafting the provision, care has been taken to ensure compatibility with the provisions of the International Convention on Standards of Training, Certification and Watchkeeping for Seafarers, 1978, as amended, the International Safety Management (ISM) Code and the harmonized system of survey and certification.

7 The provisions represent a significant change in the approach of the international maritime industries to the issue of security in the maritime transport sector. It is recognized that they may place a significant additional burden on certain Contracting Governments. The importance of technical co-operation to assist Contracting Governments implement the provisions is fully recognized.

8 Implementation of the provisions will require continuing effective co-operation and understanding between all those involved with, or using, ships and port facilities, including ship's personnel, port personnel, passengers, cargo interests, ship and port management and those in national and local authorities with security responsibilities. Existing practices and procedures will have to be reviewed and changed if they do not provide an adequate level of security. In the interests of enhanced maritime security, additional responsibilities will have to be carried by the shipping and port industries and by national and local authorities.

9 The guidance given in part B of this Code should be taken into account when implementing the security provisions set out in chapter XI-2 of SOLAS 74 and in part A of this Code. However, it is recognized that the extent to which the guidance applies may vary depending on the nature of the port facility and of the ship, its trade and/or cargo.

10 Nothing in this Code shall be interpreted or applied in a manner inconsistent with the proper respect of fundamental rights and freedoms as set out in international instruments, particularly those relating to maritime workers and refugees, including the International Labour Organization Declaration of Fundamental Principles and Rights at Work as well as international standards concerning maritime and port workers.

11 Recognizing that the Convention on the Facilitation of Maritime Traffic, 1965, as amended, provides that foreign crew members shall be allowed ashore by the public authorities while the ship on which they arrive is in port, provided that the formalities on arrival of the ship have been fulfilled and the public authorities have no reason to refuse permission to come ashore for reasons of public health, public safety or public order, Contracting Governments, when approving ship and port facility security plans, should pay due cognisance to the fact that ship's personnel live and work on the vessel and need shore leave and access to shore-based seafarer welfare facilities, including medical care.

ISPS Code

International Ship and Port Facility Security Code

Foreword

The International Ship and Port Facility Security Code (ISPS Code) represents the culmination of just over a year's intense work by IMO's Maritime Safety Committee and its Maritime Security Working Group since the twenty-second session of the Assembly adopted resolution A.924(22), on the review of measures and procedures to prevent acts of terrorism which threaten the security of passengers and crews and the safety of ships, in November 2001. The ISPS Code was adopted by one of the resolutions that were adopted on 12 December 2002 by the Conference of Contracting Governments to the International Convention for the Safety of Life at Sea, 1974 (London, 9 to 13 December 2002). Another resolution also includes the necessary amendments to chapters V and XI of SOLAS by which compliance with the Code became mandatory on 1 July 2004. The existing chapter XI of SOLAS was amended and re-identified as chapter XI-1 and a new chapter XI-2 was adopted on special measures to enhance maritime security. The ISPS Code and these amendments to SOLAS are set out in this publication, as are other resolutions (relating to the work that needed to be completed before the Code could be implemented in 2004 and the revision of the Code, technical co-operation, and co-operative work with the International Labour Organization and the World Customs Organization) that were adopted by the Conference.

The objectives of this Code are to establish an international framework involving co-operation between Contracting Governments, Government agencies, local administrations and the shipping and port industries to detect/assess security threats and take preventive measures against security incidents affecting ships or port facilities used in international trade; to establish the respective roles and responsibilities of all these parties concerned, at the national and international level, for ensuring maritime security; to ensure the early and efficient collation and exchange of security-related information; to provide a methodology for security assessments so as to have in place plans and procedures to react to changing security levels; and to ensure confidence that adequate and proportionate maritime security measures are in place. The objectives are to be achieved by the designation of appropriate officers/personnel on each ship, in each port facility and in each shipping company to prepare and to put into effect the security plans that will be approved for each ship and port facility. Parts A and B of the Code are, respectively, the mandatory requirements regarding the provisions of chapter XI-2 of SOLAS, 1974, as amended, and guidance regarding the provisions of chapter XI-2 of SOLAS, 1974, as amended, and part A of the Code.

Part A
Mandatory requirements regarding the provisions of chapter XI-2 of the Annex to the International Convention for the Safety of Life At Sea, 1974, as amended

1 General

1.1 Introduction

This part of the International Code for the Security of Ships and of Port Facilities contains mandatory provisions to which reference is made in chapter XI-2 of the International Convention for the Safety of Life at Sea, 1974, as amended.

1.2 Objectives

The objectives of this Code are:

.1 to establish an international framework involving co-operation between Contracting Governments, Government agencies, local administrations and the shipping and port industries to detect security threats and take preventive measures against security incidents affecting ships or port facilities used in international trade;

.2 to establish the respective roles and responsibilities of the Contracting Governments, Government agencies, local administrations and the shipping and port industries, at the national and international level, for ensuring maritime security;

.3 to ensure the early and efficient collection and exchange of security-related information;

.4 to provide a methodology for security assessments so as to have in place plans and procedures to react to changing security levels; and

.5 to ensure confidence that adequate and proportionate maritime security measures are in place.

1.3 Functional requirements

In order to achieve its objectives, this Code embodies a number of functional requirements. These include, but are not limited to:

.1 gathering and assessing information with respect to security threats and exchanging such information with appropriate Contracting Governments;

.2 requiring the maintenance of communication protocols for ships and port facilities;

.3 preventing unauthorized access to ships, port facilities and their restricted areas;

.4 preventing the introduction of unauthorized weapons, incendiary devices or explosives to ships or port facilities;

.5 providing means for raising the alarm in reaction to security threats or security incidents;

.6 requiring ship and port facility security plans based upon security assessments; and

.7 requiring training, drills and exercises to ensure familiarity with security plans and procedures.

2 Definitions

2.1 For the purpose of this part, unless expressly provided otherwise:

.1 *Convention* means the International Convention for the Safety of Life at Sea, 1974, as amended.

.2 *Regulation* means a regulation of the Convention.

.3 *Chapter* means a chapter of the Convention.

.4 *Ship security plan* means a plan developed to ensure the application of measures on board the ship designed to protect persons on board, cargo, cargo transport units, ship's stores or the ship from the risks of a security incident.

.5 *Port facility security plan* means a plan developed to ensure the application of measures designed to protect the port facility and ships, persons, cargo, cargo transport units and ship's stores within the port facility from the risks of a security incident.

.6 *Ship security officer* means the person on board the ship, accountable to the master, designated by the Company as responsible for the security of the ship, including implementation and maintenance of the ship security plan, and for liaison with the company security officer and port facility security officers.

.7 *Company security officer* means the person designated by the Company for ensuring that a ship security assessment is carried out; that a ship security plan is developed, submitted for approval, and thereafter implemented and maintained, and for liaison with port facility security officers and the ship security officer.

.8 *Port facility security officer* means the person designated as responsible for the development, implementation, revision and maintenance of the port facility security plan and for liaison with the ship security officers and company security officers.

.9 *Security level 1* means the level for which minimum appropriate protective security measures shall be maintained at all times.

.10 *Security level 2* means the level for which appropriate additional protective security measures shall be maintained for a period of time as a result of heightened risk of a security incident.

.11 *Security level 3* means the level for which further specific protective security measures shall be maintained for a limited period of time when a security incident is probable or imminent, although it may not be possible to identify the specific target.

2.2 The term "ship", when used in this Code, includes mobile offshore drilling units and high-speed craft as defined in regulation XI-2/1.

2.3 The term "Contracting Government" in connection with any reference to a port facility, when used in sections 14 to 18, includes a reference to the Designated Authority.

2.4 Terms not otherwise defined in this part shall have the same meaning as the meaning attributed to them in chapters I and XI-2.

3 Application

3.1 This Code applies to:

.1 the following types of ships engaged on international voyages:

  .1 passenger ships, including high-speed passenger craft;

  .2 cargo ships, including high-speed craft, of 500 gross tonnage and upwards; and

  .3 mobile offshore drilling units; and

.2 port facilities serving such ships engaged on international voyages.

3.2 Notwithstanding the provisions of section 3.1.2, Contracting Governments shall decide the extent of application of this part of the Code to those port facilities within their territory which, although used primarily by ships not engaged on international voyages, are required, occasionally, to serve ships arriving or departing on an international voyage.

3.2.1 Contracting Governments shall base their decisions, under section 3.2, on a port facility security assessment carried out in accordance with this part of the Code.

3.2.2 Any decision which a Contracting Government makes, under section 3.2, shall not compromise the level of security intended to be achieved by chapter XI-2 or by this part of the Code.

3.3 This Code does not apply to warships, naval auxiliaries or other ships owned or operated by a Contracting Government and used only on Government non-commercial service.

3.4 Sections 5 to 13 and 19 of this part apply to companies and ships as specified in regulation XI-2/4.

3.5 Sections 5 and 14 to 18 of this part apply to port facilities as specified in regulation XI-2/10.

3.6 Nothing in this Code shall prejudice the rights or obligations of States under international law.

4 Responsibilities of Contracting Governments

4.1 Subject to the provisions of regulations XI-2/3 and XI-2/7, Contracting Governments shall set security levels and provide guidance for protection from security incidents. Higher security levels indicate greater likelihood of occurrence of a security incident. Factors to be considered in setting the appropriate security level include:

- **.1** the degree that the threat information is credible;
- **.2** the degree that the threat information is corroborated;
- **.3** the degree that the threat information is specific or imminent; and
- **.4** the potential consequences of such a security incident.

4.2 Contracting Governments, when they set security level 3, shall issue, as necessary, appropriate instructions and shall provide security-related information to the ships and port facilities that may be affected.

4.3 Contracting Governments may delegate to a recognized security organization certain of their security-related duties under chapter XI-2 and this part of the Code with the exception of:

- **.1** setting of the applicable security level;
- **.2** approving a port facility security assessment and subsequent amendments to an approved assessment;
- **.3** determining the port facilities which will be required to designate a port facility security officer;
- **.4** approving a port facility security plan and subsequent amendments to an approved plan;
- **.5** exercising control and compliance measures pursuant to regulation XI-2/9; and
- **.6** establishing the requirements for a Declaration of Security.

4.4 Contracting Governments shall, to the extent they consider appropriate, test the effectiveness of the ship security plans or the port facility security plans, or of amendments to such plans, they have approved, or, in the case of ships, of plans which have been approved on their behalf.

5 Declaration of Security

5.1 Contracting Governments shall determine when a Declaration of Security is required by assessing the risk the ship/port interface or ship-to-ship activity poses to persons, property or the environment.

5.2 A ship can request completion of a Declaration of Security when:

- **.1** the ship is operating at a higher security level than the port facility or another ship it is interfacing with;
- **.2** there is an agreement on a Declaration of Security between Contracting Governments covering certain international voyages or specific ships on those voyages;

.3 there has been a security threat or a security incident involving the ship or involving the port facility, as applicable;

.4 the ship is at a port which is not required to have and implement an approved port facility security plan; or

.5 the ship is conducting ship-to-ship activities with another ship not required to have and implement an approved ship security plan.

5.3 Requests for the completion of a Declaration of Security, under this section, shall be acknowledged by the applicable port facility or ship.

5.4 The Declaration of Security shall be completed by:

.1 the master or the ship security officer on behalf of the ship(s); and, if appropriate,

.2 the port facility security officer or, if the Contracting Government determines otherwise, by any other body responsible for shore-side security, on behalf of the port facility.

5.5 The Declaration of Security shall address the security requirements that could be shared between a port facility and a ship (or between ships) and shall state the responsibility for each.

5.6 Contracting Governments shall specify, bearing in mind the provisions of regulation XI-2/9.2.3, the minimum period for which Declarations of Security shall be kept by the port facilities located within their territory.

5.7 Administrations shall specify, bearing in mind the provisions of regulation XI-2/9.2.3, the minimum period for which Declarations of Security shall be kept by ships entitled to fly their flag.

6 Obligations of the Company

6.1 The Company shall ensure that the ship security plan contains a clear statement emphasizing the master's authority. The Company shall establish in the ship security plan that the master has the overriding authority and responsibility to make decisions with respect to the safety and security of the ship and to request the assistance of the Company or of any Contracting Government as may be necessary.

6.2 The Company shall ensure that the company security officer, the master and the ship security officer are given the necessary support to fulfil their duties and responsibilities in accordance with chapter XI-2 and this part of the Code.

7 Ship security

7.1 A ship is required to act upon the security levels set by Contracting Governments as set out below.

7.2 At security level 1, the following activities shall be carried out, through appropriate measures, on all ships, taking into account the guidance given in part B of this Code, in order to identify and take preventive measures against security incidents:

.1 ensuring the performance of all ship security duties;

.2 controlling access to the ship;

.3 controlling the embarkation of persons and their effects;

.4 monitoring restricted areas to ensure that only authorized persons have access;

.5 monitoring of deck areas and areas surrounding the ship;

.6 supervising the handling of cargo and ship's stores; and

.7 ensuring that security communication is readily available.

7.3 At security level 2, additional protective measures, specified in the ship security plan, shall be implemented for each activity detailed in section 7.2, taking into account the guidance given in part B of this Code.

7.4 At security level 3, further specific protective measures, specified in the ship security plan, shall be implemented for each activity detailed in section 7.2, taking into account the guidance given in part B of this Code.

7.5 Whenever security level 2 or 3 is set by the Administration, the ship shall acknowledge receipt of the instructions on change of the security level.

7.6 Prior to entering a port or whilst in a port within the territory of a Contracting Government that has set security level 2 or 3, the ship shall acknowledge receipt of this instruction and shall confirm to the port facility security officer the initiation of the implementation of the appropriate measures and procedures as detailed in the ship security plan, and in the case of security level 3, in instructions issued by the Contracting Government which has set security level 3. The ship shall report any difficulties in implementation. In such cases, the port facility security officer and ship security officer shall liaise and co-ordinate the appropriate actions.

7.7 If a ship is required by the Administration to set, or is already at, a higher security level than that set for the port it intends to enter or in which it is already located, then the ship shall advise, without delay, the competent authority of the Contracting Government within whose territory the port facility is located and the port facility security officer of the situation.

7.7.1 In such cases, the ship security officer shall liaise with the port facility security officer and co-ordinate appropriate actions, if necessary.

7.8 An Administration requiring ships entitled to fly its flag to set security level 2 or 3 in a port of another Contracting Government shall inform that Contracting Government without delay.

7.9 When Contracting Governments set security levels and ensure the provision of security-level information to ships operating in their territorial sea, or having communicated an intention to enter their territorial sea, such ships shall be advised to maintain vigilance and report immediately to their Administration and any nearby coastal States any information that comes to their attention that might affect maritime security in the area.

7.9.1 When advising such ships of the applicable security level, a Contracting Government shall, taking into account the guidance given in part B of this Code, also advise those ships of any security measure that they should take and, if appropriate, of measures that have been taken by the Contracting Government to provide protection against the threat.

8 Ship security assessment

8.1 The ship security assessment is an essential and integral part of the process of developing and updating the ship security plan.

8.2 The company security officer shall ensure that the ship security assessment is carried out by persons with appropriate skills to evaluate the security of a ship, in accordance with this section, taking into account the guidance given in part B of this Code.

8.3 Subject to the provisions of section 9.2.1, a recognized security organization may carry out the ship security assessment of a specific ship.

8.4 The ship security assessment shall include an on-scene security survey and, at least, the following elements:

.1 identification of existing security measures, procedures and operations;

.2 identification and evaluation of key shipboard operations that it is important to protect;

.3 identification of possible threats to the key shipboard operations and the likelihood of their occurrence, in order to establish and prioritize security measures; and

.4 identification of weaknesses, including human factors, in the infrastructure, policies and procedures.

8.5 The ship security assessment shall be documented, reviewed, accepted and retained by the Company.

9 Ship security plan

9.1 Each ship shall carry on board a ship security plan approved by the Administration. The plan shall make provisions for the three security levels as defined in this part of the Code.

9.1.1 Subject to the provisions of section 9.2.1, a recognized security organization may prepare the ship security plan for a specific ship.

9.2 The Administration may entrust the review and approval of ship security plans, or of amendments to a previously approved plan, to recognized security organizations.

9.2.1 In such cases, the recognized security organization undertaking the review and approval of a ship security plan, or its amendments, for a specific ship shall not have been involved in either the preparation of the ship security assessment or of the ship security plan, or of the amendments, under review.

9.3 The submission of a ship security plan, or of amendments to a previously approved plan, for approval shall be accompanied by the security assessment on the basis of which the plan, or the amendments, has been developed.

9.4 Such a plan shall be developed, taking into account the guidance given in part B of this Code, and shall be written in the working language or languages of the ship. If the language or languages used is not English, French or Spanish, a translation into one of these languages shall be included. The plan shall address, at least, the following:

- **.1** measures designed to prevent weapons, dangerous substances and devices intended for use against persons, ships or ports and the carriage of which is not authorized from being taken on board the ship;
- **.2** identification of the restricted areas and measures for the prevention of unauthorized access to them;
- **.3** measures for the prevention of unauthorized access to the ship;
- **.4** procedures for responding to security threats or breaches of security, including provisions for maintaining critical operations of the ship or ship/port interface;
- **.5** procedures for responding to any security instructions Contracting Governments may give at security level 3;
- **.6** procedures for evacuation in case of security threats or breaches of security;
- **.7** duties of shipboard personnel assigned security responsibilities and of other shipboard personnel on security aspects;
- **.8** procedures for auditing the security activities;
- **.9** procedures for training, drills and exercises associated with the plan;
- **.10** procedures for interfacing with port facility security activities;
- **.11** procedures for the periodic review of the plan and for updating;
- **.12** procedures for reporting security incidents;
- **.13** identification of the ship security officer;
- **.14** identification of the company security officer, including 24-hour contact details;
- **.15** procedures to ensure the inspection, testing, calibration, and maintenance of any security equipment provided on board;
- **.16** frequency for testing or calibration of any security equipment provided on board;

.17 identification of the locations where the ship security alert system activation points are provided;* and

.18 procedures, instructions and guidance on the use of the ship security alert system, including the testing, activation, deactivation and resetting and to limit false alerts.*

9.4.1 Personnel conducting internal audits of the security activities specified in the plan or evaluating its implementation shall be independent of the activities being audited unless this is impracticable due to the size and the nature of the Company or of the ship.

9.5 The Administration shall determine which changes to an approved ship security plan or to any security equipment specified in an approved plan shall not be implemented unless the relevant amendments to the plan are approved by the Administration. Any such changes shall be at least as effective as those measures prescribed in chapter XI-2 and this part of the Code.

9.5.1 The nature of the changes to the ship security plan or the security equipment that have been specifically approved by the Administration, pursuant to section 9.5, shall be documented in a manner that clearly indicates such approval. This approval shall be available on board and shall be presented together with the International Ship Security Certificate (or the Interim International Ship Security Certificate). If these changes are temporary, once the original approved measures or equipment are reinstated, this documentation no longer needs to be retained by the ship.

9.6 The plan may be kept in an electronic format. In such a case, it shall be protected by procedures aimed at preventing its unauthorized deletion, destruction or amendment.

9.7 The plan shall be protected from unauthorized access or disclosure.

9.8 Ship security plans are not subject to inspection by officers duly authorized by a Contracting Government to carry out control and compliance measures in accordance with regulation XI-2/9, save in circumstances specified in section 9.8.1.

9.8.1 If the officers duly authorized by a Contracting Government have clear grounds to believe that the ship is not in compliance with the requirements of chapter XI-2 or part A of this Code, and the only means to verify or rectify the non-compliance is to review the relevant requirements of the ship security plan, limited access to the specific sections of the plan relating to the non-compliance is exceptionally allowed, but only with the consent of the Contracting Government of, or the master of, the ship concerned. Nevertheless, the provisions in the plan relating to section 9.4 subsections .2, .4, .5, .7, .15, .17 and .18 of this part of the Code are considered as confidential information, and cannot be subject to inspection unless otherwise agreed by the Contracting Governments concerned.

10 Records

10.1 Records of the following activities addressed in the ship security plan shall be kept on board for at least the minimum period specified by the Administration, bearing in mind the provisions of regulation XI-2/9.2.3:

.1 training, drills and exercises;

.2 security threats and security incidents;

.3 breaches of security;

.4 changes in security level;

.5 communications relating to the direct security of the ship such as specific threats to the ship or to port facilities the ship is, or has been, in;

* Administrations may allow, in order to avoid compromising in any way the objective of providing on board the ship security alert system, this information to be kept elsewhere on board in a document known to the master, the ship security officer and other senior shipboard personnel as may be decided by the Company.

.6 internal audits and reviews of security activities;

.7 periodic review of the ship security assessment;

.8 periodic review of the ship security plan;

.9 implementation of any amendments to the plan; and

.10 maintenance, calibration and testing of any security equipment provided on board, including testing of the ship security alert system.

10.2 The records shall be kept in the working language or languages of the ship. If the language or languages used are not English, French or Spanish, a translation into one of these languages shall be included.

10.3 The records may be kept in an electronic format. In such a case, they shall be protected by procedures aimed at preventing their unauthorized deletion, destruction or amendment.

10.4 The records shall be protected from unauthorized access or disclosure.

11 Company security officer

11.1 The Company shall designate a company security officer. A person designated as the company security officer may act as the company security officer for one or more ships, depending on the number or types of ships the Company operates, provided it is clearly identified for which ships this person is responsible. A Company may, depending on the number or types of ships they operate, designate several persons as company security officers provided it is clearly identified for which ships each person is responsible.

11.2 In addition to those specified elsewhere in this part of the Code, the duties and responsibilities of the company security officer shall include, but are not limited to:

.1 advising the level of threats likely to be encountered by the ship, using appropriate security assessments and other relevant information;

.2 ensuring that ship security assessments are carried out;

.3 ensuring the development, the submission for approval, and thereafter the implementation and maintenance of the ship security plan;

.4 ensuring that the ship security plan is modified, as appropriate, to correct deficiencies and satisfy the security requirements of the individual ship;

.5 arranging for internal audits and reviews of security activities;

.6 arranging for the initial and subsequent verifications of the ship by the Administration or the recognized security organization;

.7 ensuring that deficiencies and non-conformities identified during internal audits, periodic reviews, security inspections and verifications of compliance are promptly addressed and dealt with;

.8 enhancing security awareness and vigilance;

.9 ensuring adequate training for personnel responsible for the security of the ship;

.10 ensuring effective communication and co-operation between the ship security officer and the relevant port facility security officers;

.11 ensuring consistency between security requirements and safety requirements;

.12 ensuring that, if sister-ship or fleet security plans are used, the plan for each ship reflects the ship-specific information accurately; and

.13 ensuring that any alternative or equivalent arrangements approved for a particular ship or group of ships are implemented and maintained.

12 Ship security officer

12.1 A ship security officer shall be designated on each ship.

12.2 In addition to those specified elsewhere in this part of the Code, the duties and responsibilities of the ship security officer shall include, but are not limited to:

.1 undertaking regular security inspections of the ship to ensure that appropriate security measures are maintained;

.2 maintaining and supervising the implementation of the ship security plan, including any amendments to the plan;

.3 co-ordinating the security aspects of the handling of cargo and ship's stores with other shipboard personnel and with the relevant port facility security officers;

.4 proposing modifications to the ship security plan;

.5 reporting to the company security officer any deficiencies and non-conformities identified during internal audits, periodic reviews, security inspections and verifications of compliance and implementing any corrective actions;

.6 enhancing security awareness and vigilance on board;

.7 ensuring that adequate training has been provided to shipboard personnel, as appropriate;

.8 reporting all security incidents;

.9 co-ordinating implementation of the ship security plan with the company security officer and the relevant port facility security officer; and

.10 ensuring that security equipment is properly operated, tested, calibrated and maintained, if any.

13 Training, drills and exercises on ship security

13.1 The company security officer and appropriate shore-based personnel shall have knowledge and have received training, taking into account the guidance given in part B of this Code.

13.2 The ship security officer shall have knowledge and have received training, taking into account the guidance given in part B of this Code.

13.3 Shipboard personnel having specific security duties and responsibilities shall understand their responsibilities for ship security as described in the ship security plan and shall have sufficient knowledge and ability to perform their assigned duties, taking into account the guidance given in part B of this Code.

13.4 To ensure the effective implementation of the ship security plan, drills shall be carried out at appropriate intervals taking into account the ship type, ship personnel changes, port facilities to be visited and other relevant circumstances, taking into account the guidance given in part B of this Code.

13.5 The company security officer shall ensure the effective co-ordination and implementation of ship security plans by participating in exercises at appropriate intervals, taking into account the guidance given in part B of this Code.

14 Port facility security

14.1 A port facility is required to act upon the security levels set by the Contracting Government within whose territory it is located. Security measures and procedures shall be applied at the port facility in such a manner as to cause a minimum of interference with, or delay to, passengers, ship, ship's personnel and visitors, goods and services.

14.2 At security level 1, the following activities shall be carried out through appropriate measures in all port facilities, taking into account the guidance given in part B of this Code, in order to identify and take preventive measures against security incidents:

- **.1** ensuring the performance of all port facility security duties;
- **.2** controlling access to the port facility;
- **.3** monitoring of the port facility, including anchoring and berthing area(s);
- **.4** monitoring restricted areas to ensure that only authorized persons have access;
- **.5** supervising the handling of cargo;
- **.6** supervising the handling of ship's stores; and
- **.7** ensuring that security communication is readily available.

14.3 At security level 2, additional protective measures, specified in the port facility security plan, shall be implemented for each activity detailed in section 14.2, taking into account the guidance given in part B of this Code.

14.4 At security level 3, further specific protective measures, specified in the port facility security plan, shall be implemented for each activity detailed in section 14.2, taking into account the guidance given in part B of this Code.

14.4.1 In addition, at security level 3, port facilities are required to respond to and implement any security instructions given by the Contracting Government within whose territory the port facility is located.

14.5 When a port facility security officer is advised that a ship encounters difficulties in complying with the requirements of chapter XI-2 or this part or in implementing the appropriate measures and procedures as detailed in the ship security plan, and in the case of security level 3 following any security instructions given by the Contracting Government within whose territory the port facility is located, the port facility security officer and the ship security officer shall liaise and co-ordinate appropriate actions.

14.6 When a port facility security officer is advised that a ship is at a security level which is higher than that of the port facility, the port facility security officer shall report the matter to the competent authority and shall liaise with the ship security officer and co-ordinate appropriate actions, if necessary.

15 Port facility security assessment

15.1 The port facility security assessment is an essential and integral part of the process of developing and updating the port facility security plan.

15.2 The port facility security assessment shall be carried out by the Contracting Government within whose territory the port facility is located. A Contracting Government may authorize a recognized security organization to carry out the port facility security assessment of a specific port facility located within its territory.

15.2.1 When the port facility security assessment has been carried out by a recognized security organization, the security assessment shall be reviewed and approved for compliance with this section by the Contracting Government within whose territory the port facility is located.

15.3 The persons carrying out the assessment shall have appropriate skills to evaluate the security of the port facility in accordance with this section, taking into account the guidance given in part B of this Code.

15.4 The port facility security assessments shall periodically be reviewed and updated, taking account of changing threats and/or minor changes in the port facility, and shall always be reviewed and updated when major changes to the port facility take place.

15.5 The port facility security assessment shall include, at least, the following elements:

- **.1** identification and evaluation of important assets and infrastructure it is important to protect;

.2 identification of possible threats to the assets and infrastructure and the likelihood of their occurrence, in order to establish and prioritize security measures;

.3 identification, selection and prioritization of countermeasures and procedural changes and their level of effectiveness in reducing vulnerability; and

.4 identification of weaknesses, including human factors, in the infrastructure, policies and procedures.

15.6 The Contracting Government may allow a port facility security assessment to cover more than one port facility if the operator, location, operation, equipment, and design of these port facilities are similar. Any Contracting Government which allows such an arrangement shall communicate to the Organization particulars thereof.

15.7 Upon completion of the port facility security assessment, a report shall be prepared, consisting of a summary of how the assessment was conducted, a description of each vulnerability found during the assessment and a description of countermeasures that could be used to address each vulnerability. The report shall be protected from unauthorized access or disclosure.

16 Port facility security plan

16.1 A port facility security plan shall be developed and maintained, on the basis of a port facility security assessment for each port facility, adequate for the ship/port interface. The plan shall make provisions for the three security levels, as defined in this part of the Code.

16.1.1 Subject to the provisions of section 16.2, a recognized security organization may prepare the port facility security plan of a specific port facility.

16.2 The port facility security plan shall be approved by the Contracting Government in whose territory the port facility is located.

16.3 Such a plan shall be developed taking into account the guidance given in part B of this Code and shall be in the working language of the port facility. The plan shall address, at least, the following:

.1 measures designed to prevent weapons or any other dangerous substances and devices intended for use against persons, ships or ports, and the carriage of which is not authorized, from being introduced into the port facility or on board a ship;

.2 measures designed to prevent unauthorized access to the port facility, to ships moored at the facility, and to restricted areas of the facility;

.3 procedures for responding to security threats or breaches of security, including provisions for maintaining critical operations of the port facility or ship/port interface;

.4 procedures for responding to any security instructions the Contracting Government in whose territory the port facility is located may give at security level 3;

.5 procedures for evacuation in case of security threats or breaches of security;

.6 duties of port facility personnel assigned security responsibilities and of other facility personnel on security aspects;

.7 procedures for interfacing with ship security activities;

.8 procedures for the periodic review of the plan and updating;

.9 procedures for reporting security incidents;

.10 identification of the port facility security officer, including 24-hour contact details;

.11 measures to ensure the security of the information contained in the plan;

.12 measures designed to ensure effective security of cargo and the cargo handling equipment at the port facility;

.13 procedures for auditing the port facility security plan;

.14 procedures for responding in case the ship security alert system of a ship at the port facility has been activated; and

.15 procedures for facilitating shore leave for ship's personnel or personnel changes, as well as access of visitors to the ship, including representatives of seafarers' welfare and labour organizations.

16.4 Personnel conducting internal audits of the security activities specified in the plan or evaluating its implementation shall be independent of the activities being audited unless this is impracticable due to the size and the nature of the port facility.

16.5 The port facility security plan may be combined with, or be part of, the port security plan or any other port emergency plan or plans.

16.6 The Contracting Government in whose territory the port facility is located shall determine which changes to the port facility security plan shall not be implemented unless the relevant amendments to the plan are approved by them.

16.7 The plan may be kept in an electronic format. In such a case, it shall be protected by procedures aimed at preventing its unauthorized deletion, destruction or amendment.

16.8 The plan shall be protected from unauthorized access or disclosure.

16.9 Contracting Governments may allow a port facility security plan to cover more than one port facility if the operator, location, operation, equipment, and design of these port facilities are similar. Any Contracting Government which allows such an alternative arrangement shall communicate to the Organization particulars thereof.

17 Port facility security officer

17.1 A port facility security officer shall be designated for each port facility. A person may be designated as the port facility security officer for one or more port facilities.

17.2 In addition to those specified elsewhere in this part of the Code, the duties and responsibilities of the port facility security officer shall include, but are not limited to:

.1 conducting an initial comprehensive security survey of the port facility, taking into account the relevant port facility security assessment;

.2 ensuring the development and maintenance of the port facility security plan;

.3 implementing and exercising the port facility security plan;

.4 undertaking regular security inspections of the port facility to ensure the continuation of appropriate security measures;

.5 recommending and incorporating, as appropriate, modifications to the port facility security plan in order to correct deficiencies and to update the plan to take into account relevant changes to the port facility;

.6 enhancing security awareness and vigilance of the port facility personnel;

.7 ensuring adequate training has been provided to personnel responsible for the security of the port facility;

.8 reporting to the relevant authorities and maintaining records of occurrences which threaten the security of the port facility;

.9 co-ordinating implementation of the port facility security plan with the appropriate Company and ship security officer(s);

.10 co-ordinating with security services, as appropriate;

.11 ensuring that standards for personnel responsible for security of the port facility are met;

.12 ensuring that security equipment is properly operated, tested, calibrated and maintained, if any; and

.13 assisting ship security officers in confirming the identity of those seeking to board the ship when requested.

17.3 The port facility security officer shall be given the necessary support to fulfil the duties and responsibilities imposed by chapter XI-2 and this part of the Code.

18 Training, drills and exercises on port facility security

18.1 The port facility security officer and appropriate port facility security personnel shall have knowledge and have received training, taking into account the guidance given in part B of this Code.

18.2 Port facility personnel having specific security duties shall understand their duties and responsibilities for port facility security, as described in the port facility security plan, and shall have sufficient knowledge and ability to perform their assigned duties, taking into account the guidance given in part B of this Code.

18.3 To ensure the effective implementation of the port facility security plan, drills shall be carried out at appropriate intervals, taking into account the types of operation of the port facility, port facility personnel changes, the type of ship the port facility is serving and other relevant circumstances, taking into account guidance given in part B of this Code.

18.4 The port facility security officer shall ensure the effective co-ordination and implementation of the port facility security plan by participating in exercises at appropriate intervals, taking into account the guidance given in part B of this Code.

19 Verification and certification for ships

19.1 Verifications

19.1.1 Each ship to which this part of the Code applies shall be subject to the verifications specified below:

.1 an initial verification before the ship is put in service or before the certificate required under section 19.2 is issued for the first time, which shall include a complete verification of its security system and any associated security equipment covered by the relevant provisions of chapter XI-2, of this part of the Code and of the approved ship security plan. This verification shall ensure that the security system and any associated security equipment of the ship fully complies with the applicable requirements of chapter XI-2 and this part of the Code, is in satisfactory condition and fit for the service for which the ship is intended;

.2 a renewal verification at intervals specified by the Administration, but not exceeding five years, except where section 19.3 is applicable. This verification shall ensure that the security system and any associated security equipment of the ship fully complies with the applicable requirements of chapter XI-2, this part of the Code and the approved ship security plan, is in satisfactory condition and fit for the service for which the ship is intended;

.3 at least one intermediate verification. If only one intermediate verification is carried out it shall take place between the second and third anniversary date of the certificate as defined in regulation I/2(n). The intermediate verification shall include inspection of the security system and any associated security equipment of the ship to ensure that it remains satisfactory for the service for which the ship is intended. Such intermediate verification shall be endorsed on the certificate;

.4 any additional verifications as determined by the Administration.

19.1.2 The verifications of ships shall be carried out by officers of the Administration. The Administration may, however, entrust the verifications to a recognized security organization referred to in regulation XI-2/1.

19.1.3 In every case, the Administration concerned shall fully guarantee the completeness and efficiency of the verification and shall undertake to ensure the necessary arrangements to satisfy this obligation.

19.1.4 The security system and any associated security equipment of the ship after verification shall be maintained to conform with the provisions of regulations XI-2/4.2 and XI-2/6, of this part of the Code and of the approved ship security plan. After any verification under section 19.1.1 has been completed, no changes shall be made in the security system and in any associated security equipment or the approved ship security plan without the sanction of the Administration.

19.2 Issue or endorsement of Certificate

19.2.1 An International Ship Security Certificate shall be issued after the initial or renewal verification in accordance with the provisions of section 19.1.

19.2.2 Such Certificate shall be issued or endorsed either by the Administration or by a recognized security organization acting on behalf of the Administration.

19.2.3 Another Contracting Government may, at the request of the Administration, cause the ship to be verified and, if satisfied that the provisions of section 19.1.1 are complied with, shall issue or authorize the issue of an International Ship Security Certificate to the ship and, where appropriate, endorse or authorize the endorsement of that Certificate on the ship, in accordance with this Code.

19.2.3.1 A copy of the Certificate and a copy of the verification report shall be transmitted as soon as possible to the requesting Administration.

19.2.3.2 A Certificate so issued shall contain a statement to the effect that it has been issued at the request of the Administration and it shall have the same force and receive the same recognition as the Certificate issued under section 19.2.2.

19.2.4 The International Ship Security Certificate shall be drawn up in a form corresponding to the model given in the appendix to this Code. If the language used is not English, French or Spanish, the text shall include a translation into one of these languages.

19.3 Duration and validity of Certificate

19.3.1 An International Ship Security Certificate shall be issued for a period specified by the Administration, which shall not exceed five years.

19.3.2 When the renewal verification is completed within three months before the expiry date of the existing Certificate, the new Certificate shall be valid from the date of completion of the renewal verification to a date not exceeding five years from the date of expiry of the existing Certificate.

19.3.2.1 When the renewal verification is completed after the expiry date of the existing Certificate, the new Certificate shall be valid from the date of completion of the renewal verification to a date not exceeding five years from the date of expiry of the existing Certificate.

19.3.2.2 When the renewal verification is completed more than three months before the expiry date of the existing Certificate, the new Certificate shall be valid from the date of completion of the renewal verification to a date not exceeding five years from the date of completion of the renewal verification.

19.3.3 If a Certificate is issued for a period of less than five years, the Administration may extend the validity of the Certificate beyond the expiry date to the maximum period specified in section 19.3.1, provided that the verifications referred to in section 19.1.1 applicable when a Certificate is issued for a period of five years are carried out as appropriate.

19.3.4 If a renewal verification has been completed and a new Certificate cannot be issued or placed on board the ship before the expiry date of the existing Certificate, the Administration or recognized security organization acting on behalf of the Administration may endorse the existing Certificate and such a Certificate shall be accepted as valid for a further period which shall not exceed five months from the expiry date.

19.3.5 If a ship, at the time when a Certificate expires, is not in a port in which it is to be verified, the Administration may extend the period of validity of the Certificate but this extension shall be granted only for the purpose of allowing the ship to complete its voyage to the port in which it is to be verified, and then only in cases where it appears proper and reasonable to do so. No Certificate shall be extended for a period longer than three months, and the ship to which an extension is granted shall not, on its arrival in the port in which it is to be verified, be entitled by virtue of such extension to leave that port without having a new Certificate. When the renewal verification is completed, the new Certificate shall be valid to a date not exceeding five years from the expiry date of the existing Certificate before the extension was granted.

19.3.6 A Certificate issued to a ship engaged on short voyages which has not been extended under the foregoing provisions of this section may be extended by the Administration for a period of grace of up to one month from the date of expiry stated on it. When the renewal verification is completed, the new Certificate shall be valid to a date not exceeding five years from the date of expiry of the existing Certificate before the extension was granted.

19.3.7 If an intermediate verification is completed before the period specified in section 19.1.1, then:

.1 the expiry date shown on the Certificate shall be amended by endorsement to a date which shall not be more than three years later than the date on which the intermediate verification was completed;

.2 the expiry date may remain unchanged provided one or more additional verifications are carried out so that the maximum intervals between the verifications prescribed by section 19.1.1 are not exceeded.

19.3.8 A Certificate issued under section 19.2 shall cease to be valid in any of the following cases:

.1 if the relevant verifications are not completed within the periods specified under section 19.1.1;

.2 if the Certificate is not endorsed in accordance with sections 19.1.1.3 and 19.3.7.1, if applicable;

.3 when a Company assumes the responsibility for the operation of a ship not previously operated by that Company; and

.4 upon transfer of the ship to the flag of another State.

19.3.9 In the case of:

.1 a transfer of a ship to the flag of another Contracting Government, the Contracting Government whose flag the ship was formerly entitled to fly shall, as soon as possible, transmit to the receiving Administration copies of, or all information relating to, the International Ship Security Certificate carried by the ship before the transfer and copies of available verification reports, or

.2 a Company that assumes responsibility for the operation of a ship not previously operated by that Company, the previous Company shall, as soon as possible, transmit to the receiving Company copies of any information related to the International Ship Security Certificate or to facilitate the verifications described in section 19.4.2.

19.4 Interim certification

19.4.1 The Certificates specified in section 19.2 shall be issued only when the Administration issuing the Certificate is fully satisfied that the ship complies with the requirements of section 19.1. However, after 1 July 2004, for the purposes of:

.1 a ship without a Certificate, on delivery or prior to its entry or re-entry into service;

.2 transfer of a ship from the flag of a Contracting Government to the flag of another Contracting Government;

.3 transfer of a ship to the flag of a Contracting Government from a State which is not a Contracting Government; or

.4 a Company assuming the responsibility for the operation of a ship not previously operated by that Company

until the Certificate referred to in section 19.2 is issued, the Administration may cause an Interim International Ship Security Certificate to be issued, in a form corresponding to the model given in the appendix to this part of the Code.

19.4.2 An Interim International Ship Security Certificate shall only be issued when the Administration or recognized security organization, on behalf of the Administration, has verified that:

.1 the ship security assessment required by this part of the Code has been completed;

.2 a copy of the ship security plan meeting the requirements of chapter XI-2 and part A of this Code is provided on board, has been submitted for review and approval, and is being implemented on the ship;

.3 the ship is provided with a ship security alert system meeting the requirements of regulation XI-2/6, if required;

.4 the company security officer:

.1 has ensured:

.1 the review of the ship security plan for compliance with this part of the Code;

.2 that the plan has been submitted for approval; and

.3 that the plan is being implemented on the ship; and

.2 has established the necessary arrangements, including arrangements for drills, exercises and internal audits, through which the company security officer is satisfied that the ship will successfully complete the required verification in accordance with section 19.1.1.1, within 6 months;

.5 arrangements have been made for carrying out the required verifications under section 19.1.1.1;

.6 the master, the ship security officer and other ship's personnel with specific security duties are familiar with their duties and responsibilities as specified in this part of the Code; and with the relevant provisions of the ship security plan placed on board; and have been provided such information in the working language of the ship's personnel or languages understood by them; and

.7 the ship security officer meets the requirements of this part of the Code.

19.4.3 An Interim International Ship Security Certificate may be issued by the Administration or by a recognized security organization authorized to act on its behalf.

19.4.4 An Interim International Ship Security Certificate shall be valid for 6 months, or until the Certificate required by section 19.2 is issued, whichever comes first, and may not be extended.

19.4.5 No Contracting Government shall cause a subsequent, consecutive Interim International Ship Security Certificate to be issued to a ship if, in the judgement of the Administration or the recognized security organization, one of the purposes of the ship or a Company in requesting such Certificate is to avoid full compliance with chapter XI-2 and this part of the Code beyond the period of the initial Interim Certificate as specified in section 19.4.4.

19.4.6 For the purposes of regulation XI-2/9, Contracting Governments may, prior to accepting an Interim International Ship Security Certificate as a valid Certificate, ensure that the requirements of sections 19.4.2.4 to 19.4.2.6 have been met.

Appendix to part A

Appendix 1
Form of the International Ship Security Certificate

INTERNATIONAL SHIP SECURITY CERTIFICATE

(Official seal) *(State)*

Certificate number . . . .

Issued under the provisions of the

INTERNATIONAL CODE FOR THE SECURITY OF SHIPS AND OF PORT FACILITIES
(ISPS CODE)

under the authority of the Government of

. . . .

(name of State)

by . . . .

(person(s) or organization authorized)

Name of ship. . . . .

Distinctive number or letters. . . . .

Port of registry . . . .

Type of ship. . . . .

Gross tonnage. . . . .

IMO Number . . . .

Name and address of the Company. . . . .

Company Identification Number . . . .

THIS IS TO CERTIFY:

1 that the security system and any associated security equipment of the ship has been verified in accordance with section 19.1 of part A of the ISPS Code;

2 that the verification showed that the security system and any associated security equipment of the ship is in all respects satisfactory and that the ship complies with the applicable requirements of chapter XI-2 of the Convention and part A of the ISPS Code;

3 that the ship is provided with an approved ship security plan.

Date of initial/renewal verification on which this Certificate is based . . . .

This Certificate is valid until . . . . subject to verifications in accordance with section 19.1.1 of part A of the ISPS Code.

Issued at . . . .

(Place of issue of certificate)

Date of issue . . . .

(Signature of the duly authorized official issuing the Certificate)

(Seal or stamp of issuing authority, as appropriate)

Endorsement for intermediate verification

THIS IS TO CERTIFY that at an intermediate verification required by section 19.1.1 of part A of the ISPS Code the ship was found to comply with the relevant provisions of chapter XI-2 of the Convention and part A of the ISPS Code.

Intermediate verification | Signed: . . . .
(Signature of duly authorized official)

Place: . . . .

Date: . . . .

(Seal or stamp of the authority, as appropriate)

Endorsement for additional verifications*

Additional verification | Signed: . . . .
(Signature of duly authorized official)

Place: . . . .

Date: . . . .

(Seal or stamp of the authority, as appropriate)

Additional verification | Signed: . . . .
(Signature of duly authorized official)

Place: . . . .

Date: . . . .

(Seal or stamp of the authority, as appropriate)

Additional verification | Signed: . . . .
(Signature of duly authorized official)

Place: . . . .

Date: . . . .

(Seal or stamp of the authority, as appropriate)

* This part of the certificate shall be adapted by the Administration to indicate whether it has established additional verifications as provided for in section 19.1.1.4.

Additional verification in accordance with section A/19.3.7.2 of the ISPS Code

THIS IS TO CERTIFY that at an additional verification required by section 19.3.7.2 of part A of the ISPS Code the ship was found to comply with the relevant provisions of chapter XI-2 of the Convention and part A of the ISPS Code.

Signed: . . . .
(Signature of duly authorized official)

Place: . . . .

Date: . . . .

(Seal or stamp of the authority, as appropriate)

Endorsement to extend the certificate if valid for less than 5 years where section A/19.3.3 of the ISPS Code applies

The ship complies with the relevant provisions of part A of the ISPS Code, and the Certificate shall, in accordance with section 19.3.3 of part A of the ISPS Code, be accepted as valid until . . . .

Signed: . . . .
(Signature of duly authorized official)

Place: . . . .

Date: . . . .

(Seal or stamp of the authority, as appropriate)

Endorsement where the renewal verification has been completed and section A/19.3.4 of the ISPS Code applies

The ship complies with the relevant provisions of part A of the ISPS Code, and the Certificate shall, in accordance with section 19.3.4 of part A of the ISPS Code, be accepted as valid until . . . .

Signed: . . . .
(Signature of duly authorized official)

Place: . . . .

Date: . . . .

(Seal or stamp of the authority, as appropriate)

Endorsement to extend the validity of the certificate until reaching the port of verification where section A/19.3.5 of the ISPS Code applies or for a period of grace where section A/19.3.6 of the ISPS Code applies

This Certificate shall, in accordance with section 19.3.5/19.3.6* of part A of the ISPS Code, be accepted as valid until

Signed: . . . .
(Signature of duly authorized official)

Place: . . . .

Date: . . . .

(Seal or stamp of the authority, as appropriate)

Endorsement for advancement of expiry date where section A/19.3.7.1 of the ISPS Code applies

In accordance with section 19.3.7.1 of part A of the ISPS Code, the new expiry date† is . . . .

Signed: . . . .
(Signature of duly authorized official)

Place: . . . .

Date: . . . .

(Seal or stamp of the authority, as appropriate)

* Delete as appropriate.

† This part of the certificate shall be adapted by the Administration to indicate whether it has established additional verifications as provided for in section 19.1.1.4.

Appendix 2
Form of the Interim International Ship Security Certificate

INTERIM INTERNATIONAL SHIP SECURITY CERTIFICATE

(Official seal) *(State)*

Certificate number . . . .

Issued under the provisions of the

INTERNATIONAL CODE FOR THE SECURITY OF SHIPS AND OF PORT FACILITIES
(ISPS CODE)

under the authority of the Government of

. . . .

(name of State)

by . . . .

(person(s) or organization authorized)

Name of ship. . . . .

Distinctive number or letters. . . . .

Port of registry . . . .

Type of ship. . . . .

Gross tonnage. . . . .

IMO Number . . . .

Name and address of Company. . . . .

Company Identification Number . . . .

Is this a subsequent, consecutive Interim Certificate? Yes/No*

If Yes, date of issue of initial Interim Certificate . . . .

THIS IS TO CERTIFY THAT the requirements of section A/19.4.2 of the ISPS Code have been complied with.

This Certificate is issued pursuant to section A/19.4 of the ISPS Code.

This Certificate is valid until . . . .

Issued at . . . .

(Place of issue of the Certificate)

(Date of issue) . . . .

(Signature of the duly authorized official issuing the Certificate)

(Seal or stamp of issuing authority, as appropriate)

* Delete as appropriate.

Part B
Guidance regarding the provisions of chapter XI-2 of the Annex to the International Convention for the Safety of Life at Sea, 1974 as amended and part A of this Code

1 Introduction

General

1.1 The preamble of this Code indicates that chapter XI-2 and part A of this Code establish the new international framework of measures to enhance maritime security and through which ships and port facilities can co-operate to detect and deter acts which threaten security in the maritime transport sector.

1.2 This introduction outlines, in a concise manner, the processes envisaged in establishing and implementing the measures and arrangements needed to achieve and maintain compliance with the provisions of chapter XI-2 and of part A of this Code and identifies the main elements on which guidance is offered. The guidance is provided in paragraphs 2 through to 19. It also sets down essential considerations which should be taken into account when considering the application of the guidance relating to ships and port facilities.

1.3 If the reader's interest relates to ships alone, it is strongly recommended that this part of the Code is still read as a whole, particularly the paragraphs relating to port facilities. The same applies to those whose primary interest is port facilities; they should also read the paragraphs relating to ships.

1.4 The guidance provided in the following paragraphs relates primarily to protection of the ship when it is at a port facility. There could, however, be situations when a ship may pose a threat to the port facility, e.g., because, once within the port facility, it could be used as a base from which to launch an attack. When considering the appropriate security measures to respond to ship-based security threats, those completing the port facility security assessment or preparing the port facility security plan should consider making appropriate adaptations to the guidance offered in the following paragraphs.

1.5 The reader is advised that nothing in this part of the Code should be read or interpreted in conflict with any of the provisions of either chapter XI-2 or part A of this Code and that the aforesaid provisions always prevail and override any unintended inconsistency which may have been inadvertently expressed in this part of the Code. The guidance provided in this part of the Code should always be read, interpreted and applied in a manner which is consistent with the aims, objectives and principles established in chapter XI-2 and part A of this Code.

Responsibilities of Contracting Governments

1.6 Contracting Governments have, under the provisions of chapter XI-2 and part A of this Code, various responsibilities, which, amongst others, include:

- setting the applicable security level;
- approving the ship security plan (SSP) and relevant amendments to a previously approved plan;
- verifying the compliance of ships with the provisions of chapter XI-2 and part A of this Code and issuing to ships the International Ship Security Certificate;
- determining which of the port facilities located within their territory are required to designate a port facility security officer (PFSO) who will be responsible for the preparation of the port facility security plan;
- ensuring completion and approval of the port facility security assessment (PFSA) and of any subsequent amendments to a previously approved assessment;

- approving the port facility security plan (PFSP) and any subsequent amendments to a previously approved plan;
- exercising control and compliance measures;
- testing approved plans; and
- communicating information to the International Maritime Organization and to the shipping and port industries.

1.7 Contracting Governments can designate, or establish, Designated Authorities within Government to undertake, with respect to port facilities, their security duties under chapter XI-2 and part A of this Code and allow recognized security organizations to carry out certain work with respect to port facilities, but the final decision on the acceptance and approval of this work should be given by the Contracting Government or the Designated Authority. Administrations may also delegate the undertaking of certain security duties, relating to ships, to recognized security organizations. The following duties or activities cannot be delegated to a recognized security organization:

- setting of the applicable security level;
- determining which of the port facilities located within the territory of a Contracting Government are required to designate a PFSO and to prepare a PFSP;
- approving a PFSA or any subsequent amendments to a previously approved assessment;
- approving a PFSP or any subsequent amendments to a previously approved plan;
- exercising control and compliance measures; and
- establishing the requirements for a Declaration of Security.

Setting the security level

1.8 The setting of the security level applying at any particular time is the responsibility of Contracting Governments and can apply to ships and port facilities. Part A of this Code defines three security levels for international use. These are:

- Security level 1, normal; the level at which ships and port facilities normally operate;
- Security level 2, heightened; the level applying for as long as there is a heightened risk of a security incident; and
- Security level 3, exceptional; the level applying for the period of time when there is the probable or imminent risk of a security incident.

The Company and the ship

1.9 Any Company operating ships to which chapter XI-2 and part A of this Code apply has to designate a CSO for the Company and an SSO for each of its ships. The duties, responsibilities and training requirements of these officers and requirements for drills and exercises are defined in part A of this Code.

1.10 The company security officer's responsibilities include, in brief amongst others, ensuring that a ship security assessment (SSA) is properly carried out, that an SSP is prepared and submitted for approval by, or on behalf of, the Administration and thereafter is placed on board each ship to which part A of this Code applies and in respect of which that person has been appointed as the CSO.

1.11 The SSP should indicate the operational and physical security measures the ship itself should take to ensure it always operates at security level 1. The plan should also indicate the additional, or intensified, security measures the ship itself can take to move to and operate at security level 2 when instructed to do so. Furthermore, the plan should indicate the possible preparatory actions the ship could take to allow prompt response to the instructions that may be issued to the ship by those responding at security level 3 to a security incident or threat thereof.

1.12 The ships to which the requirements of chapter XI-2 and part A of this Code apply are required to have, and be operated in accordance with, an SSP approved by, or on behalf of, the Administration. The CSO and SSO should monitor the continuing relevance and effectiveness of the plan, including the undertaking of internal audits. Amendments to any of the elements of an approved plan, for which the Administration has determined that approval is required, have to be submitted for review and approval before their incorporation into the approved plan and their implementation by the ship.

1.13 The ship has to carry an International Ship Security Certificate indicating that it complies with the requirements of chapter XI-2 and part A of this Code. Part A of this Code includes provisions relating to the verification and certification of the ship's compliance with the requirements on an initial, renewal and intermediate verification basis.

1.14 When a ship is at a port or is proceeding to a port of a Contracting Government, the Contracting Government has the right, under the provisions of regulation XI-2/9, to exercise various control and compliance measures with respect to that ship. The ship is subject to port State control inspections but such inspections will not normally extend to examination of the SSP itself except in specific circumstances. The ship may also be subject to additional control measures if the Contracting Government exercising the control and compliance measures has reason to believe that the security of the ship has, or the port facilities it has served have, been compromised.

1.15 The ship is also required to have on board information, to be made available to Contracting Governments upon request, indicating who is responsible for deciding the employment of the ship's personnel and for deciding various aspects relating to the employment of the ship.

The port facility

1.16 Each Contracting Government has to ensure completion of a PFSA for each of the port facilities, located within its territory, serving ships engaged on international voyages. The Contracting Government, a Designated Authority or a recognized security organization may carry out this assessment. The completed PFSA has to be approved by the Contracting Government or the Designated Authority concerned. This approval cannot be delegated. Port facility security assessments should be periodically reviewed.

1.17 The PFSA is fundamentally a risk analysis of all aspects of a port facility's operation in order to determine which part(s) of it are more susceptible, and/or more likely, to be the subject of attack. Security risk is a function of the threat of an attack coupled with the vulnerability of the target and the consequences of an attack.

The assessment must include the following components:

- the determination of the perceived threat to port installations and infrastructure;
- identification of the potential vulnerabilities; and
- calculation of the consequences of incidents calculated.

On completion of the analysis, it will be possible to produce an overall assessment of the level of risk. The PFSA will help determine which port facilities are required to appoint a PFSO and prepare a PFSP.

1.18 The port facilities which have to comply with the requirements of chapter XI-2 and part A of this Code are required to designate a PFSO. The duties, responsibilities and training requirements of these officers and requirements for drills and exercises are defined in part A of this Code.

1.19 The PFSP should indicate the operational and physical security measures the port facility should take to ensure that it always operates at security level 1. The plan should also indicate the additional, or intensified, security measures the port facility can take to move to and operate at security level 2 when instructed to do so. Furthermore, the plan should indicate the possible preparatory actions the port facility could take to allow prompt response to the instructions that may be issued by those responding at security level 3 to a security incident or threat thereof.

1.20 The port facilities which have to comply with the requirements of chapter XI-2 and part A of this Code are required to have, and operate in accordance with, a PFSP approved by the Contracting Government or by the Designated Authority concerned. The PFSO should implement its provisions and monitor the continuing effectiveness and relevance of the plan, including commissioning internal audits of the application of the plan. Amendments to any of the elements of an approved plan for which the Contracting Government or the Designated Authority concerned has determined that approval is required, have to be submitted for review and approval before their incorporation into the approved plan and their implementation at the port facility. The Contracting Government or the Designated Authority concerned may test the effectiveness of the plan. The PFSA covering the port facility or on which the development of the plan has been based should be regularly reviewed. All these activities may lead to amendment of the approved plan. Any amendments to specified elements of an approved plan will have to be submitted for approval by the Contracting Government or by the Designated Authority concerned.

1.21 Ships using port facilities may be subject to the port State control inspections and additional control measures outlined in regulation XI-2/9. The relevant authorities may request the provision of information regarding the ship, its cargo, passengers and ship's personnel prior to the ship's entry into port. There may be circumstances in which entry into port could be denied.

Information and communication

1.22 Chapter XI-2 and part A of this Code require Contracting Governments to provide certain information to the International Maritime Organization and for information to be made available to allow effective communication between Contracting Governments and between company security officers/ship security officers and the port facility security officers.

2 Definitions

2.1 No guidance is provided with respect to the definitions set out in chapter XI-2 or part A of this Code.

2.2 For the purpose of this part of the Code:

.1 *section* means a section of part A of the Code and is indicated as "section A/[*followed by the number of the section*]";

.2 *paragraph* means a paragraph of this part of the Code and is indicated as "paragraph [*followed by the number of the paragraph*]"; and

.3 *Contracting Government*, when used in paragraphs 14 to 18, means the Contracting Government within whose territory the port facility is located, and includes a reference to the Designated Authority.

3 Application

General

3.1 The guidance given in this part of the Code should be taken into account when implementing the requirements of chapter XI-2 and part A of this Code.

3.2 However, it should be recognized that the extent to which the guidance on ships applies will depend on the type of ship, its cargoes and/or passengers, its trading pattern and the characteristics of the port facilities visited by the ship.

3.3 Similarly, in relation to the guidance on port facilities, the extent to which this guidance applies will depend on the port facilities, the types of ships using the port facility, the types of cargo and/or passengers and the trading patterns of visiting ships.

3.4 The provisions of chapter XI-2 and part A of this Code are not intended to apply to port facilities designed and used primarily for military purposes.

4 Responsibilities of Contracting Governments

Security of assessments and plans

4.1 Contracting Governments should ensure that appropriate measures are in place to avoid unauthorized disclosure of, or access to, security-sensitive material relating to ship security assessments, ship security plans, port facility security assessments and port facility security plans, and to individual assessments or plans.

Designated Authorities

4.2 Contracting Governments may identify a Designated Authority within Government to undertake their security duties relating to port facilities as set out in chapter XI-2 or part A of this Code.

Recognized security organizations

4.3 Contracting Governments may authorize a recognized security organization (RSO) to undertake certain security-related activities, including:

.1 approval of ship security plans, or amendments thereto, on behalf of the Administration;

.2 verification and certification of compliance of ships with the requirements of chapter XI-2 and part A of this Code on behalf of the Administration; and

.3 conducting port facility security assessments required by the Contracting Government.

4.4 An RSO may also advise or provide assistance to Companies or port facilities on security matters, including ship security assessments, ship security plans, port facility security assessments and port facility security plans. This can include completion of an SSA or SSP or PFSA or PFSP. If an RSO has done so in respect of an SSA or SSP, that RSO should not be authorized to approve that SSP.

4.5 When authorizing an RSO, Contracting Governments should give consideration to the competency of such an organization. An RSO should be able to demonstrate:

.1 expertise in relevant aspects of security;

.2 appropriate knowledge of ship and port operations, including knowledge of ship design and construction if providing services in respect of ships and of port design and construction if providing services in respect of port facilities;

.3 their capability to assess the likely security risks that could occur during ship and port facility operations, including the ship/port interface, and how to minimize such risks;

.4 their ability to maintain and improve the expertise of their personnel;

.5 their ability to monitor the continuing trustworthiness of their personnel;

.6 their ability to maintain appropriate measures to avoid unauthorized disclosure of, or access to, security-sensitive material;

.7 their knowledge of the requirements of chapter XI-2 and part A of this Code and relevant national and international legislation and security requirements;

.8 their knowledge of current security threats and patterns;

.9 their knowledge of recognition and detection of weapons, dangerous substances and devices;

.10 their knowledge of recognition, on a non-discriminatory basis, of characteristics and behavioural patterns of persons who are likely to threaten security;

.11 their knowledge of techniques used to circumvent security measures; and

.12 their knowledge of security and surveillance equipment and systems and their operational limitations.

When delegating specific duties to an RSO, Contracting Governments, including Administrations, should ensure that the RSO has the competencies needed to undertake the task.

4.6 A recognized organization, as referred to in regulation I/6 and fulfilling the requirements of regulation XI-1/1, may be appointed as an RSO provided it has the appropriate security-related expertise listed in paragraph 4.5.

4.7 A port or harbour authority or port facility operator may be appointed as an RSO provided it has the appropriate security-related expertise listed in paragraph 4.5.

Setting the security level

4.8 In setting the security level, Contracting Governments should take account of general and specific threat information. Contracting Governments should set the security level applying to ships or port facilities at one of three levels:

- Security level 1, normal; the level at which the ship or port facility normally operates;
- Security level 2, heightened; the level applying for as long as there is a heightened risk of a security incident; and
- Security level 3, exceptional; the level applying for the period of time when there is the probable or imminent risk of a security incident.

4.9 Setting security level 3 should be an exceptional measure applying only when there is credible information that a security incident is probable or imminent. Security level 3 should only be set for the duration of the identified security threat or actual security incident. While the security levels may change from security level 1, through security level 2 to security level 3, it is also possible that the security levels will change directly from security level 1 to security level 3.

4.10 At all times the master of a ship has the ultimate responsibility for the safety and security of the ship. Even at security level 3 a master may seek clarification or amendment of instructions issued by those responding to a security incident, or threat thereof, if there are reasons to believe that compliance with any instruction may imperil the safety of the ship.

4.11 The CSO or the SSO should liaise at the earliest opportunity with the PFSO of the port facility the ship is intended to visit to establish the security level applying for that ship at the port facility. Having established contact with a ship, the PFSO should advise the ship of any subsequent change in the port facility's security level and should provide the ship with any relevant security information.

4.12 While there may be circumstances when an individual ship may be operating at a higher security level than the port facility it is visiting, there will be no circumstances when a ship can have a lower security level than the port facility it is visiting. If a ship has a higher security level than the port facility it intends to use, the CSO or SSO should advise the PFSO without delay. The PFSO should undertake an assessment of the particular situation in consultation with the CSO or SSO and agree on appropriate security measures with the ship, which may include completion and signing of a Declaration of Security.

4.13 Contracting Governments should consider how information on changes in security levels should be promulgated rapidly. Administrations may wish to use NAVTEX messages or Notices to Mariners as the method for notifying such changes in security levels to the ship and to the CSO and SSO. Or, they may wish to consider other methods of communication that provide equivalent or better speed and coverage. Contracting Governments should establish means of notifying PFSOs of changes in security levels. Contracting Governments should compile and maintain the contact details for a list of those who need to be informed of changes in security levels. Whereas the security level need not be regarded as being particularly sensitive, the underlying threat information may be highly sensitive. Contracting Governments should give careful consideration to the type and detail of the information conveyed and the method by which it is conveyed to SSOs, CSOs and PFSOs.

Contact points and information on port facility security plans

4.14 Where a port facility has a PFSP, that fact has to be communicated to the Organization and that information must also be made available to CSOs and SSOs. No further details of the PFSP have to be published other than that it is in place. Contracting Governments should consider establishing either central or regional points of contact, or other means of providing up-to-date information on the locations where PFSPs are in place, together with contact details for the relevant PFSO. The existence of such contact points should be publicized. They could also provide information on the recognized security organizations appointed to act on behalf of the Contracting Government, together with details of the specific responsibility and conditions of authority delegated to such recognized security organizations.

4.15 In the case of a port that does not have a PFSP (and therefore does not have a PFSO), the central or regional point of contact should be able to identify a suitably qualified person ashore who can arrange for appropriate security measures to be in place, if needed, for the duration of the ship's visit.

4.16 Contracting Governments should also provide the contact details of Government officers to whom an SSO, a CSO and a PFSO can report security concerns. These Government officers should assess such reports before taking appropriate action. Such reported concerns may have a bearing on the security measures falling under the jurisdiction of another Contracting Government. In that case, the Contracting Governments should consider contacting their counterpart in the other Contracting Government to discuss whether remedial action is appropriate. For this purpose, the contact details of the Government officers should be communicated to the International Maritime Organization.

4.17 Contracting Governments should also make the information indicated in paragraphs 4.14 to 4.16 available to other Contracting Governments on request.

Identification documents

4.18 Contracting Governments are encouraged to issue appropriate identification documents to Government officials entitled to board ships or enter port facilities when performing their official duties and to establish procedures whereby the authenticity of such documents might be verified.

Fixed and floating platforms and mobile offshore drilling units on location

4.19 Contracting Governments should consider establishing appropriate security measures for fixed and floating platforms and mobile offshore drilling units on location to allow interaction with ships which are required to comply with the provisions of chapter XI-2 and part A of this Code.*

Ships which are not required to comply with part A of this Code

4.20 Contracting Governments should consider establishing appropriate security measures to enhance the security of ships to which chapter XI-2 and part A of this Code do not apply and to ensure that any security provisions applying to such ships allow interaction with ships to which part A of this Code applies.

Threats to ships and other incidents at sea

4.21 Contracting Governments should provide general guidance on the measures considered appropriate to reduce the security risk to ships flying their flag when at sea. They should provide specific advice on the action to be taken in accordance with security levels 1 to 3, if:

.1 there is a change in the security level applying to the ship while it is at sea, e.g., because of the geographical area in which it is operating or relating to the ship itself; and

.2 there is a security incident or threat thereof involving the ship while at sea.

---

* Refer to Establishment of appropriate measures to enhance the security of ships, port facilities, mobile offshore drilling units on location and fixed and floating platforms not covered by chapter XI-2 of the 1974 SOLAS Convention, adopted by the 2002 SOLAS Conference by resolution 7.

Contracting Governments should establish the best methods and procedures for these purposes. In the case of an imminent attack, the ship should seek to establish direct communication with those responsible in the flag State for responding to security incidents.

4.22 Contracting Governments should also establish a point of contact for advice on security for any ship:

.1 entitled to fly their flag; or

.2 operating in their territorial sea or having communicated an intention to enter their territorial sea.

4.23 Contracting Governments should offer advice to ships operating in their territorial sea or having communicated an intention to enter their territorial sea, which could include advice:

.1 to alter or delay their intended passage;

.2 to navigate on a particular course or proceed to a specific location;

.3 on the availability of any personnel or equipment that could be placed on the ship;

.4 to co-ordinate the passage, arrival into port or departure from port, to allow escort by patrol craft or aircraft (fixed-wing or helicopter).

Contracting Governments should remind ships operating in their territorial sea, or having communicated an intention to enter their territorial sea, of any temporary restricted areas that they have published.

4.24 Contracting Governments should recommend that ships operating in their territorial sea, or having communicated an intention to enter their territorial sea, implement expeditiously, for the ship's protection and for the protection of other ships in the vicinity, any security measure the Contracting Government may have advised.

4.25 The plans prepared by the Contracting Governments for the purposes given in paragraph 4.22 should include information on an appropriate point of contact, available on a 24-hour basis, within the Contracting Government including the Administration. These plans should also include information on the circumstances in which the Administration considers assistance should be sought from nearby coastal States, and a procedure for liaison between PFSOs and SSOs.

Alternative security agreements

4.26 Contracting Governments, in considering how to implement chapter XI-2 and part A of this Code, may conclude one or more agreements with one or more Contracting Governments. The scope of an agreement is limited to short international voyages on fixed routes between port facilities in the territory of the parties to the agreement. When concluding an agreement, and thereafter, the Contracting Governments should consult other Contracting Governments and Administrations with an interest in the effects of the agreement. Ships flying the flag of a State that is not party to the agreement should only be allowed to operate on the fixed routes covered by the agreement if their Administration agrees that the ship should comply with the provisions of the agreement and requires the ship to do so. In no case can such an agreement compromise the level of security of other ships and port facilities not covered by it, and specifically, all ships covered by such an agreement may not conduct ship-to-ship activities with ships not so covered. Any operational interface undertaken by ships covered by the agreement should be covered by it. The operation of each agreement must be continually monitored and amended when the need arises and in any event should be reviewed every five years.

Equivalent arrangements for port facilities

4.27 For certain specific port facilities with limited or special operations but with more than occasional traffic, it may be appropriate to ensure compliance by security measures equivalent to those prescribed in chapter XI-2 and in part A of this Code. This can, in particular, be the case for terminals such as those attached to factories, or quaysides with no frequent operations.

Manning level

4.28 In establishing the minimum safe manning of a ship, the Administration should take into account[*] that the minimum safe manning provisions established by regulation V/14[†] only address the safe navigation of the ship. The Administration should also take into account any additional workload which may result from the implementation of the SSP and ensure that the ship is sufficiently and effectively manned. In doing so, the Administration should verify that ships are able to implement the hours of rest and other measures to address fatigue which have been promulgated by national law, in the context of all shipboard duties assigned to the various shipboard personnel.

Control and compliance measures[‡]

General

4.29 Regulation XI-2/9 describes the Control and compliance measures applicable to ships under chapter XI-2. It is divided into three distinct sections; control of ships already in a port, control of ships intending to enter a port of another Contracting Government, and additional provisions applicable to both situations.

4.30 Regulation XI-2/9.1, Control of ships in port, implements a system for the control of ships while in the port of a foreign country where duly authorized officers of the Contracting Government ("duly authorized officers") have the right to go on board the ship to verify that the required certificates are in proper order. Then, if there are clear grounds to believe the ship does not comply, control measures such as additional inspections or detention may be taken. This reflects current control systems.[§] Regulation XI-2/9.1 builds on such systems and allows for additional measures (including expulsion of a ship from a port to be taken as a control measure) when duly authorized officers have clear grounds for believing that a ship is in non-compliance with the requirements of chapter XI-2 or part A of this Code. Regulation XI-2/9.3 describes the safeguards that promote fair and proportionate implementation of these additional measures.

4.31 Regulation XI-2/9.2 applies control measures to ensure compliance to ships intending to enter a port of another Contracting Government and introduces an entirely different concept of control within chapter XI-2, applying to security only. Under this regulation, measures may be implemented prior to the ship entering port, to better ensure security. Just as in regulation XI-2/9.1, this additional control system is based on the concept of clear grounds for believing the ship does not comply with chapter XI-2 or part A of this Code, and includes significant safeguards in regulations XI-2/9.2.2 and XI-2/9.2.5 as well as in regulation XI-2/9.3.

4.32 Clear grounds that the ship is not in compliance means evidence or reliable information that the ship does not correspond with the requirements of chapter XI-2 or part A of this Code, taking into account the guidance given in this part of the Code. Such evidence or reliable information may arise from the duly authorized officer's professional judgement or observations gained while verifying the ship's International Ship Security Certificate or Interim International Ship Security Certificate issued in accordance with part A of this Code ("Certificate") or from other sources. Even if a valid Certificate is on board the ship, the duly authorized officers may still have clear grounds for believing that the ship is not in compliance based on their professional judgement.

---

[*] Refer to Further work by the International Maritime Organization pertaining to enhancement of maritime security, adopted by the 2002 SOLAS Conference by resolution 3, inviting, amongst others, the Organization to review Assembly resolution A.890(21) on Principles of safe manning. This review may also lead to amendments of regulation V/14.

[†] As was in force on the date of adoption of this Code.

[‡] Refer to Further work by the International Maritime Organization pertaining to enhancement of maritime security, adopted by the 2002 SOLAS Conference by resolution 3, inviting, amongst others, the Organization to review Assembly resolutions A.787(19) and A.882(21) on Procedures for port State control.

[§] See regulation I/19 and regulation IX/6.2 of SOLAS 74 as amended, article 21 of Load Line 66 as modified by the 1988 Load Line Protocol, articles 5 and 6 and regulation 8A of Annex I and regulation 15 of Annex II of MARPOL 73/78 as amended, article X of STCW 78 as amended and IMO Assembly resolutions A.787(19) and A.882(21).

4.33 Examples of possible clear grounds under regulations XI-2/9.1 and XI-2/9.2 may include, when relevant:

.1 evidence from a review of the Certificate that it is not valid or it has expired;

.2 evidence or reliable information that serious deficiencies exist in the security equipment, documentation or arrangements required by chapter XI-2 and part A of this Code;

.3 receipt of a report or complaint which, in the professional judgement of the duly authorized officer, contains reliable information clearly indicating that the ship does not comply with the requirements of chapter XI-2 or part A of this Code;

.4 evidence or observation gained by a duly authorized officer using professional judgement that the master or ship's personnel is not familiar with essential shipboard security procedures or cannot carry out drills related to the security of the ship or that such procedures or drills have not been carried out;

.5 evidence or observation gained by a duly authorized officer using professional judgement that key members of ship's personnel are not able to establish proper communication with any other key members of ship's personnel with security responsibilities on board the ship;

.6 evidence or reliable information that the ship has embarked persons or loaded stores or goods at a port facility or from another ship where either the port facility or the other ship is in violation of chapter XI-2 or part A of this Code, and the ship in question has not completed a Declaration of Security, nor taken appropriate, special or additional security measures or has not maintained appropriate ship security procedures;

.7 evidence or reliable information that the ship has embarked persons or loaded stores or goods at a port facility or from another source (e.g., another ship or helicopter transfer) where either the port facility or the other source is not required to comply with chapter XI-2 or part A of this Code, and the ship has not taken appropriate, special or additional security measures or has not maintained appropriate security procedures; and

.8 if the ship is holding a subsequent, consecutively issued Interim International Ship Security Certificate as described in section A/19.4, and if, in the professional judgement of an officer duly authorized, one of the purposes of the ship or a Company in requesting such a Certificate is to avoid full compliance with chapter XI-2 and part A of this Code beyond the period of the initial Interim Certificate as described in section A/19.4.4.

4.34 The international law implications of regulation XI-2/9 are particularly relevant, and the regulation should be implemented with regulation XI-2/2.4 in mind, as the potential exists for situations where either measures will be taken which fall outside the scope of chapter XI-2, or where rights of affected ships, outside chapter XI-2, should be considered. Thus, regulation XI-2/9 does not prejudice the Contracting Government from taking measures having a basis in, and consistent with, international law to ensure the safety or security of persons, ships, port facilities and other property in cases where the ship, although in compliance with chapter XI-2 and part A of this Code, is still considered to present a security risk.

4.35 When a Contracting Government imposes control measures on a ship, the Administration should, without delay, be contacted with sufficient information to enable the Administration to fully liaise with the Contracting Government.

Control of ships in port

4.36 Where the non-compliance is either a defective item of equipment or faulty documentation leading to the ship's detention and the non-compliance cannot be remedied in the port of inspection, the Contracting Government may allow the ship to sail to another port provided that any conditions agreed between the port States and the Administration or master are met.

Ships intending to enter the port of another Contracting Government

4.37 Regulation XI-2/9.2.1 lists the information Contracting Governments may require from a ship as a condition of entry into port. One item of information listed is confirmation of any special or additional measures taken by the ship during its last 10 calls at a port facility. Examples could include:

.1 records of the measures taken while visiting a port facility located in the territory of a State which is not a Contracting Government, especially those measures that would normally have been provided by port facilities located in the territories of Contracting Governments; and

.2 any Declarations of Security that were entered into with port facilities or other ships.

4.38 Another item of information listed, that may be required as a condition of entry into port, is confirmation that appropriate ship security procedures were maintained during ship-to-ship activity conducted within the period of the last 10 calls at a port facility. It would not normally be required to include records of transfers of pilots or of customs, immigration or security officials nor bunkering, lightering, loading of supplies and unloading of waste by ship within port facilities as these would normally fall within the auspices of the PFSP. Examples of information that might be given include:

.1 records of the measures taken while engaged in a ship-to-ship activity with a ship flying the flag of a State which is not a Contracting Government, especially those measures that would normally have been provided by ships flying the flag of Contracting Governments;

.2 records of the measures taken while engaged in a ship-to-ship activity with a ship that is flying the flag of a Contracting Government but is not required to comply with the provisions of chapter XI-2 and part A of this Code, such as a copy of any security certificate issued to that ship under other provisions; and

.3 in the event that persons or goods rescued at sea are on board, all known information about such persons or goods, including their identities when known and the results of any checks run on behalf of the ship to establish the security status of those rescued. It is not the intention of chapter XI-2 or part A of this Code to delay or prevent the delivery of those in distress at sea to a place of safety. It is the sole intention of chapter XI-2 and part A of this Code to provide States with enough appropriate information to maintain their security integrity.

4.39 Examples of other practical security-related information that may be required as a condition of entry into port in order to assist with ensuring the safety and security of persons, port facilities, ships and other property include:

.1 information contained in the Continuous Synopsis Record;

.2 location of the ship at the time the report is made;

.3 expected time of arrival of the ship in port;

.4 crew list;

.5 general description of cargo aboard the ship;

.6 passenger list; and

.7 information required to be carried under regulation XI-2/5.

4.40 Regulation XI-2/9.2.5 allows the master of a ship, upon being informed that the coastal or port State will implement control measures under regulation XI-2/9.2, to withdraw the intention for the ship to enter port. If the master withdraws that intention, regulation XI-2/9 no longer applies, and any other steps that are taken must be based on, and consistent with, international law.

Additional provisions

4.41 In all cases where a ship is denied entry or is expelled from a port, all known facts should be communicated to the authorities of relevant States. This communication should consist of the following, when known:

.1 name of ship, its flag, the Ship Identification Number, call sign, ship type and cargo;

.2 reason for denying entry or for expulsion from port or port areas;

.3 if relevant, the nature of any security non-compliance;

.4 if relevant, details of any attempts made to rectify any non-compliance, including any conditions imposed on the ship for the voyage;

.5 past port(s) of call and next declared port of call;

.6 time of departure and likely estimated time of arrival at those ports;

.7 any instructions given to the ship, e.g., reporting on its route;

.8 available information on the security level at which the ship is currently operating;

.9 information regarding any communications the port State has had with the Administration;

.10 contact point within the port State making the report for the purpose of obtaining further information;

.11 crew list; and

.12 any other relevant information.

4.42 Relevant States to contact should include those along the ship's intended passage to its next port, particularly if the ship intends to enter the territorial sea of that coastal State. Other relevant States could include previous ports of call, so that further information might be obtained and security issues relating to the previous ports resolved.

4.43 In exercising control and compliance measures, the duly authorized officers should ensure that any measures or steps imposed are proportionate. Such measures or steps should be reasonable and of the minimum severity and duration necessary to rectify or mitigate the non-compliance.

4.44 The word "delay" in regulation XI-2/9.3.5.1 also refers to situations where, pursuant to actions taken under this regulation, the ship is unduly denied entry into port or the ship is unduly expelled from port.

Non-Party ships and ships below Convention size

4.45 With respect to ships flying the flag of a State which is not a Contracting Government to the Convention and not a Party to the 1988 SOLAS Protocol,* Contracting Governments should not give more favourable treatment to such ships. Accordingly, the requirements of regulation XI-2/9 and the guidance provided in this part of the Code should be applied to those ships.

4.46 Ships below Convention size are subject to measures by which States maintain security. Such measures should be taken with due regard to the requirements in chapter XI-2 and the guidance provided in this part of the Code.

5 Declaration of Security

General

5.1 A Declaration of Security (DoS) should be completed when the Contracting Government of the port facility deems it to be necessary or when a ship deems it necessary.

5.1.1 The need for a DoS may be indicated by the results of the port facility security assessment (PFSA) and the reasons and circumstances in which a DoS is required should be set out in the port facility security plan (PFSP).

5.1.2 The need for a DoS may be indicated by an Administration for ships entitled to fly its flag or as a result of a ship security assessment (SSA) and should be set out in the ship security plan (SSP).

* Protocol of 1988 relating to the International Convention for the Safety of Life at Sea, 1974.

5.2 It is likely that a DoS will be requested at higher security levels, when a ship has a higher security level than the port facility, or another ship with which it interfaces, and for ship/port interface or ship-to-ship activities that pose a higher risk to persons, property or the environment for reasons specific to that ship, including its cargo or passengers, or the circumstances at the port facility or a combination of these factors.

5.2.1 In the case that a ship or an Administration, on behalf of ships entitled to fly its flag, requests completion of a DoS, the PFSO or SSO should acknowledge the request and discuss appropriate security measures.

5.3 A PFSO may also initiate a DoS prior to ship/port interfaces that are identified in the approved PFSA as being of particular concern. Examples may include embarking or disembarking passengers and the transfer, loading or unloading of dangerous goods or hazardous substances. The PFSA may also identify facilities at or near highly populated areas or economically significant operations that warrant a DoS.

5.4 The main purpose of a DoS is to ensure agreement is reached between the ship and the port facility or with other ships with which it interfaces as to the respective security measures each will undertake in accordance with the provisions of their respective approved security plans.

5.4.1 The agreed DoS should be signed and dated by both the port facility and the ship(s), as applicable, to indicate compliance with chapter XI-2 and part A of this Code and should include its duration, the relevant security level or levels and the relevant contact details.

5.4.2 A change in the security level may require that a new or revised DoS be completed.

5.5 The DoS should be completed in English, French or Spanish or in a language common to both the port facility and the ship or the ships, as applicable.

5.6 A model DoS is included in appendix 1 to this part of the Code. This model is for a DoS between a ship and a port facility. If the DoS is to cover two ships this model should be appropriately adjusted.

6 Obligations of the Company

General

6.1 Regulation XI-2/5 requires the Company to provide the master of the ship with information to meet the requirements of the Company under the provisions of this regulation. This information should include items such as:

- **.1** parties responsible for appointing shipboard personnel, such as ship management companies, manning agents, contractors, concessionaries (for example, retail sales outlets, casinos, etc.);
- **.2** parties responsible for deciding the employment of the ship, including time or bareboat charterer(s) or any other entity acting in such capacity; and
- **.3** in cases when the ship is employed under the terms of a charter party, the contact details of those parties, including time or voyage charterers.

6.2 In accordance with regulation XI-2/5, the Company is obliged to update and keep this information current as and when changes occur.

6.3 This information should be in English, French or Spanish language.

6.4 With respect to ships constructed before 1 July 2004, this information should reflect the actual condition on that date.

6.5 With respect to ships constructed on or after 1 July 2004 and for ships constructed before 1 July 2004 which were out of service on 1 July 2004, the information should be provided as from the date of entry of the ship into service and should reflect the actual condition on that date.

6.6 After 1 July 2004, when a ship is withdrawn from service, the information should be provided as from the date of re-entry of the ship into service and should reflect the actual condition on that date.

6.7 Previously provided information that does not relate to the actual condition on that date need not be retained on board.

6.8 When the responsibility for the operation of the ship is assumed by another Company, the information relating to the Company which operated the ship is not required to be left on board.

In addition, other relevant guidance is provided under sections 8, 9 and 13.

7 Ship security

Relevant guidance is provided under sections 8, 9 and 13.

8 Ship security assessment

Security assessment

8.1 The company security officer (CSO) is responsible for ensuring that a ship security assessment (SSA) is carried out for each of the ships in the Company's fleet which is required to comply with the provisions of chapter XI-2 and part A of this Code for which the CSO is responsible. While the CSO need not necessarily personally undertake all the duties associated with the post, the ultimate responsibility for ensuring that they are properly performed remains with the individual CSO.

8.2 Prior to commencing the SSA, the CSO should ensure that advantage is taken of information available on the assessment of threat for the ports at which the ship will call or at which passengers embark or disembark and about the port facilities and their protective measures. The CSO should study previous reports on similar security needs. Where feasible, the CSO should meet with appropriate persons on the ship and in the port facilities to discuss the purpose and methodology of the assessment. The CSO should follow any specific guidance offered by the Contracting Governments.

8.3 An SSA should address the following elements on board or within the ship:

- **.1** physical security;
- **.2** structural integrity;
- **.3** personnel protection systems;
- **.4** procedural policies;
- **.5** radio and telecommunication systems, including computer systems and networks; and
- **.6** other areas that may, if damaged or used for illicit observation, pose a risk to persons, property, or operations on board the ship or within a port facility.

8.4 Those involved in conducting an SSA should be able to draw upon expert assistance in relation to:

- **.1** knowledge of current security threats and patterns;
- **.2** recognition and detection of weapons, dangerous substances and devices;
- **.3** recognition, on a non-discriminatory basis, of characteristics and behavioural patterns of persons who are likely to threaten security;
- **.4** techniques used to circumvent security measures;
- **.5** methods used to cause a security incident;
- **.6** effects of explosives on ship's structures and equipment;
- **.7** ship security;
- **.8** ship/port interface business practices;
- **.9** contingency planning, emergency preparedness and response;

.10 physical security;

.11 radio and telecommunications systems, including computer systems and networks;

.12 marine engineering; and

.13 ship and port operations.

8.5 The CSO should obtain and record the information required to conduct an assessment, including:

.1 the general layout of the ship;

.2 the location of areas which should have restricted access, such as navigation bridge, machinery spaces of category A and other control stations as defined in chapter II-2, etc.;

.3 the location and function of each actual or potential access point to the ship;

.4 changes in the tide which may have an impact on the vulnerability or security of the ship;

.5 the cargo spaces and stowage arrangements;

.6 the locations where the ship's stores and essential maintenance equipment is stored;

.7 the locations where unaccompanied baggage is stored;

.8 the emergency and stand-by equipment available to maintain essential services;

.9 the number of ship's personnel, any existing security duties and any existing training requirement practices of the Company;

.10 existing security and safety equipment for the protection of passengers and ship's personnel;

.11 escape and evacuation routes and assembly stations which have to be maintained to ensure the orderly and safe emergency evacuation of the ship;

.12 existing agreements with private security companies providing ship/water-side security services; and

.13 existing security measures and procedures in effect, including inspection and control procedures, identification systems, surveillance and monitoring equipment, personnel identification documents and communication, alarms, lighting, access control and other appropriate systems.

8.6 The SSA should examine each identified point of access, including open weather decks, and evaluate its potential for use by individuals who might seek to breach security. This includes points of access available to individuals having legitimate access as well as those who seek to obtain unauthorized entry.

8.7 The SSA should consider the continuing relevance of the existing security measures and guidance, procedures and operations, under both routine and emergency conditions, and should determine security guidance including:

.1 the restricted areas;

.2 the response procedures to fire or other emergency conditions;

.3 the level of supervision of the ship's personnel, passengers, visitors, vendors, repair technicians, dock workers, etc.;

.4 the frequency and effectiveness of security patrols;

.5 the access control systems, including identification systems;

.6 the security communications systems and procedures;

.7 the security doors, barriers and lighting; and

.8 the security and surveillance equipment and systems, if any.

8.8 The SSA should consider the persons, activities, services and operations that it is important to protect. This includes:

- **.1** the ship's personnel;
- **.2** passengers, visitors, vendors, repair technicians, port facility personnel, etc.;
- **.3** the capacity to maintain safe navigation and emergency response;
- **.4** the cargo, particularly dangerous goods or hazardous substances;
- **.5** the ship's stores;
- **.6** the ship security communication equipment and systems, if any; and
- **.7** the ship's security surveillance equipment and systems, if any.

8.9 The SSA should consider all possible threats, which may include the following types of security incidents:

- **.1** damage to, or destruction of, the ship or of a port facility, e.g., by explosive devices, arson, sabotage or vandalism;
- **.2** hijacking or seizure of the ship or of persons on board;
- **.3** tampering with cargo, essential ship equipment or systems or ship's stores;
- **.4** unauthorized access or use, including presence of stowaways;
- **.5** smuggling weapons or equipment, including weapons of mass destruction;
- **.6** use of the ship to carry those intending to cause a security incident and/or their equipment;
- **.7** use of the ship itself as a weapon or as a means to cause damage or destruction;
- **.8** attacks from seaward whilst at berth or at anchor; and
- **.9** attacks whilst at sea.

8.10 The SSA should take into account all possible vulnerabilities, which may include:

- **.1** conflicts between safety and security measures;
- **.2** conflicts between shipboard duties and security assignments;
- **.3** watchkeeping duties, number of ship's personnel, particularly with implications on crew fatigue, alertness and performance;
- **.4** any identified security training deficiencies; and
- **.5** any security equipment and systems, including communication systems.

8.11 The CSO and ship security officer (SSO) should always have regard to the effect that security measures may have on ship's personnel who will remain on the ship for long periods. When developing security measures, particular consideration should be given to the convenience, comfort and personal privacy of the ship's personnel and their ability to maintain their effectiveness over long periods.

8.12 Upon completion of the SSA, a report shall be prepared, consisting of a summary of how the assessment was conducted, a description of each vulnerability found during the assessment and a description of countermeasures that could be used to address each vulnerability. The report shall be protected from unauthorized access or disclosure.

8.13 If the SSA has not been carried out by the Company, the report of the SSA should be reviewed and accepted by the CSO.

On-scene security survey

8.14 The on-scene security survey is an integral part of any SSA. The on-scene security survey should examine and evaluate existing shipboard protective measures, procedures and operations for:

.1 ensuring the performance of all ship security duties;

.2 monitoring restricted areas to ensure that only authorized persons have access;

.3 controlling access to the ship, including any identification systems;

.4 monitoring of deck areas and areas surrounding the ship;

.5 controlling the embarkation of persons and their effects (accompanied and unaccompanied baggage and ship's personnel personal effects);

.6 supervising the handling of cargo and the delivery of ship's stores; and

.7 ensuring that ship security communication, information, and equipment are readily available.

9 Ship security plan

General

9.1 The company security officer (CSO) has the responsibility of ensuring that a ship security plan (SSP) is prepared and submitted for approval. The content of each individual SSP should vary depending on the particular ship it covers. The ship security assessment (SSA) will have identified the particular features of the ship and the potential threats and vulnerabilities. The preparation of the SSP will require these features to be addressed in detail. Administrations may prepare advice on the preparation and content of an SSP.

9.2 All SSPs should:

.1 detail the organizational structure of security for the ship;

.2 detail the ship's relationships with the Company, port facilities, other ships and relevant authorities with security responsibility;

.3 detail the communication systems to allow effective continuous communication within the ship and between the ship and others, including port facilities;

.4 detail the basic security measures for security level 1, both operational and physical, that will always be in place;

.5 detail the additional security measures that will allow the ship to progress without delay to security level 2 and, when necessary, to security level 3;

.6 provide for regular review, or audit, of the SSP and for its amendment in response to experience or changing circumstances; and

.7 detail reporting procedures to the appropriate Contracting Government's contact points.

9.3 Preparation of an effective SSP should rest on a thorough assessment of all issues that relate to the security of the ship, including, in particular, a thorough appreciation of the physical and operational characteristics, including the voyage pattern, of the individual ship.

9.4 All SSPs should be approved by, or on behalf of, the Administration. If an Administration uses a recognized security organization (RSO) to review or approve the SSP, that RSO should not be associated with any other RSO that prepared, or assisted in the preparation of, the plan.

9.5 CSOs and SSOs should develop procedures to:

.1 assess the continuing effectiveness of the SSP; and

.2 prepare amendments of the plan subsequent to its approval.

9.6 The security measures included in the SSP should be in place when the initial verification for compliance with the requirements of chapter XI-2 and part A of this Code will be carried out. Otherwise the process of issue to the ship of the required International Ship Security Certificate cannot be carried out. If there is any subsequent failure of security equipment or systems, or suspension of a security measure for whatever reason, equivalent temporary security measures should be adopted, notified to, and agreed by the Administration.

Organization and performance of ship security duties

9.7 In addition to the guidance given in paragraph 9.2, the SSP should establish the following, which relate to all security levels:

.1 the duties and responsibilities of all shipboard personnel with a security role;

.2 the procedures or safeguards necessary to allow such continuous communications to be maintained at all times;

.3 the procedures needed to assess the continuing effectiveness of security procedures and any security and surveillance equipment and systems, including procedures for identifying and responding to equipment or systems failure or malfunction;

.4 the procedures and practices to protect security-sensitive information held in paper or electronic format;

.5 the type and maintenance requirements of security and surveillance equipment and systems, if any;

.6 the procedures to ensure the timely submission, and assessment, of reports relating to possible breaches of security or security concerns; and

.7 procedures to establish, maintain and update an inventory of any dangerous goods or hazardous substances carried on board, including their location.

9.8 The remainder of section 9 addresses specifically the security measures that could be taken at each security level covering:

.1 access to the ship by ship's personnel, passengers, visitors, etc.;

.2 restricted areas on the ship;

.3 handling of cargo;

.4 delivery of ship's stores;

.5 handling unaccompanied baggage; and

.6 monitoring the security of the ship.

Access to the ship

9.9 The SSP should establish the security measures covering all means of access to the ship identified in the SSA. This should include any:

.1 access ladders;

.2 access gangways;

.3 access ramps;

.4 access doors, sidescuttles, windows and ports;

.5 mooring lines and anchor chains; and

.6 cranes and hoisting gear.

9.10 For each of these the SSP should identify the appropriate locations where access restrictions or prohibitions should be applied for each of the security levels. For each security level the SSP should establish the type of restriction or prohibition to be applied and the means of enforcing them.

9.11 The SSP should establish for each security level the means of identification required to allow access to the ship and for individuals to remain on the ship without challenge. This may involve developing an appropriate identification system, allowing for permanent and temporary identifications for ship's personnel and for visitors respectively. Any ship identification system should, when it is practicable to do so, be co-ordinated with that applying to the port facility. Passengers should be able to prove their identity by boarding passes, tickets, etc., but should not be permitted access to restricted areas unless supervised. The SSP should establish provisions to ensure that the identification systems are regularly updated, and that abuse of procedures should be subject to disciplinary action.

9.12 Those unwilling or unable to establish their identity and/or to confirm the purpose of their visit when requested to do so should be denied access to the ship and their attempt to obtain access should be reported, as appropriate, to the SSO, the CSO, the PFSO and to the national or local authorities with security responsibilities.

9.13 The SSP should establish the frequency of application of any access controls, particularly if they are to be applied on a random, or occasional, basis.

Security level 1

9.14 At security level 1, the SSP should establish the security measures to control access to the ship, where the following may be applied:

.1 checking the identity of all persons seeking to board the ship and confirming their reasons for doing so by checking, for example, joining instructions, passenger tickets, boarding passes, work orders, etc.;

.2 in liaison with the port facility, the ship should ensure that designated secure areas are established in which inspections and searching of persons, baggage (including carry-on items), personal effects, vehicles and their contents can take place;

.3 in liaison with the port facility, the ship should ensure that vehicles destined to be loaded on board car carriers, ro–ro and other passenger ships are subjected to search prior to loading, in accordance with the frequency required in the SSP;

.4 segregating checked persons and their personal effects from unchecked persons and their personal effects;

.5 segregating embarking from disembarking passengers;

.6 identifying access points that should be secured or attended to prevent unauthorized access;

.7 securing, by locking or other means, access to unattended spaces adjoining areas to which passengers and visitors have access; and

.8 providing security briefings to all ship personnel on possible threats, the procedures for reporting suspicious persons, objects or activities and the need for vigilance.

9.15 At security level 1, all those seeking to board a ship should be liable to search. The frequency of such searches, including random searches, should be specified in the approved SSP and should be specifically approved by the Administration. Such searches may best be undertaken by the port facility in close co-operation with the ship and in close proximity to it. Unless there are clear security grounds for doing so, members of the ship's personnel should not be required to search their colleagues or their personal effects. Any such search shall be undertaken in a manner which fully takes into account the human rights of the individual and preserves their basic human dignity.

Security level 2

9.16 At security level 2, the SSP should establish the security measures to be applied to protect against a heightened risk of a security incident to ensure higher vigilance and tighter control, which may include:

.1 assigning additional personnel to patrol deck areas during silent hours to deter unauthorized access;

.2 limiting the number of access points to the ship, identifying those to be closed and the means of adequately securing them;

.3 deterring waterside access to the ship, including, for example, in liaison with the port facility, provision of boat patrols;

.4 establishing a restricted area on the shore side of the ship, in close co-operation with the port facility;

.5 increasing the frequency and detail of searches of persons, personal effects, and vehicles being embarked or loaded onto the ship;

.6 escorting visitors on the ship;

.7 providing additional specific security briefings to all ship personnel on any identified threats, re-emphasizing the procedures for reporting suspicious persons, objects, or activities and stressing the need for increased vigilance; and

.8 carrying out a full or partial search of the ship.

Security level 3

9.17 At security level 3, the ship should comply with the instructions issued by those responding to the security incident or threat thereof. The SSP should detail the security measures which could be taken by the ship, in close co-operation with those responding and the port facility, which may include:

.1 limiting access to a single, controlled, access point;

.2 granting access only to those responding to the security incident or threat thereof;

.3 directing persons on board;

.4 suspension of embarkation or disembarkation;

.5 suspension of cargo handling operations, deliveries, etc.;

.6 evacuation of the ship;

.7 movement of the ship; and

.8 preparing for a full or partial search of the ship.

Restricted areas on the ship

9.18 The SSP should identify the restricted areas to be established on the ship, specify their extent, times of application, the security measures to be taken to control access to them and those to be taken to control activities within them. The purposes of restricted areas are to:

.1 prevent unauthorized access;

.2 protect passengers, ship's personnel, and personnel from port facilities or other agencies authorized to be on board the ship;

.3 protect security-sensitive areas within the ship; and

.4 protect cargo and ship's stores from tampering.

9.19 The SSP should ensure that there are clearly established policies and practices to control access to all restricted areas.

9.20 The SSP should provide that all restricted areas should be clearly marked, indicating that access to the area is restricted and that unauthorized presence within the area constitutes a breach of security.

9.21 Restricted areas may include:

.1 navigation bridge, machinery spaces of category A and other control stations as defined in chapter II-2;

.2 spaces containing security and surveillance equipment and systems and their controls and lighting system controls;

.3 ventilation and air-conditioning systems and other similar spaces;

.4 spaces with access to potable water tanks, pumps, or manifolds;

.5 spaces containing dangerous goods or hazardous substances;

.6 spaces containing cargo pumps and their controls;

.7 cargo spaces and spaces containing ship's stores;

.8 crew accommodation; and

.9 any other areas as determined by the CSO, through the SSA, to which access must be restricted to maintain the security of the ship.

Security level 1

9.22 At security level 1, the SSP should establish the security measures to be applied to restricted areas, which may include:

.1 locking or securing access points;

.2 using surveillance equipment to monitor the areas;

.3 using guards or patrols; and

.4 using automatic intrusion-detection devices to alert the ship's personnel of unauthorized access.

Security level 2

9.23 At security level 2, the frequency and intensity of the monitoring of, and control of access to, restricted areas should be increased to ensure that only authorized persons have access. The SSP should establish the additional security measures to be applied, which may include:

.1 establishing restricted areas adjacent to access points;

.2 continuously monitoring surveillance equipment; and

.3 dedicating additional personnel to guard and patrol restricted areas.

Security level 3

9.24 At security level 3, the ship should comply with the instructions issued by those responding to the security incident or threat thereof. The SSP should detail the security measures which could be taken by the ship, in close co-operation with those responding and the port facility, which may include:

.1 setting up of additional restricted areas on the ship in proximity to the security incident, or the believed location of the security threat, to which access is denied; and

.2 searching of restricted areas as part of a search of the ship.

Handling of cargo

9.25 The security measures relating to cargo handling should:

.1 prevent tampering; and

.2 prevent cargo that is not meant for carriage from being accepted and stored on board the ship.

9.26 The security measures, some of which may have to be applied in liaison with the port facility, should include inventory control procedures at access points to the ship. Once on board the ship, cargo should be capable of being identified as having been approved for loading onto the ship. In addition, security measures should be developed to ensure that cargo, once on board, is not tampered with.

Security level 1

9.27 At security level 1, the SSP should establish the security measures to be applied during cargo handling, which may include:

.1 routine checking of cargo, cargo transport units and cargo spaces prior to, and during, cargo handling operations;

.2 checks to ensure that cargo being loaded matches the cargo documentation;

.3 ensuring, in liaison with the port facility, that vehicles to be loaded on board car carriers, ro–ro and passenger ships are subjected to search prior to loading, in accordance with the frequency required in the SSP; and

.4 checking of seals or other methods used to prevent tampering.

9.28 Checking of cargo may be accomplished by the following means:

.1 visual and physical examination; and

.2 using scanning/detection equipment, mechanical devices, or dogs.

9.29 When there are regular or repeated cargo movements, the CSO or SSO may, in consultation with the port facility, agree arrangements with shippers or others responsible for such cargo covering off-site checking, sealing, scheduling, supporting documentation, etc. Such arrangements should be communicated to and agreed with the PFSO concerned.

Security level 2

9.30 At security level 2, the SSP should establish the additional security measures to be applied during cargo handling, which may include:

.1 detailed checking of cargo, cargo transport units and cargo spaces;

.2 intensified checks to ensure that only the intended cargo is loaded;

.3 intensified searching of vehicles to be loaded on car carriers, ro–ro and passenger ships; and

.4 increased frequency and detail in checking of seals or other methods used to prevent tampering.

9.31 Detailed checking of cargo may be accomplished by the following means:

.1 increasing the frequency and detail of visual and physical examination;

.2 increasing the frequency of the use of scanning/detection equipment, mechanical devices, or dogs; and

.3 co-ordinating enhanced security measures with the shipper or other responsible party in accordance with an established agreement and procedures.

Security level 3

9.32 At security level 3, the ship should comply with the instructions issued by those responding to the security incident or threat thereof. The SSP should detail the security measures which could be taken by the ship, in close co-operation with those responding and the port facility, which may include:

.1 suspending the loading or unloading of cargo; and

.2 verifying the inventory of dangerous goods and hazardous substances carried on board, if any, and their location.

Delivery of ship's stores

9.33 The security measures relating to the delivery of ship's stores should:

.1 ensure checking of ship's stores and package integrity;

.2 prevent ship's stores from being accepted without inspection;

.3 prevent tampering; and

.4 prevent ship's stores from being accepted unless ordered.

9.34 For ships regularly using the port facility, it may be appropriate to establish procedures involving the ship, its suppliers and the port facility covering notification and timing of deliveries and their documentation. There should always be some way of confirming that stores presented for delivery are accompanied by evidence that they have been ordered by the ship.

Security level 1

9.35 At security level 1, the SSP should establish the security measures to be applied during delivery of ship's stores, which may include:

.1 checking to ensure stores match the order prior to being loaded on board; and

.2 ensuring immediate secure stowage of ship's stores.

Security level 2

9.36 At security level 2, the SSP should establish the additional security measures to be applied during delivery of ship's stores by exercising checks prior to receiving stores on board and intensifying inspections.

Security level 3

9.37 At security level 3, the ship should comply with the instructions issued by those responding to the security incident or threat thereof. The SSP should detail the security measures which could be taken by the ship, in close co-operation with those responding and the port facility, which may include:

.1 subjecting ship's stores to more extensive checking;

.2 preparation for restriction or suspension of handling of ship's stores; and

.3 refusal to accept ship's stores on board the ship.

Handling unaccompanied baggage

9.38 The SSP should establish the security measures to be applied to ensure that unaccompanied baggage (i.e., any baggage, including personal effects, which is not with the passenger or member of ship's personnel at the point of inspection or search) is identified and subjected to appropriate screening, including searching, before it is accepted on board the ship. It is not envisaged that such baggage will be subjected to screening by both the ship and the port facility, and in cases where both are suitably equipped, the responsibility for screening should rest with the port facility. Close co-operation with the port facility is essential and steps should be taken to ensure that unaccompanied baggage is handled securely after screening.

Security level 1

9.39 At security level 1, the SSP should establish the security measures to be applied when handling unaccompanied baggage to ensure that unaccompanied baggage is screened or searched up to and including 100%, which may include use of x-ray screening.

Security level 2

9.40 At security level 2, the SSP should establish the additional security measures to be applied when handling unaccompanied baggage, which should include 100% x-ray screening of all unaccompanied baggage.

Security level 3

9.41 At security level 3, the ship should comply with the instructions issued by those responding to the security incident or threat thereof. The SSP should detail the security measures which could be taken by the ship, in close co-operation with those responding and the port facility, which may include:

- **.1** subjecting such baggage to more extensive screening, for example x-raying it from at least two different angles;
- **.2** preparation for restriction or suspension of handling of unaccompanied baggage; and
- **.3** refusal to accept unaccompanied baggage on board the ship.

Monitoring the security of the ship

9.42 The ship should have the capability to monitor the ship, the restricted areas on board and areas surrounding the ship. Such monitoring capabilities may include use of:

- **.1** lighting;
- **.2** watchkeepers, security guards and deck watches, including patrols; and
- **.3** automatic intrusion-detection devices and surveillance equipment.

9.43 When used, automatic intrusion-detection devices should activate an audible and/or visual alarm at a location that is continuously attended or monitored.

9.44 The SSP should establish the procedures and equipment needed at each security level and the means of ensuring that monitoring equipment will be able to perform continually, including consideration of the possible effects of weather conditions or of power disruptions.

Security level 1

9.45 At security level 1, the SSP should establish the security measures to be applied, which may be a combination of lighting, watchkeepers, security guards or use of security and surveillance equipment to allow ship's security personnel to observe the ship in general, and barriers and restricted areas in particular.

9.46 The ship's deck and access points to the ship should be illuminated during hours of darkness and periods of low visibility while conducting ship/port interface activities or at a port facility or anchorage when necessary. While under way, when necessary, ships should use the maximum lighting available consistent with safe navigation, having regard to the provisions of the International Regulations for the Prevention of Collisions at Sea in force. The following should be considered when establishing the appropriate level and location of lighting:

- **.1** the ship's personnel should be able to detect activities beyond the ship, on both the shore side and the water side;
- **.2** coverage should include the area on and around the ship;
- **.3** coverage should facilitate personnel identification at access points; and
- **.4** coverage may be provided through co-ordination with the port facility.

Security level 2

9.47 At security level 2, the SSP should establish the additional security measures to be applied to enhance the monitoring and surveillance capabilities, which may include:

.1 increasing the frequency and detail of security patrols;

.2 increasing the coverage and intensity of lighting or the use of security and surveillance equipment;

.3 assigning additional personnel as security look-outs; and

.4 ensuring co-ordination with water-side boat patrols, and foot or vehicle patrols on the shore side, when provided.

9.48 Additional lighting may be necessary to protect against a heightened risk of a security incident. When necessary, the additional lighting requirements may be accomplished by co-ordinating with the port facility to provide additional shoreside lighting.

Security level 3

9.49 At security level 3, the ship should comply with the instructions issued by those responding to the security incident or threat thereof. The SSP should detail the security measures which could be taken by the ship, in close co-operation with those responding and the port facility, which may include:

.1 switching on of all lighting on, or illuminating the vicinity of, the ship;

.2 switching on of all on-board surveillance equipment capable of recording activities on, or in the vicinity of, the ship;

.3 maximizing the length of time such surveillance equipment can continue to record;

.4 preparation for underwater inspection of the hull of the ship; and

.5 initiation of measures, including the slow revolution of the ship's propellers, if practicable, to deter underwater access to the hull of the ship.

Differing security levels

9.50 The SSP should establish details of the procedures and security measures the ship could adopt if the ship is at a higher security level than that applying to a port facility.

Activities not covered by the Code

9.51 The SSP should establish details of the procedures and security measures the ship should apply when:

.1 it is at a port of a State which is not a Contracting Government;

.2 it is interfacing with a ship to which this Code does not apply;*

.3 it is interfacing with fixed or floating platforms or a mobile drilling unit on location; or

.4 it is interfacing with a port or port facility which is not required to comply with chapter XI-2 and part A of this Code.

Declarations of Security

9.52 The SSP should detail how requests for Declarations of Security from a port facility will be handled and the circumstances under which the ship itself should request a DoS.

---

* Refer to Further work by the International Maritime Organization pertaining to the enhancement of maritime security and to Establishment of appropriate measures to enhance the security of ships, port facilities, mobile offshore drilling units on location and fixed and floating platforms not covered by chapter XI-2 of the 1974 SOLAS Convention, adopted by the 2002 SOLAS Conference by resolutions 3 and 7 respectively.

Audit and review

9.53 The SSP should establish how the CSO and the SSO intend to audit the continued effectiveness of the SSP and the procedure to be followed to review, update or amend the SSP.

10 Records

General

10.1 Records should be available to duly authorized officers of Contracting Governments to verify that the provisions of ship security plans are being implemented.

10.2 Records may be kept in any format but should be protected from unauthorized access or disclosure.

11 Company security officer

Relevant guidance is provided under sections 8, 9 and 13.

12 Ship security officer

Relevant guidance is provided under sections 8, 9 and 13.

13 Training, drills and exercises on ship security

Training

13.1 The company security officer (CSO) and appropriate shore-based Company personnel, and the ship security officer (SSO), should have knowledge of, and receive training, in some or all of the following, as appropriate:

- **.1** security administration;
- **.2** relevant international conventions, codes and recommendations;
- **.3** relevant Government legislation and regulations;
- **.4** responsibilities and functions of other security organizations;
- **.5** methodology of ship security assessment;
- **.6** methods of ship security surveys and inspections;
- **.7** ship and port operations and conditions;
- **.8** ship and port facility security measures;
- **.9** emergency preparedness and response and contingency planning;
- **.10** instruction techniques for security training and education, including security measures and procedures;
- **.11** handling sensitive security-related information and security-related communications;
- **.12** knowledge of current security threats and patterns;
- **.13** recognition and detection of weapons, dangerous substances and devices;
- **.14** recognition, on a non-discriminatory basis, of characteristics and behavioural patterns of persons who are likely to threaten security;
- **.15** techniques used to circumvent security measures;
- **.16** security equipment and systems and their operational limitations;

.17 methods of conducting audits, inspection, control and monitoring;

.18 methods of physical searches and non-intrusive inspections;

.19 security drills and exercises, including drills and exercises with port facilities; and

.20 assessment of security drills and exercises.

13.2 In addition, the SSO should have adequate knowledge of, and receive training in, some or all of the following, as appropriate:

.1 the layout of the ship;

.2 the ship security plan (SSP) and related procedures (including scenario-based training on how to respond);

.3 crowd management and control techniques;

.4 operations of security equipment and systems; and

.5 testing, calibration and at-sea maintenance of security equipment and systems.

13.3 Shipboard personnel having specific security duties should have sufficient knowledge and ability to perform their assigned duties, including, as appropriate:

.1 knowledge of current security threats and patterns;

.2 recognition and detection of weapons, dangerous substances and devices;

.3 recognition of characteristics and behavioural patterns of persons who are likely to threaten security;

.4 techniques used to circumvent security measures;

.5 crowd management and control techniques;

.6 security-related communications;

.7 knowledge of the emergency procedures and contingency plans;

.8 operations of security equipment and systems;

.9 testing, calibration and at-sea maintenance of security equipment and systems;

.10 inspection, control, and monitoring techniques; and

.11 methods of physical searches of persons, personal effects, baggage, cargo, and ship's stores.

13.4 All other shipboard personnel should have sufficient knowledge of and be familiar with relevant provisions of the SSP, including:

.1 the meaning and the consequential requirements of the different security levels;

.2 knowledge of the emergency procedures and contingency plans;

.3 recognition and detection of weapons, dangerous substances and devices;

.4 recognition, on a non-discriminatory basis, of characteristics and behavioural patterns of persons who are likely to threaten security; and

.5 techniques used to circumvent security measures.

Drills and exercises

13.5 The objective of drills and exercises is to ensure that shipboard personnel are proficient in all assigned security duties at all security levels and the identification of any security-related deficiencies which need to be addressed.

13.6 To ensure the effective implementation of the provisions of the ship security plan, drills should be conducted at least once every three months. In addition, in cases where more than 25% of the ship's personnel has been changed, at any one time, with personnel that has not previously participated in any drill on that ship within the last 3 months, a drill should be conducted within one week of the change. These drills should test individual elements of the plan such as those security threats listed in paragraph 8.9.

13.7 Various types of exercises, which may include participation of company security officers, port facility security officers, relevant authorities of Contracting Governments as well as ship security officers, if available, should be carried out at least once each calendar year with no more than 18 months between the exercises. These exercises should test communications, co-ordination, resource availability, and response. These exercises may be:

.1 full-scale or live;

.2 tabletop simulation or seminar; or

.3 combined with other exercises held, such as search and rescue or emergency response exercises.

13.8 Company participation in an exercise with another Contracting Government should be recognized by the Administration.

14 Port facility security

Relevant guidance is provided under sections 15, 16 and 18.

15 Port facility security assessment

General

15.1 The port facility security assessment (PFSA) may be conducted by a recognized security organization (RSO). However, approval of a completed PFSA should only be given by the relevant Contracting Government.

15.2 If a Contracting Government uses an RSO to review or verify compliance of the PFSA, the RSO should not be associated with any other RSO that prepared or assisted in the preparation of that assessment.

15.3 A PFSA should address the following elements within a port facility:

.1 physical security;

.2 structural integrity;

.3 personnel protection systems;

.4 procedural policies;

.5 radio and telecommunication systems, including computer systems and networks;

.6 relevant transportation infrastructure;

.7 utilities; and

.8 other areas that may, if damaged or used for illicit observation, pose a risk to persons, property, or operations within the port facility.

15.4 Those involved in a PFSA should be able to draw upon expert assistance in relation to:

.1 knowledge of current security threats and patterns;

.2 recognition and detection of weapons, dangerous substances and devices;

.3 recognition, on a non-discriminatory basis, of characteristics and behavioural patterns of persons who are likely to threaten security;

.4 techniques used to circumvent security measures;

.5 methods used to cause a security incident;

.6 effects of explosives on structures and port facility services;

.7 port facility security;

.8 port business practices;

.9 contingency planning, emergency preparedness and response;

.10 physical security measures, e.g., fences;

.11 radio and telecommunications systems, including computer systems and networks;

.12 transport and civil engineering; and

.13 ship and port operations.

Identification and evaluation of important assets and infrastructure it is important to protect

15.5 The identification and evaluation of important assets and infrastructure is a process through which the relative importance of structures and installations to the functioning of the port facility can be established. This identification and evaluation process is important because it provides a basis for focusing mitigation strategies on those assets and structures which it is more important to protect from a security incident. This process should take into account potential loss of life, the economic significance of the port, symbolic value, and the presence of Government installations.

15.6 Identification and evaluation of assets and infrastructure should be used to prioritize their relative importance for protection. The primary concern should be avoidance of death or injury. It is also important to consider whether the port facility, structure or installation can continue to function without the asset, and the extent to which rapid re-establishment of normal functioning is possible.

15.7 Assets and infrastructure that should be considered important to protect may include:

.1 accesses, entrances, approaches, and anchorages, manoeuvring and berthing areas;

.2 cargo facilities, terminals, storage areas, and cargo handling equipment;

.3 systems such as electrical distribution systems, radio and telecommunication systems and computer systems and networks;

.4 port vessel traffic management systems and aids to navigation;

.5 power plants, cargo transfer piping, and water supplies;

.6 bridges, railways, roads;

.7 port service vessels, including pilot boats, tugs, lighters, etc.;

.8 security and surveillance equipment and systems; and

.9 the waters adjacent to the port facility.

15.8 The clear identification of assets and infrastructure is essential to the evaluation of the port facility's security requirements, the prioritization of protective measures, and decisions concerning the allocation of resources to better protect the port facility. The process may involve consultation with the relevant authorities relating to structures adjacent to the port facility which could cause damage within the facility or be used for the purpose of causing damage to the facility or for illicit observation of the facility or for diverting attention.

Identification of the possible threats to the assets and infrastructure and the likelihood of their occurrence, in order to establish and prioritize security measures

15.9 Possible acts that could threaten the security of assets and infrastructure, and the methods of carrying out those acts, should be identified to evaluate the vulnerability of a given asset or location to a security incident, and to establish and prioritize security requirements to enable planning and resource allocations. Identification and evaluation of each potential act and its method should be based on various factors, including threat assessments by Government agencies. By identifying and assessing threats, those conducting the assessment do not have to rely on worst-case scenarios to guide planning and resource allocations.

15.10 The PFSA should include an assessment undertaken in consultation with the relevant national security organizations to determine:

.1 any particular aspects of the port facility, including the vessel traffic using the facility, which make it likely to be the target of an attack;

.2 the likely consequences in terms of loss of life, damage to property and economic disruption, including disruption to transport systems, of an attack on, or at, the port facility;

.3 the capability and intent of those likely to mount such an attack; and

.4 the possible type, or types, of attack,

producing an overall assessment of the level of risk against which security measures have to be developed.

15.11 The PFSA should consider all possible threats, which may include the following types of security incidents:

.1 damage to, or destruction of, the port facility or of the ship, e.g., by explosive devices, arson, sabotage or vandalism;

.2 hijacking or seizure of the ship or of persons on board;

.3 tampering with cargo, essential ship equipment or systems or ship's stores;

.4 unauthorized access or use, including presence of stowaways;

.5 smuggling weapons or equipment, including weapons of mass destruction;

.6 use of the ship to carry those intending to cause a security incident and their equipment;

.7 use of the ship itself as a weapon or as a means to cause damage or destruction;

.8 blockage of port entrances, locks, approaches, etc.; and

.9 nuclear, biological and chemical attack.

15.12 The process should involve consultation with the relevant authorities relating to structures adjacent to the port facility which could cause damage within the facility or be used for the purpose of causing damage to the facility or for illicit observation of the facility or for diverting attention.

Identification, selection, and prioritization of countermeasures and procedural changes and their level of effectiveness in reducing vulnerability

15.13 The identification and prioritization of countermeasures is designed to ensure that the most effective security measures are employed to reduce the vulnerability of a port facility or ship/port interface to the possible threats.

15.14 Security measures should be selected on the basis of factors such as whether they reduce the probability of an attack and should be evaluated using information that includes:

.1 security surveys, inspections and audits;

.2 consultation with port facility owners and operators, and owners/operators of adjacent structures if appropriate;

.3 historical information on security incidents; and

.4 operations within the port facility.

Identification of vulnerabilities

15.15 Identification of vulnerabilities in physical structures, personnel protection systems, processes, or other areas that may lead to a security incident can be used to establish options to eliminate or mitigate those vulnerabilities. For example, an analysis might reveal vulnerabilities in a port facility's security systems or unprotected infrastructure such as water supplies, bridges, etc. that could be resolved through physical measures, e.g., permanent barriers, alarms, surveillance equipment, etc.

15.16 Identification of vulnerabilities should include consideration of:

.1 water-side and shore-side access to the port facility and ships berthing at the facility;

.2 structural integrity of the piers, facilities, and associated structures;

.3 existing security measures and procedures, including identification systems;

.4 existing security measures and procedures relating to port services and utilities;

.5 measures to protect radio and telecommunication equipment, port services and utilities, including computer systems and networks;

.6 adjacent areas that may be exploited during, or for, an attack;

.7 existing agreements with private security companies providing water-side/shore-side security services;

.8 any conflicting policies between safety and security measures and procedures;

.9 any conflicting port facility and security duty assignments;

.10 any enforcement and personnel constraints;

.11 any deficiencies identified during training and drills; and

.12 any deficiencies identified during daily operation, following incidents or alerts, the report of security concerns, the exercise of control measures, audits, etc.

16 Port facility security plan

General

16.1 Preparation of the port facility security plan (PFSP) is the responsibility of the port facility security officer (PFSO). While the PFSO need not necessarily personally undertake all the duties associated with the post, the ultimate responsibility for ensuring that they are properly performed remains with the individual PFSO.

16.2 The content of each individual PFSP should vary depending on the particular circumstances of the port facility, or facilities, it covers. The port facility security assessment (PFSA) will have identified the particular features of the port facility, and of the potential security risks, that have led to the need to appoint a PFSO and to prepare a PFSP. The preparation of the PFSP will require these features, and other local or national security considerations, to be addressed in the PFSP and for appropriate security measures to be established so as to minimize the likelihood of a breach of security and the consequences of potential risks. Contracting Governments may prepare advice on the preparation and content of a PFSP.

16.3 All PFSPs should:

.1 detail the security organization of the port facility;

.2 detail the organization's links with other relevant authorities and the necessary communication systems to allow the effective continuous operation of the organization and its links with others, including ships in port;

.3 detail the basic security level 1 measures, both operational and physical, that will be in place;

.4 detail the additional security measures that will allow the port facility to progress without delay to security level 2 and, when necessary, to security level 3;

.5 provide for regular review, or audit, of the PFSP and for its amendment in response to experience or changing circumstances; and

.6 detail reporting procedures to the appropriate Contracting Government's contact points.

16.4 Preparation of an effective PFSP will rest on a thorough assessment of all issues that relate to the security of the port facility, including, in particular, a thorough appreciation of the physical and operational characteristics of the individual port facility.

16.5 Contracting Governments should approve the PFSPs of the port facilities under their jurisdiction. Contracting Governments should develop procedures to assess the continuing effectiveness of each PFSP and may require amendment of the PFSP prior to its initial approval or subsequent to its approval. The PFSP should make provision for the retention of records of security incidents and threats, reviews, audits, training, drills and exercises as evidence of compliance with those requirements.

16.6 The security measures included in the PFSP should be in place within a reasonable period of the PFSP's approval and the PFSP should establish when each measure will be in place. If there is likely to be any delay in their provision, this should be discussed with the Contracting Government responsible for approval of the PFSP and satisfactory alternative temporary security measures that provide an equivalent level of security should be agreed to cover any interim period.

16.7 The use of firearms on or near ships and in port facilities may pose particular and significant safety risks, in particular in connection with certain dangerous or hazardous substances, and should be considered very carefully. In the event that a Contracting Government decides that it is necessary to use armed personnel in these areas, that Contracting Government should ensure that these personnel are duly authorized and trained in the use of their weapons and that they are aware of the specific risks to safety that are present in these areas. If a Contracting Government authorizes the use of firearms they should issue specific safety guidelines on their use. The PFSP should contain specific guidance on this matter, in particular with regard to its application to ships carrying dangerous goods or hazardous substances.

Organization and performance of port facility security duties

16.8 In addition to the guidance given under paragraph 16.3, the PFSP should establish the following, which relate to all security levels:

.1 the role and structure of the port facility security organization;

.2 the duties, responsibilities and training requirements of all port facility personnel with a security role and the performance measures needed to allow their individual effectiveness to be assessed;

.3 the port facility security organization's links with other national or local authorities with security responsibilities;

.4 the communication systems provided to allow effective and continuous communication between port facility security personnel, ships in port and, when appropriate, with national or local authorities with security responsibilities;

.5 the procedures or safeguards necessary to allow such continuous communications to be maintained at all times;

.6 the procedures and practices to protect security-sensitive information held in paper or electronic format;

.7 the procedures to assess the continuing effectiveness of security measures, procedures and equipment, including identification of, and response to, equipment failure or malfunction;

.8 the procedures to allow the submission, and assessment, of reports relating to possible breaches of security or security concerns;

.9 procedures relating to cargo handling;

.10 procedures covering the delivery of ship's stores;

.11 the procedures to maintain, and update, records of dangerous goods and hazardous substances and their location within the port facility;

.12 the means of alerting and obtaining the services of waterside patrols and specialist search teams, including bomb searches and underwater searches;

.13 the procedures for assisting ship security officers in confirming the identity of those seeking to board the ship when requested; and

.14 the procedures for facilitating shore leave for ship's personnel or personnel changes, as well as access of visitors to the ship, including representatives of seafarers' welfare and labour organizations.

16.9 The remainder of section 16 addresses specifically the security measures that could be taken at each security level covering:

.1 access to the port facility;

.2 restricted areas within the port facility;

.3 handling of cargo;

.4 delivery of ship's stores;

.5 handling unaccompanied baggage; and

.6 monitoring the security of the port facility.

Access to the port facility

16.10 The PFSP should establish the security measures covering all means of access to the port facility identified in the PFSA.

16.11 For each of these the PFSP should identify the appropriate locations where access restrictions or prohibitions should be applied for each of the security levels. For each security level the PFSP should specify the type of restriction or prohibition to be applied and the means of enforcing them.

16.12 The PFSP should establish for each security level the means of identification required to allow access to the port facility and for individuals to remain within the port facility without challenge. This may involve developing an appropriate identification system, allowing for permanent and temporary identifications for port facility personnel and for visitors respectively. Any port facility identification system should, when it is practicable to do so, be co-ordinated with that applying to ships that regularly use the port facility. Passengers should be able to prove their identity by boarding passes, tickets, etc., but should not be permitted access to restricted areas unless supervised. The PFSP should establish provisions to ensure that the identification systems are regularly updated, and that abuse of procedures should be subject to disciplinary action.

16.13 Those unwilling or unable to establish their identity and/or to confirm the purpose of their visit when requested to do so should be denied access to the port facility and their attempt to obtain access should be reported to the PFSO and to the national or local authorities with security responsibilities.

16.14 The PFSP should identify the locations where persons, personal effects, and vehicle searches are to be undertaken. Such locations should be covered to facilitate continuous operation, regardless of prevailing weather conditions, in accordance with the frequency laid down in the PFSP. Once subjected to search, persons, personal effects and vehicles should proceed directly to the restricted holding, embarkation or car loading areas.

16.15 The PFSP should establish separate locations for checked and unchecked persons and their effects and if possible separate areas for embarking/disembarking passengers, ship's personnel and their effects to ensure that unchecked persons are not able to come in contact with checked persons.

16.16 The PFSP should establish the frequency of application of any access controls, particularly if they are to be applied on a random, or occasional, basis.

Security level 1

16.17 At security level 1, the PFSP should establish the control points where the following security measures may be applied:

.1 restricted areas, which should be bounded by fencing or other barriers to a standard which should be approved by the Contracting Government;

.2 checking identity of all persons seeking entry to the port facility in connection with a ship, including passengers, ship's personnel and visitors, and confirming their reasons for doing so by checking, for example, joining instructions, passenger tickets, boarding passes, work orders, etc.;

.3 checking vehicles used by those seeking entry to the port facility in connection with a ship;

.4 verification of the identity of port facility personnel and those employed within the port facility and their vehicles;

.5 restricting access to exclude those not employed by the port facility or working within it, if they are unable to establish their identity;

.6 undertaking searches of persons, personal effects, vehicles and their contents; and

.7 identification of any access points not in regular use, which should be permanently closed and locked.

16.18 At security level 1, all those seeking access to the port facility should be liable to search. The frequency of such searches, including random searches, should be specified in the approved PFSP and should be specifically approved by the Contracting Government. Unless there are clear security grounds for doing so, members of the ship's personnel should not be required to search their colleagues or their personal effects. Any such search shall be undertaken in a manner which fully takes into account the human rights of the individual and preserves their basic human dignity.

Security level 2

16.19 At security level 2, the PFSP should establish the additional security measures to be applied, which may include:

.1 assigning additional personnel to guard access points and patrol perimeter barriers;

.2 limiting the number of access points to the port facility, and identifying those to be closed and the means of adequately securing them;

.3 providing for means of impeding movement through the remaining access points, e.g., security barriers;

.4 increasing the frequency of searches of persons, personal effects, and vehicles;

.5 denying access to visitors who are unable to provide a verifiable justification for seeking access to the port facility; and

.6 using patrol vessels to enhance water-side security.

Security level 3

16.20 At security level 3, the port facility should comply with instructions issued by those responding to the security incident or threat thereof. The PFSP should detail the security measures which could be taken by the port facility, in close co-operation with those responding and the ships at the port facility, which may include:

.1 suspension of access to all, or part, of the port facility;

.2 granting access only to those responding to the security incident or threat thereof;

.3 suspension of pedestrian or vehicular movement within all, or part, of the port facility;

.4 increased security patrols within the port facility, if appropriate;

.5 suspension of port operations within all, or part, of the port facility;

.6 direction of vessel movements relating to all, or part, of the port facility; and

.7 evacuation of all, or part, of the port facility.

Restricted areas within the port facility

16.21 The PFSP should identify the restricted areas to be established within the port facility and specify their extent, times of application, the security measures to be taken to control access to them and those to be taken to control activities within them. This should also include, in appropriate circumstances, measures to ensure that temporary restricted areas are security swept both before and after that area is established. The purpose of restricted areas is to:

.1 protect passengers, ship's personnel, port facility personnel and visitors, including those visiting in connection with a ship;

.2 protect the port facility;

.3 protect ships using, and serving, the port facility;

.4 protect security-sensitive locations and areas within the port facility;

.5 protect security and surveillance equipment and systems; and

.6 protect cargo and ship's stores from tampering.

16.22 The PFSP should ensure that all restricted areas have clearly established security measures to control:

.1 access by individuals;

.2 the entry, parking, loading and unloading of vehicles;

.3 movement and storage of cargo and ship's stores; and

.4 unaccompanied baggage or personal effects.

16.23 The PFSP should provide that all restricted areas should be clearly marked, indicating that access to the area is restricted and that unauthorized presence within the area constitutes a breach of security.

16.24 When automatic intrusion-detection devices are installed they should alert a control centre which can respond to the triggering of an alarm.

16.25 Restricted areas may include:

.1 shore- and water-side areas immediately adjacent to the ship;

.2 embarkation and disembarkation areas, passenger and ship's personnel holding and processing areas, including search points;

.3 areas where loading, unloading or storage of cargo and stores is undertaken;

.4 locations where security-sensitive information, including cargo documentation, is held;

.5 areas where dangerous goods and hazardous substances are held;

.6 vessel traffic management system control rooms, aids to navigation and port control buildings, including security and surveillance control rooms;

.7 areas where security and surveillance equipment are stored or located;

.8 essential electrical, radio and telecommunication, water and other utility installations; and

.9 other locations in the port facility where access by vessels, vehicles and individuals should be restricted.

16.26 The security measures may extend, with the agreement of the relevant authorities, to restrictions on unauthorized access to structures from which the port facility can be observed.

Security level 1

16.27 At security level 1, the PFSP should establish the security measures to be applied to restricted areas, which may include:

.1 provision of permanent or temporary barriers to surround the restricted area, whose standard should be accepted by the Contracting Government;

.2 provision of access points where access can be controlled by security guards when in operation and which can be effectively locked or barred when not in use;

.3 providing passes which must be displayed to identify individual's entitlement to be within the restricted area;

.4 clearly marking vehicles allowed access to restricted areas;

.5 providing guards and patrols;

.6 providing automatic intrusion-detection devices, or surveillance equipment or systems to detect unauthorized access into, or movement within, restricted areas; and

.7 control of the movement of vessels in the vicinity of ships using the port facility.

Security level 2

16.28 At security level 2, the PFSP should establish the enhancement of the frequency and intensity of the monitoring of, and control of access to, restricted areas. The PFSP should establish the additional security measures, which may include:

.1 enhancing the effectiveness of the barriers or fencing surrounding restricted areas, including the use of patrols or automatic intrusion-detection devices;

.2 reducing the number of access points to restricted areas and enhancing the controls applied at the remaining accesses;

.3 restrictions on parking adjacent to berthed ships;

.4 further restricting access to the restricted areas and movements and storage within them;

.5 use of continuously monitored and recording surveillance equipment;

.6 enhancing the number and frequency of patrols, including water-side patrols, undertaken on the boundaries of the restricted areas and within the areas;

.7 establishing and restricting access to areas adjacent to the restricted areas; and

.8 enforcing restrictions on access by unauthorized craft to the waters adjacent to ships using the port facility.

Security level 3

16.29 At security level 3, the port facility should comply with the instructions issued by those responding to the security incident or threat thereof. The PFSP should detail the security measures which could be taken by the port facility in close co-operation with those responding and the ships at the port facility, which may include:

.1 setting up of additional restricted areas within the port facility in proximity to the security incident, or the believed location of the security threat, to which access is denied; and

.2 preparing for the searching of restricted areas as part of a search of all, or part, of the port facility.

Handling of cargo

16.30 The security measures relating to cargo handling should:

.1 prevent tampering; and

.2 prevent cargo that is not meant for carriage from being accepted and stored within the port facility.

16.31 The security measures should include inventory control procedures at access points to the port facility. Once within the port facility, cargo should be capable of being identified as having been checked and accepted for loading onto a ship or for temporary storage in a restricted area while awaiting loading. It may be appropriate to restrict the entry of cargo to the port facility that does not have a confirmed date for loading.

Security level 1

16.32 At security level 1, the PFSP should establish the security measures to be applied during cargo handling, which may include:

.1 routine checking of cargo, cargo transport units and cargo storage areas within the port facility prior to, and during, cargo handling operations;

.2 checks to ensure that cargo entering the port facility matches the delivery note or equivalent cargo documentation;

.3 searches of vehicles; and

.4 checking of seals and other methods used to prevent tampering upon entering the port facility and upon storage within the port facility.

16.33 Checking of cargo may be accomplished by some or all of the following means:

.1 visual and physical examination; and

.2 using scanning/detection equipment, mechanical devices, or dogs.

16.34 When there are regular or repeated cargo movements, the CSO or the SSO may, in consultation with the port facility, agree arrangements with shippers or others responsible for such cargo covering off-site checking, sealing, scheduling, supporting documentation, etc. Such arrangements should be communicated to and agreed with the PFSO concerned.

Security level 2

16.35 At security level 2, the PFSP should establish the additional security measures to be applied during cargo handling to enhance control, which may include:

.1 detailed checking of cargo, cargo transport units and cargo storage areas within the port facility;

.2 intensified checks, as appropriate, to ensure that only the documented cargo enters the port facility, is temporarily stored there and is then loaded onto the ship;

.3 intensified searches of vehicles; and

.4 increased frequency and detail in checking of seals and other methods used to prevent tampering.

16.36 Detailed checking of cargo may be accomplished by some or all of the following means:

.1 increasing the frequency and detail of checking of cargo, cargo transport units and cargo storage areas within the port facility (visual and physical examination);

.2 increasing the frequency of the use of scanning/detection equipment, mechanical devices, or dogs; and

.3 co-ordinating enhanced security measures with the shipper or other responsible party in addition to an established agreement and procedures.

Security level 3

16.37 At security level 3, the port facility should comply with the instructions issued by those responding to the security incident or threat thereof. The PFSP should detail the security measures which could be taken by the port facility in close co-operation with those responding and the ships at the port facility, which may include:

.1 restriction or suspension of cargo movements or operations within all, or part, of the port facility or specific ships; and

.2 verifying the inventory of dangerous goods and hazardous substances held within the port facility and their location.

Delivery of ship's stores

16.38 The security measures relating to the delivery of ship's stores should:

.1 ensure checking of ship's stores and package integrity;

.2 prevent ship's stores from being accepted without inspection;

.3 prevent tampering;

.4 prevent ship's stores from being accepted unless ordered;

.5 ensure searching the delivery vehicle; and

.6 ensure escorting delivery vehicles within the port facility.

16.39 For ships regularly using the port facility it may be appropriate to establish procedures involving the ship, its suppliers and the port facility covering notification and timing of deliveries and their documentation. There should always be some way of confirming that stores presented for delivery are accompanied by evidence that they have been ordered by the ship.

Security level 1

16.40 At security level 1, the PFSP should establish the security measures to be applied to control the delivery of ship's stores, which may include:

.1 checking of ship's stores;

.2 advance notification as to composition of load, driver details and vehicle registration; and

.3 searching the delivery vehicle.

16.41 Checking of ship's stores may be accomplished by some or all of the following means:

.1 visual and physical examination; and

.2 using scanning/detection equipment, mechanical devices or dogs.

Security level 2

16.42 At security level 2, the PFSP should establish the additional security measures to be applied to enhance the control of the delivery of ship's stores, which may include:

.1 detailed checking of ship's stores;

.2 detailed searches of the delivery vehicles;

.3 co-ordination with ship personnel to check the order against the delivery note prior to entry to the port facility; and

.4 escorting the delivery vehicle within the port facility.

16.43 Detailed checking of ship's stores may be accomplished by some or all of the following means:

.1 increasing the frequency and detail of searches of delivery vehicles;

.2 increasing the use of scanning/detection equipment, mechanical devices, or dogs; and

.3 restricting, or prohibiting, entry of stores that will not leave the port facility within a specified period.

Security level 3

16.44 At security level 3, the port facility should comply with the instructions issued by those responding to the security incident or threat thereof. The PFSP should detail the security measures which could be taken by the port facility, in close co-operation with those responding and the ships at the port facility, which may include preparation for restriction, or suspension, of the delivery of ship's stores within all, or part, of the port facility.

Handling unaccompanied baggage

16.45 The PFSP should establish the security measures to be applied to ensure that unaccompanied baggage (i.e., any baggage, including personal effects, which is not with the passenger or member of ship's personnel at the point of inspection or search) is identified and subjected to appropriate screening, including searching, before it is allowed in the port facility and, depending on the storage arrangements, before it is transferred between the port facility and the ship. It is not envisaged that such baggage will be subjected to screening by both the port facility and the ship, and in cases where both are suitably equipped, the responsibility for screening should rest with the port facility. Close co-operation with the ship is essential and steps should be taken to ensure that unaccompanied baggage is handled securely after screening.

Security level 1

16.46 At security level 1, the PFSP should establish the security measures to be applied when handling unaccompanied baggage to ensure that unaccompanied baggage is screened or searched up to and including 100%, which may include use of x-ray screening.

Security level 2

16.47 At security level 2, the PFSP should establish the additional security measures to be applied when handling unaccompanied baggage which should include 100% x-ray screening of all unaccompanied baggage.

Security level 3

16.48 At security level 3, the port facility should comply with the instructions issued by those responding to the security incident or threat thereof. The PFSP should detail the security measures which could be taken by the port facility in close co-operation with those responding and the ships at the port facility, which may include:

.1 subjecting such baggage to more extensive screening, for example x-raying it from at least two different angles;

.2 preparations for restriction or suspension of handling of unaccompanied baggage; and

.3 refusal to accept unaccompanied baggage into the port facility.

Monitoring the security of the port facility

16.49 The port facility security organization should have the capability to monitor the port facility and its nearby approaches, on land and water, at all times, including the night hours and periods of limited visibility, the restricted areas within the port facility, the ships at the port facility and areas surrounding ships. Such monitoring can include use of:

.1 lighting;

.2 security guards, including foot, vehicle and waterborne patrols; and

.3 automatic intrusion-detection devices and surveillance equipment.

16.50 When used, automatic intrusion-detection devices should activate an audible and/or visual alarm at a location that is continuously attended or monitored.

16.51 The PFSP should establish the procedures and equipment needed at each security level and the means of ensuring that monitoring equipment will be able to perform continually, including consideration of the possible effects of weather or of power disruptions.

Security level 1

16.52 At security level 1, the PFSP should establish the security measures to be applied, which may be a combination of lighting, security guards or use of security and surveillance equipment to allow port facility security personnel to:

.1 observe the general port facility area, including shore- and water-side accesses to it;

.2 observe access points, barriers and restricted areas; and

.3 allow port facility security personnel to monitor areas and movements adjacent to ships using the port facility, including augmentation of lighting provided by the ship itself.

Security level 2

16.53 At security level 2, the PFSP should establish the additional security measures to be applied, to enhance the monitoring and surveillance capability, which may include:

.1 increasing the coverage and intensity of lighting and surveillance equipment, including the provision of additional lighting and surveillance coverage;

.2 increasing the frequency of foot, vehicle or waterborne patrols; and

.3 assigning additional security personnel to monitor and patrol.

Security level 3

16.54 At security level 3, the port facility should comply with the instructions issued by those responding to the security incident or threat thereof. The PFSP should detail the security measures which could be taken by the port facility in close co-operation with those responding and the ships at the port facility, which may include:

.1 switching on all lighting within, or illuminating the vicinity of, the port facility;

.2 switching on all surveillance equipment capable of recording activities within, or adjacent to, the port facility; and

.3 maximizing the length of time such surveillance equipment can continue to record.

Differing security levels

16.55 The PFSP should establish details of the procedures and security measures the port facility could adopt if the port facility is at a lower security level than that applying to a ship.

Activities not covered by the Code

16.56 The PFSP should establish details of the procedures and security measures the port facility should apply when:

.1 it is interfacing with a ship which has been at a port of a State which is not a Contracting Government;

.2 it is interfacing with a ship to which this Code does not apply; and

.3 it is interfacing with fixed or floating platforms or mobile offshore drilling units on location.

Declarations of Security

16.57 The PFSP should establish the procedures to be followed when, on the instructions of the Contracting Government, the PFSO requests a DoS or when a DoS is requested by a ship.

Audit, review and amendment

16.58 The PFSP should establish how the PFSO intends to audit the continued effectiveness of the PFSP and the procedure to be followed to review, update or amend the PFSP.

16.59 The PFSP should be reviewed at the discretion of the PFSO. In addition it should be reviewed:

.1 if the PFSA relating to the port facility is altered;

.2 if an independent audit of the PFSP or the Contracting Government's testing of the port facility security organization identifies failings in the organization or questions the continuing relevance of significant elements of the approved PFSP;

.3 following security incidents or threats thereof involving the port facility; and

.4 following changes in ownership or operational control of the port facility.

16.60 The PFSO can recommend appropriate amendments to the approved plan following any review of the plan. Amendments to the PFSP relating to:

.1 proposed changes which could fundamentally alter the approach adopted to maintaining the security of the port facility; and

.2 the removal, alteration or replacement of permanent barriers, security and surveillance equipment and systems, etc., previously considered essential in maintaining the security of the port facility

should be submitted to the Contracting Government that approved the original PFSP for their consideration and approval. Such approval can be given by, or on behalf of, the Contracting Government with, or without, amendments to the proposed changes. On approval of the PFSP, the Contracting Government should indicate which procedural or physical alterations have to be submitted to it for approval.

Approval of port facility security plans

16.61 PFSPs have to be approved by the relevant Contracting Government, which should establish appropriate procedures to provide for:

.1 the submission of PFSPs to them;

.2 the consideration of PFSPs;

.3 the approval of PFSPs, with or without amendments;

.4 consideration of amendments submitted after approval; and

.5 procedures for inspecting or auditing the continuing relevance of the approved PFSP.

At all stages, steps should be taken to ensure that the contents of the PFSP remain confidential.

Statement of Compliance of a Port Facility

16.62 The Contracting Government within whose territory a port facility is located may issue an appropriate Statement of Compliance of a Port Facility (SoCPF) indicating:

.1 the port facility;

.2 that the port facility complies with the provisions of chapter XI-2 and part A of the Code;

.3 the period of validity of the SoCPF, which should be specified by the Contracting Governments but should not exceed five years; and

.4 the subsequent verification arrangements established by the Contracting Government and a confirmation when these are carried out.

16.63 The Statement of Compliance of a Port Facility should be in the form set out in the appendix to this part of the Code. If the language used is not Spanish, French or English, the Contracting Government, if it considers it appropriate, may also include a translation into one of these languages.

17 Port facility security officer

General

17.1 In those exceptional instances where the ship security officer has questions about the validity of identification documents of those seeking to board the ship for official purposes, the port facility security officer should assist.

17.2 The port facility security officer should not be responsible for routine confirmation of the identity of those seeking to board the ship.

In addition, other relevant guidance is provided under sections 15, 16 and 18.

18 Training, drills and exercises on port facility security

Training

18.1 The port facility security officer should have knowledge and receive training, in some or all of the following, as appropriate:

.1 security administration;

.2 relevant international conventions, codes and recommendations;

.3 relevant Government legislation and regulations;

.4 responsibilities and functions of other security organizations;

.5 methodology of port facility security assessment;

.6 methods of ship and port facility security surveys and inspections;

.7 ship and port operations and conditions;

.8 ship and port facility security measures;

.9 emergency preparedness and response and contingency planning;

.10 instruction techniques for security training and education, including security measures and procedures;

.11 handling sensitive security-related information and security-related communications;

.12 knowledge of current security threats and patterns;

.13 recognition and detection of weapons, dangerous substances and devices;

.14 recognition, on a non-discriminatory basis, of characteristics and behavioural patterns of persons who are likely to threaten the security;

.15 techniques used to circumvent security measures;

.16 security equipment and systems, and their operational limitations;

.17 methods of conducting audits, inspection, control and monitoring;

.18 methods of physical searches and non-intrusive inspections;

.19 security drills and exercises, including drills and exercises with ships; and

.20 assessment of security drills and exercises.

18.2 Port facility personnel having specific security duties should have knowledge and receive training in some or all of the following, as appropriate:

- **.1** knowledge of current security threats and patterns;
- **.2** recognition and detection of weapons, dangerous substances and devices;
- **.3** recognition of characteristics and behavioural patterns of persons who are likely to threaten security;
- **.4** techniques used to circumvent security measures;
- **.5** crowd management and control techniques;
- **.6** security-related communications;
- **.7** operation of security equipment and systems;
- **.8** testing, calibration and maintenance of security equipment and systems;
- **.9** inspection, control, and monitoring techniques; and
- **.10** methods of physical searches of persons, personal effects, baggage, cargo, and ship's stores.

18.3 All other port facility personnel should have knowledge of and be familiar with relevant provisions of the PFSP in some or all of the following, as appropriate:

- **.1** the meaning and the consequential requirements of the different security levels;
- **.2** recognition and detection of weapons, dangerous substances and devices;
- **.3** recognition of characteristics and behavioural patterns of persons who are likely to threaten the security; and
- **.4** techniques used to circumvent security measures.

Drills and exercises

18.4 The objective of drills and exercises is to ensure that port facility personnel are proficient in all assigned security duties, at all security levels, and to identify any security-related deficiencies which need to be addressed.

18.5 To ensure the effective implementation of the provisions of the port facility security plan, drills should be conducted at least every three months unless the specific circumstances dictate otherwise. These drills should test individual elements of the plan such as those security threats listed in paragraph 15.11.

18.6 Various types of exercises, which may include participation of port facility security officers, in conjunction with relevant authorities of Contracting Governments, company security officers, or ship security officers, if available, should be carried out at least once each calendar year with no more than 18 months between the exercises. Requests for the participation of company security officers or ship security officers in joint exercises should be made, bearing in mind the security and work implications for the ship. These exercises should test communication, co-ordination, resource availability and response. These exercises may be:

- **.1** full-scale or live;
- **.2** tabletop simulation or seminar; or
- **.3** combined with other exercises held, such as emergency response or other port State authority exercises.

19 Verification and certification for ships

No additional guidance.

Appendix to part B

Appendix 1
Form of a Declaration of Security between a ship and a port facility*

DECLARATION OF SECURITY

Name of ship: ....

Port of registry: ....

IMO Number: ....

Name of port facility: ....

This Declaration of Security is valid from .... until .... for the following activities:

....

(list the activities with relevant details)

under the following security levels

Security level(s) for the ship:	
Security level(s) for the port facility:	

The port facility and ship agree to the following security measures and responsibilities to ensure compliance with the requirements of part A of the International Code for the Security of Ships and of Port Facilities.

* This form of Declaration of Security is for use between a ship and a port facility. If the Declaration of Security is to cover two ships, this model should be appropriately modified.

Activity	The affixing of the initials of the SSO or PFSO under these columns indicates that the activity will be done, in accordance with the relevant approved plan, by	
	The port facility:	The ship:
Ensuring the performance of all security duties		
Monitoring restricted areas to ensure that only authorized personnel have access		
Controlling access to the port facility		
Controlling access to the ship		
Monitoring of the port facility, including berthing areas and areas surrounding the ship		
Monitoring of the ship, including berthing areas and areas surrounding the ship		
Handling of cargo		
Delivery of ship's stores		
Handling unaccompanied baggage		
Controlling the embarkation of persons and their effects		
Ensuring that security communication is readily available between the ship and the port facility		

The signatories to this agreement certify that security measures and arrangements for both the port facility and the ship during the specified activities meet the provisions of chapter XI-2 and part A of the Code that will be implemented in accordance with the provisions already stipulated in their approved plan or the specific arrangements agreed to and set out in the attached annex.

Dated at .... on the ....

Signed for and on behalf of

the port facility:	the ship:
....	....
(Signature of port facility security officer)	*(Signature of master or ship security officer)*

Name and title of person who signed

Name:	Name: ....
Title:...	Title:...

Contact Details

(to be completed as appropriate)
(indicate the telephone numbers or the radio channels or frequencies to be used)

for the port facility:

Port facility

. . . .

Port facility security officer

. . . .

for the ship:

Master

. . . .

Ship security officer

. . . .

Company

. . . .

Company security officer

. . . .

Appendix 2
Form of a Statement of Compliance of a Port Facility

STATEMENT OF COMPLIANCE OF A PORT FACILITY

(Official seal) *(State)*

Statement Number . . . .

Issued under the provisions of part B of the

INTERNATIONAL CODE FOR THE SECURITY OF SHIPS AND OF PORT FACILITIES
(ISPS CODE)

The Government of . . . .
(name of the State)

Name of the port facility . . . .

Address of the port facility . . . .

THIS IS TO CERTIFY that the compliance of this port facility with the provisions of chapter XI-2 and part A of the International Code for the Security of Ships and of Port Facilities (ISPS Code) has been verified and that this port facility operates in accordance with the approved port facility security plan. This plan has been approved for the following [*specify the types of operations, types of ship or activities or other relevant information*] (delete as appropriate):

Passenger ship
Passenger high-speed craft
Cargo high-speed craft
Bulk carrier
Oil tanker
Chemical tanker
Gas carrier
Mobile offshore drilling units
Cargo ships other than those referred to above

This Statement of Compliance is valid until . . . ., subject to verifications (as indicated overleaf)

Issued at . . . .
(place of issue of the statement)

Date of issue . . . .
(Signature of the duly authorized official issuing the document)

(Seal or stamp of the issuing authority, as appropriate)

Endorsement for verifications

The Government of [*insert name of the State*] has established that the validity of this Statement of Compliance is subject to [*insert relevant details of the verifications (e.g., mandatory annual or unscheduled).*

THIS IS TO CERTIFY that, during a verification carried out in accordance with paragraph B/16.62.4 of the ISPS Code, the port facility was found to comply with the relevant provisions of chapter XI-2 of the Convention and part A of the ISPS Code.

First verification

Signed: . . . .
(Signature of authorized official)

Place: . . . .

Date: . . . .

(Seal or stamp of authority, as appropriate)

Second verification

Signed: . . . .
(Signature of authorized official)

Place: . . . .

Date: . . . .

(Seal or stamp of authority, as appropriate)

Third verification

Signed: . . . .
(Signature of authorized official)

Place: . . . .

Date: . . . .

(Seal or stamp of authority, as appropriate)

Fourth verification

Signed: . . . .
(Signature of authorized official)

Place: . . . .

Date: . . . .

(Seal or stamp of authority, as appropriate)

Conference resolution 3

(adopted on 12 December 2002)

Further work by the International Maritime Organization pertaining to the enhancement of maritime security

THE CONFERENCE,

HAVING ADOPTED amendments to the International Convention for the Safety of Life at Sea, 1974, as amended (hereinafter referred to as "the Convention"), concerning special measures to enhance maritime safety and security,

RECOGNIZING the need for further work in the area of enhancement of maritime security and in order to ensure the global and uniform application and implementation of the special measures to enhance maritime security adopted by the Conference,

1. INVITES the International Maritime Organization (hereinafter referred to as "the Organization"), bearing in mind the provisions of chapter XI-2 of the Convention and the International Ship and Port Facility Security (ISPS) Code (hereinafter referred to as "the ISPS Code"), to:

(a) develop training guidance such as model courses for ship security officers, company security officers, port facility security officers and company, ship and port security personnel;

(b) review the Organization's Assembly resolution A.787(19) as amended by resolution A.882(21) on Procedures for port State control and, if found necessary, develop appropriate amendments thereto;

(c) consider the need and, if necessary, develop further guidance on control and compliance measures on aspects other than those already addressed in part B of the ISPS Code;

(d) consider the need and, if necessary, develop guidelines on recognized security organizations;

(e) review the Organization's Assembly resolution A.890(21) on Principles of safe manning and, if found necessary, develop appropriate amendments thereto;

(f) review the aspect of security of ships to which chapter XI-2 of the Convention applies when interfacing with floating production storage units and floating storage units and take action as appropriate;

(g) consider, in the context of security, relevant aspects of facilitation of maritime traffic such as, for example, port arrivals and departures, standardized forms of reporting and electronic data interchange and take action as appropriate;

(h) review the Organization's Assembly resolution A.872(20) on Guidelines for the prevention and suppression of the smuggling of drugs, psychotropic substances and precursor chemicals on ships engaged in international maritime traffic and, if necessary, develop appropriate amendments thereto; and

(i) consider the need and, if necessary, develop any other guidance or guidelines to ensure the global, uniform and consistent implementation of the provisions of chapter XI-2 of the Convention or part A of the ISPS Code

and to adopt them in time before the entry into force of the amendments to the Convention adopted by the Conference or as and when the Organization considers appropriate;

2. INVITES ALSO the Organization to carry out, as a matter of urgency, an impact assessment of the proposals to implement the long-range identification and tracking of ships and, if found necessary, develop and adopt appropriate performance standards and guidelines for long-range ship identification and tracking systems.

Conference resolution 4

(adopted on 12 December 2002)

Future amendments to chapters XI-1 and XI-2 of the 1974 SOLAS Convention on special measures to enhance maritime safety and security

THE CONFERENCE,

HAVING ADOPTED amendments to the International Convention for the Safety of Life at Sea (SOLAS), 1974, as amended (hereinafter referred to as "the Convention"), concerning special measures to enhance maritime safety and security,

NOTING the special nature of the measures now included in the new chapter XI-2 of the Convention aimed at enhancing maritime security,

RECOGNIZING the need for urgent and special measures to enhance maritime security and the desire of Contracting Governments to bring these measures into force as soon as possible,

NOTING ALSO that it may be necessary, due to the special nature of the issues involved, to frequently amend, in the future, the provisions of chapter XI-2 of the Convention in order to respond, in a proactive manner, to new or emerging security risks and threats,

RECALLING Resolution 5 entitled "Future amendments to chapter XI of the 1974 SOLAS Convention on special measures to enhance maritime safety", adopted by the 1994 Conference of Contracting Governments to the International Convention for the Safety of Life at Sea, 1974,

DESIRING that future amendments to chapters XI-1 and XI-2 of the Convention are adopted, brought into force and given effect in the shortest possible time,

RECOMMENDS that future amendments to the provisions of chapters XI-1 and XI-2 of the Convention should be adopted by either the Maritime Safety Committee of the International Maritime Organization in accordance with article VIII(b) of the Convention or by a Conference of Contracting Governments to the Convention in accordance with article VIII(c) thereof.

Conference resolution 5

(adopted on 12 December 2002)

Promotion of technical co-operation and assistance

THE CONFERENCE,

HAVING ADOPTED amendments to the International Convention for the Safety of Life at Sea, 1974, as amended (hereinafter referred to as "the Convention"), concerning special measures to enhance maritime safety and security,

RECALLING operative paragraph 5 of resolution A.924(22) on Review of measures and procedures to prevent acts of terrorism which threaten the security of passengers and crews and the safety of ships, adopted on 20 November 2001 by the Assembly of the International Maritime Organization (hereinafter referred to as "the Organization"), whereby the Secretary-General of the Organization is requested to take appropriate measures within the Integrated Technical Co-operation Programme to assist Governments to assess, put in place or enhance, as the case may be, appropriate infrastructure and measures to strengthen port safety and security so as to prevent and suppress terrorist acts directed against ports and port personnel as well as ships in port areas, passengers and crew,

BEING APPRECIATIVE of the steps already taken by the Secretary-General of the Organization, in response to the request of the Assembly of the Organization, to provide assistance to States in strengthening their maritime and port security infrastructure and measures,

RECOGNIZING the need for the development of appropriate legislation and the putting in place of appropriate infrastructure for ship and port facility security and relevant training facilities in order to ensure the global and uniform application and implementation of the special measures adopted to enhance maritime security,

RECOGNIZING ALSO the importance of adequate education and training for seafarers and port facility personnel to contribute to the overall efforts to enhance maritime security,

RECOGNIZING FURTHER that, in some cases, there may be limited infrastructure, facilities and training programmes for obtaining the experience required for the purpose of preventing acts which threaten the security of ships and of port facilities, particularly in developing countries,

BELIEVING that the promotion of technical co-operation at the international level will assist those States not yet having adequate expertise or facilities for providing training and experience to assess, put in place or enhance appropriate infrastructure and, in general, implement the measures required by the adopted amendments necessary to strengthen maritime security on board ships and ashore,

EMPHASIZING, in this regard, the vital role that safe and secure shipping and port operations play in sustainable socio-economic development,

1. STRONGLY URGES Contracting Governments to the Convention and Member States of the Organization to:

(a) provide, in co-operation with the Organization, assistance to those States which have difficulty in implementing or meeting the requirements of the adopted amendments or the ISPS Code; and

(b) use the Integrated Technical Co-operation Programme of the Organization as one of the main instruments to obtain assistance in advancing effective implementation of, and compliance with, the adopted amendments and the ISPS Code;

2. REQUESTS the Secretary-General of the Organization to make adequate provision, within the Integrated Technical Co-operation Programme, to strengthen further the assistance that is already being provided and to promote, in co-operation, as appropriate, with relevant international organizations, the enhancement of the Organization's capacity to address the future needs of developing countries for continued education and training and the improvement of their maritime and port security infrastructure and measures;

3. INVITES donors, international organizations and the shipping and port industry to contribute financial, human and/or in-kind resources to the Integrated Technical Co-operation Programme of the Organization for its maritime and port security activities;

4. INVITES ALSO the Secretary General to give early consideration to establishing a Maritime Security Trust Fund for the purpose of providing a dedicated source of financial support for maritime security technical co-operation activities and, in particular, for providing support for national initiatives in developing countries to strengthen their maritime security infrastructure and measures.

Conference resolution 6

(adopted on 12 December 2002)

Early implementation of the special measures to enhance maritime security

THE CONFERENCE,

HAVING ADOPTED amendments to the International Convention for the Safety of Life at Sea, 1974, as amended (hereinafter referred to as "the Convention"), concerning special measures to enhance maritime safety and security,

RECOGNIZING the important contribution that the implementation of the special measures adopted will make towards the safe and secure operation of ships, for pollution prevention and for the safety and security of those on board and ashore,

RECOGNIZING ALSO that the task of implementing the requirements of chapter XI-2 of the Convention and of the International Ship and Port Facility Security (ISPS) Code (hereinafter referred to as "the Code") will place a significant burden on Contracting Governments, Administrations, and recognized security organizations,

RECALLING that the Code, from 1 July 2004, requires each ship to which the provisions of chapter XI-2 of the Convention and part A of the Code apply to be provided with an appropriate ship security plan,

RECALLING ALSO that each such ship is required to be provided with an International Ship Security Certificate not later than 1 July 2004,

RECOGNIZING FURTHER that the process of verifying the compliance of a ship, to which the provisions of chapter XI-2 of the Convention and part A of the Code apply, with the requirements of chapter XI-2 and of the Code cannot be undertaken until the ship security plan has been approved and its provisions have been implemented on board,

DESIRING to ensure the smooth implementation of the provisions of chapter XI-2 of the Convention and of the Code,

BEARING IN MIND the difficulties experienced during implementation of the International Safety Management (ISM) Code,

1. DRAWS the attention of Contracting Governments to the Convention and the industry to the fact that neither chapter XI-2 of the Convention nor the Code provide for any extension of the implementation dates for the introduction of the special measures concerned to enhance maritime security;

2. URGES Contracting Governments to take, as a matter of high priority, any action needed to finalize as soon as possible any legislative or administrative arrangements, which are required at the national level, to give effect to the requirements of the adopted amendments to the Convention (and the Code) relating to the certification of ships entitled to fly their flag or port facilities situated in their territory;

3. RECOMMENDS that Contracting Governments and Administrations concerned designate dates, in advance of the application date of 1 July 2004, by which requests for:

- .1 review and approval of ship security plans;
- .2 verification and certification of ships; and
- .3 review and approval of port facility security assessments and of port facility security plans

should be submitted in order to allow Contracting Governments, Administrations and recognized security organizations time to complete the review and approval and the verification and certification process and for Companies, ships and port facilities to rectify any non-compliance;

4. INVITES Contracting Governments, on and after 1 July 2004, to recognize and accept as valid and as meeting the requirements of chapter XI-2 of the Convention and part A of the Code any:

 .1 Ship security plans approved prior to 1 July 2004, pursuant to the provisions of part A of the Code, by Administrations or on their behalf; and

 .2 International Ship Security Certificates issued, prior to 1 July 2004, in accordance with the provisions of part A of the Code, by Administrations or on their behalf

as far as these relate to ships which, on 1 July 2004, were entitled to fly the flag of the State of the Administration which, or on behalf of which, the plan in question was approved or the certificate in question was issued;

5. FURTHER RECOMMENDS that Contracting Governments and the industry take early appropriate action to ensure that all necessary infrastructure is in place in time for the effective implementation of the adopted measures to enhance maritime security on board ships and ashore.

Conference resolution 7

(adopted on 12 December 2002)

Establishment of appropriate measures to enhance the security of ships, port facilities, mobile offshore drilling units on location and fixed and floating platforms not covered by chapter XI-2 of the 1974 SOLAS Convention

THE CONFERENCE,

HAVING ADOPTED amendments to the International Convention for the Safety of Life at Sea, 1974, as amended (hereinafter referred to as "the Convention"), concerning special measures to enhance maritime safety and security,

RECALLING that chapter XI-2 of the Convention applies only to:

(a) the following types of ships engaged on international voyages:

.1 passenger ships, including passenger high-speed craft; and

.2 cargo ships, including cargo high-speed craft, of 500 gross tonnage and upwards; and

.3 mobile offshore drilling units; and

(b) port facilities serving such ships engaged on international voyages,

RECOGNIZING the important contribution that the implementation of the special measures adopted will make towards the safe and secure operation of ships, for pollution prevention and for the safety and security of those on board and ashore,

RECOGNIZING ALSO the need to address and establish appropriate measures to enhance the security of ships and of port facilities other than those covered by chapter XI-2 of the Convention,

RECOGNIZING FURTHER that the establishment of such measures will further enhance and positively contribute towards the international efforts to ensure maritime security and to prevent and suppress acts threatening the security in the maritime transport sector,

1. INVITES Contracting Governments to the Convention to establish, as they may consider necessary, and to disseminate, as they deem fit, appropriate measures to enhance the security of ships and of port facilities other than those covered by chapter XI-2 of the Convention;

2. ENCOURAGES, in particular, Contracting Governments to establish, as they may consider necessary, and to disseminate, as they deem fit, information to facilitate the interactions of ships and of port facilities to which chapter XI-2 of the Convention applies with ships which are not covered by chapter XI-2 of the Convention;

3. ALSO ENCOURAGES Contracting Governments to establish, as they may consider necessary, and to disseminate, as they deem fit, information to facilitate contact and liaison between company and ship security officers and the authorities responsible for the security of port facilities not covered by chapter XI-2 of the Convention, prior to a ship entering, or anchoring off, such a port;

4. FURTHER ENCOURAGES Contracting Governments, when exercising their responsibilities for mobile offshore drilling units and for fixed and floating platforms operating on their Continental Shelf or within their Exclusive Economic Zone, to ensure that any security provisions applying to such units and platforms allow interaction with those applying to ships covered by chapter XI-2 of the Convention that serve, or operate in conjunction with, such units or platforms;

5. REQUESTS Contracting Governments to inform the Organization of any action they have taken in this respect.

Conference resolution 8

(adopted on 12 December 2002)

Enhancement of security in co-operation with the International Labour Organization

(Seafarers' Identity Documents and work on the wider issues of port security)

THE CONFERENCE,

HAVING ADOPTED amendments to the International Convention for the Safety of Life at Sea, 1974, as amended (hereinafter referred to as "the Convention"), concerning special measures to enhance maritime safety and security,

RECOGNIZING the important contribution that the implementation of the special measures adopted will make towards the safe and secure operation of ships, for pollution prevention and for the safety and security of those on board and ashore,

RECOGNIZING ALSO the need to continue the work and to establish, as the need arises, further appropriate measures to enhance the security of ships and of port facilities,

RECOGNIZING FURTHER that the development and use of a verifiable Seafarers' Identity Document will further enhance and positively contribute towards the international efforts to ensure maritime security and to prevent and suppress acts threatening the security in the maritime transport sector,

COGNIZANT of the competencies and work of the International Labour Organization (hereinafter referred to as "the ILO") in the area of development and adoption of the international labour standards,

RECALLING the Seafarers' Identity Documents Convention, 1958 (No. 108), adopted by the International Labour Conference on 13 May 1958, which entered into force on 19 February 1961,

RECALLING ALSO that the Governing Body of the ILO at its 283rd Session, in March 2002, placed the question of "Improved security for seafarers' identification" as an urgent item on the agenda of the 91st Session of the International Labour Conference, to be held in June 2003, with a view to the adoption of a Protocol to the Seafarers' Identity Documents Convention, 1958 (No. 108),

RECALLING FURTHER the long-standing co-operation between the International Maritime Organization (hereinafter referred as "the Organization") and the ILO in the area of international maritime transport,

NOTING, with satisfaction, the work undertaken, so far, by the Governing Body of the ILO and by the International Labour Office on seafarers' identity documents and on port and dock workers' security,

1. INVITES the ILO to continue the development of a Seafarers' Identity Document as a matter of urgency, which should cover, *inter alia*, a document for professional purposes, a verifiable security document and a certification information document;

2. REQUESTS the Organization to consider the results of the 91st Session of the International Labour Conference on the "Improved security for seafarers' identification" and to take appropriate action, as it deems appropriate;

3. INVITES States through their tripartite delegations to participate in the 91st Session of the International Labour Conference, in June 2003, and to give favourable consideration to the earliest possible ratification, acceptance, approval or accession to the new ILO instrument concerning seafarers' identification documents, once it is adopted;

4. INVITES the Organization and the ILO to establish a joint ILO/IMO Working Group to undertake any further work which may be required on the wider issue of port security, based on the terms of reference set out in the attached annex;

5. REQUESTS the Secretary-General of the Organization to contribute, with appropriate expertise, to the work of the ILO on the "Improved security for seafarers' identification" and to the proposed joint work on the wider issue of port security;

6. REQUESTS the Secretary-General of the Organization to transmit a copy of this resolution to the Director-General of the International Labour Office.

Annex

IMO/ILO work on port security

POSSIBLE TERMS OF REFERENCE

1 The joint IMO/ILO Working Group on Port Security, having regard to the amendments to the International Convention for the Safety of Life at Sea, 1974 and the International Ship and Port Facility Security (ISPS) Code adopted by the December 2002 Conference of Contracting Governments to the International Convention for the Safety of Life at Sea, 1974 for the purpose of introducing mandatory requirements and guidance relating to the enhancement of the safety and security of ships and of port facilities, should:

.1 consider and recommend, for the purpose of enhancing security, safety and the protection of the environment, the form and content of any further guidance which may be required on the wider issue of port security, including the relationship between ship and port security, and the wider security and safety and the protection of the environment considerations relevant to port areas, including the question of verifiable identification of those working within these areas or having access to such areas;

.2 consider the need for any mandatory requirements relating to the above and, if such a need is identified, to recommend the form and content of such requirements; and

.3 prepare and submit a report (including interim work and progress reports) on the aforesaid, together with the relevant reasons and justifications thereto, as well as an assessment of the impact, benefits and costs of the recommendations, for the consideration of the International Maritime Organization and of the International Labour Organization.

2 The International Maritime Organization and the International Labour Organization will monitor the work of the joint IMO/ILO Working Group on Port Security and, as the need arises, will issue appropriate instructions and guidance to the Working Group.

Conference resolution 9

(adopted on 12 December 2002)

Enhancement of security in co-operation with the World Customs Organization

(Closed cargo transport units)

THE CONFERENCE,

HAVING ADOPTED amendments to the International Convention for the Safety of Life at Sea, 1974, as amended (hereinafter referred to as "the Convention"), concerning special measures to enhance maritime safety and security,

RECOGNIZING the important contribution that the implementation of the special measures adopted will make towards the safe and secure operation of ships, for pollution prevention and for the safety and security of those on board and ashore,

RECOGNIZING ALSO the need to address and establish appropriate measures to enhance the security of ships and of port facilities in aspects other than those covered by chapter XI-2 of the Convention,

RECALLING that the Convention on Facilitation of International Maritime Traffic, 1965 already contains requirements related to the provision to Administrations of commercial data related to the movement of cargoes by sea,

RECOGNIZING FURTHER the need to include, in due course, in the Convention appropriate requirements to address specifically the security of closed cargo transport units (hereinafter referred to as "closed CTUs") and that such requirements will further enhance and positively contribute towards the international efforts to ensure maritime security and to prevent and suppress acts threatening security in the maritime transport sector,

FURTHERMORE RECOGNIZING the inter-modal and international nature of closed CTUs movements, the need to ensure security of the complete supply chain and the respective roles of all those involved,

RECALLING ALSO the role of frontier agencies, in particular Customs Administrations, in controlling the international movement of closed CTUs,

COGNIZANT of the competencies and work of the World Customs Organization (hereinafter referred to as "the WCO") in the area of international maritime transport,

RECALLING FURTHER the long-standing co-operation of the International Maritime Organization (hereinafter referred to as "the Organization") with the WCO in the area of international maritime transport,

NOTING with satisfaction the signing on 23 July 2002 of a Memorandum of Understanding to strengthen the co-operation between the two Organizations,

1. INVITES the WCO to consider, urgently, measures to enhance security throughout international movements of closed CTUs;

2. REQUESTS the Secretary-General of the Organization to contribute expertise relating to maritime transport and, in particular, to the carriage of closed CTUs by sea to the discussions at the WCO;

3. AGREES that the Convention should be amended, if and when appropriate, to give effect to relevant decisions taken by the WCO and endorsed by the Contracting Governments to the Convention insofar as they relate to the carriage of closed CTUs by sea;

4. REQUESTS the Secretary-General of the Organization to transmit a copy of this resolution to the Secretary-General of the WCO.

Conference resolution 10

(adopted on 12 December 2002)

Early implementation of long-range ship's identification and tracking

THE CONFERENCE,

HAVING ADOPTED amendments to the International Convention for the Safety of Life at Sea, 1974, as amended (hereinafter referred to as "the Convention"), concerning special measures to enhance safety and security,

RECALLING that long-range identification and tracking of ships at sea is a measure that fully contributes to the enhancement of the maritime and coastal States' security as a whole,

HAVING ACKNOWLEDGED that Inmarsat C polling is currently an appropriate system for long-range identification and tracking of ships,

RECOGNIZING the importance of an early implementation of long-range identification and tracking of ships,

RECOGNIZING ALSO that the equipment installed on board and ashore is available for immediate use and will allow the early implementation of such measures,

1. URGES Contracting Governments to take, as a matter of high priority, any action needed at national level to give effect to implementing and beginning the long-range identification and tracking of ships;

2. INVITES Contracting Governments to encourage ships entitled to fly the flag of their State to take the necessary measures so that they are prepared to respond automatically to Inmarsat C polling, or to other available systems;

3. REQUESTS Contracting Governments to consider all aspects related to the introduction of long-range identification and tracking of ships, including its potential for misuse as an aid to ship targeting and the need for confidentiality in respect of the information so gathered.

Conference resolution 11

(adopted on 12 December 2002)

Human-element-related aspects and shore leave for seafarers

THE CONFERENCE,

HAVING ADOPTED amendments to the International Convention for the Safety of Life at Sea, 1974, as amended (hereinafter referred to as "the Convention"), concerning special measures to enhance maritime safety and security,

RECOGNIZING that the shipping industry and the smooth transportation of goods are essential to world trade,

RECALLING that the Assembly of the International Maritime Organization (hereinafter referred to as "the Organization") adopted resolution A.907(22) on the long-term work programme of the Organization (up to 2008) and that the human element is an important item thereof,

RECALLING ALSO the provisions of the Convention on Facilitation of International Maritime Traffic, 1965, as amended, which has, *inter alia*, established a general right for foreign crew members to be entitled to shore leave while the ship on which they arrived is in port, provided that the formalities on arrival of the ship have been fulfilled and the public authorities have no reason to refuse permission to come ashore for reasons of public health, public safety or public order,

RECALLING FURTHER the generally accepted principles of international human rights applicable to all workers, including seafarers,

CONSIDERING that, given the global nature of the shipping industry, seafarers need special protection,

BEING AWARE that seafarers work and live on ships involved in international trade and that access to shore facilities and shore leave are vital elements of seafarers' general well-being and, therefore, to the realization of safer seas and cleaner oceans,

BEING AWARE ALSO that the ability to go ashore is essential for joining and leaving a ship after the agreed period of service,

1. URGES Contracting Governments to take the human element, the need to afford special protection to seafarers and the critical importance of shore leave into account when implementing the provisions of chapter XI-2 of the Convention and the International Ship and Port Facility (ISPS) Code (hereinafter referred to as "the Code");

2. ENCOURAGES Contracting Governments, Member States of the Organization and non-governmental organizations with consultative status at the Organization to report to the Organization any instances where the human element has been adversely impacted by the implementation of the provisions of chapter XI-2 of the Convention or the Code; and

3. REQUESTS the Secretary-General to bring to the attention of the Maritime Safety Committee and the Facilitation Committee of the Organization, any human-element-related problems which have been communicated to the Organization as a result of the implementation of chapter XI-2 of the Convention or the Code.

Notes

Notes

Notes

Notes

Notes

Notes

Notes

Notes

Notes

Notes

Notes

Notes

Notes

Notes

Notes